MATHEMATICAL ESSAYS ON GROWTH AND THE EMERGENCE OF FORM

MATHEMATICAL ESSAYS ON GROWTH AND THE EMERGENCE OF FORM

edited by
Peter L. Antonelli

The University of Alberta Press

First published by
The University of Alberta Press
Athabasca Hall
Edmonton, Alberta, Canada

ISBN 0–88864–089–7

Canadian Cataloguing in Publication Data

Main entry under title:

Mathematical essays on growth and the emergence of form

ISBN 0-88864–089-7

1. Biomathematics - Addresses, essays, lectures. 2. Developmental biology - Mathematics - Addresses, essays, lectures. 3. Morphogenesis - Mathematics - Addresses, essays, lectures. 4. Ecology - Mathematics - Addresses, essays, lectures. I. Antonelli, Peter L. (Peter Louis), 1941–
QH323.5.M38 1985 574'.015'1 C85–091047–1

Typeset by Vivian Spak.
Printed by Hignell Printing Limited, Winnipeg, Manitoba, Canada.

Contents

Contributors

P.L. Antonelli	Department of Mathematics, University of Alberta, Edmonton, Alberta, Canada
F.L. Bookstein	Center for Human Growth and Development, University of Michigan, Ann Arbor, Michigan, U.S.A.
B.C. Goodwin	Department of Biology, The Open University, Milton Keynes, England
N.D. Kazarinoff	Department of Mathematics, S.U.N.Y. at Buffalo, New York, U.S.A.
C.J. Lumsden	Department of Medicine, University of Toronto, Toronto, Ontario, Canada
M.A. Russell	Department of Genetics, University of Alberta, Edmonton, Alberta, Canada
L.E.H. Trainor	Department of Physics and Medicine, University of Toronto, Toronto, Ontario, Canada
B.H. Voorhees	Department of Mathematics, Athabasca University, Athabasca, Alberta, Canada

Preface

The essays in this book are meant primarily for theoretical biologists but may serve as a source book for mathematicians and physical scientists who wish to take an interesting geometric excursion into new frontiers offered by contemporary mathematical biology. They are also offered as an introduction for students who wish to pursue advanced degrees in mathematical biology. These essays on ecology and developmental biology represent a distillate accrued from two small symposia held in Edmonton, Alberta which were sponsored by the University of Alberta. The mathematical leitmotif of this book is geometry and topology. However, some knowledge of classical and statistical mechanics and elementary field theory is also required of the reader. Hidden treasures await you!

Peter L. Antonelli
Edmonton, Alberta

Acknowledgements

Many people were involved with this project who I would like to thank. My wife, Paule, deserves special mention, not only for translating Vito Volterra's "Principes de Biologie Mathematique" for this book, but for contributing in countless other ways. I would also like to thank Brian Taylor for his translation of Goethe's "Principes de Philosophie Zoologique." Another person who put many hours into this project and deserves much credit for her patience and dedication is Vivian Spak, who typed many of the original chapters of the manuscript and who also painstakingly did the typesetting for the entire book. Finally, I would like to thank the staff of the University of Alberta Press, Mrs. Norma Gutteridge, the director, Mary Mahoney-Robson, in-house editor, and Joanne Poon, the designer, for all their help.

Permission to translate into English Vito Volterra's "Principes de Biologie Mathematique," which appeared originally in French in the 1937 edition of *Acta Biotheoretica,* Vol. III, has been granted by B.J. Brill, Publishers, Leiden, Netherlands.

Introduction

For a long time, biologists and mathematicians have vehemently criticized the pioneering work of Vito Volterra on the ecology of populations. To be sure, there are difficulties with the original theory. For example, the constant coefficients are not generally justifiable. Also, the statistical properties of real populations must be related to the specific deterministic equations (first order nonlinear O.D.E.'s) in some mathematical way. E. Kerner, a Ph.D. student of physicist R. Feynman, attacked this problem for the predator/prey equations in the 1950's. Kerner, after discounting Volterra's attempt at construction of a Hamiltonian mechanics for this 2n dimensional special system via an "auxillary variable," constructed a statistical mechanics based on an unusual linear lagrangian.

This method caught on immediately and B. Goodwin adopted Kerner's formalism to questions in developmental biology, while J. Cowan did the analogous thing in neurodynamics. In the ecological context, E.G. Leigh, Jr. continued Kerner's program. The recent work of Lumsden and Trainor, included in this book, places Kerner's approach in a broader biological context to cover developmental systems as well as ecological ones. Use is made of techniques of physicist P.A.M. Dirac in the study of Kerner's unusual linear Lagrangian for the predator/prey class of systems. The reader will find it significant that they have brought the dynamical equations into canonical form and have extended the abstract theory to time-dependent nonequilibrium systems. Yet, they give a new concrete example of their statistical mechanics applied to a Volterra "encounter type" model of transport-metabolism in the cell.

However, I believe Volterra's method deserves closer scrutiny. A new translation of Volterra's paper is included as Chapter 9. I argue in my essays in this book that his "auxiliary variable" can, in the case of competition in plant and plant/herbivore communities, offer deep insight into the chemical interactions which are well-known to mediate such real systems. Volterra's variable is defined as the integral over time of the number of modular units in the sense of botanist J.L. Harper (1977) (e.g. leaves with axillary bud, stems, roots, xylem cells, etc.). These units produce allelochemicals inhibiting the growth and development of herbivores which feed on them. For the mathematical theory, it is necessary to consider only those chemicals produced so as to be *allometrically related* to the biomass produced by the modular units. Lignins and tannins are allelochemicals of this type (Whitaker and Feeny, 1971). Such systems are designed by natural selection, at least partially, for the benefit of the plant community suffering attacks by phytophagous insects (Rhoades, 1979).

The mathematical method also applies to other sorts of sessile communities. For example, coral reefs where the Volterra variable is interpreted as the total accumulated amount of aragonite calcium carbonate forming the reef skeleton (Antonelli and Voorhees, 1983). The stability (for a range of parameter values) of a limit cycle discovered by N. Kazarinoff and myself, models the well-established natural cycle for aggregating adult populations of the crown-of-thorns starfish preying on corals of the Indo-west pacific. This paper is presented in essay B in Chapter 2. The basic mathematical reference is from N. Kazarinoff's (with B. Hassard and Y-H. Wan) *Theory and Applications of Hopf Bifurcation* (1981).

The statistical problem for this growth mechanics theory is uniquely solved in the context of optimal production theory and maximum likelihood estimation. The method provides a *Gaussian white noise* perturbation theory for populations of *modular units,* whilst it simultaneously describes *sample paths* in *production space.* Effects like cloud cover variation on the growth of sessile communities can be so described. These sample paths are solutions to Stratonovich stochastic differential equations on a Reimannian manifold (Ikeda and Watanabe, 1981). It is possible to describe such solutions en masse, in the manner of R. Graham (1973). Dissipative structures, in the sense of I. Prigogine, can also occur. These may arise from Volterra's demographic potential (see Chapter 9) spanning a catastrophe in the sense of Rene Thom (1975). Furthermore, the concept of *stochastic canalization* is important for the statistical theory. Briefly, the negative curvature of biomass space (due to the ecological interaction (see 2A.1) arising from an optimal production hypothesis) causes sample paths of the stochastic theory to stay closer to deterministic trajectories than in the case of ordinary white noise. This result depends on recently discovered short time asymptotic formulas of Watanabe and Takahaski (1980) involving the curvature, **R** .

For the case of vegetative growth, − **R** may be successfully interpreted as a measure of plant vigor. Section 3.B presents the mathematical theory in a self contained form. (Large times have been studied by M. Pinsky and also D. Sullivan in 1980's issues of *J. Differential Geometry.*)

According to B. Goodwin (1976), embryological evidence strongly suggests the existence of constraints underlying conservative aspects of morphogenetic processes over large taxonomic groups such as insects and vertebrates. Evidently, embryos obey some general topological principles, such as preservation of the Euler characteristic of a developmental field or the Dirichlet principle for harmonic functions, on the surfaces over which these fields are defined. The topological constraints must somehow result in forms generated by "bifurcation" satisfying particular relations over the class of surfaces involved.

What is novel, methodologically speaking, about Goodwin's ideas is that traditional historical explanations using neodarwinian or reductionistic arguments are temporarily set aside. Thus, he is able to consider development as the emergence of form in its purest sense, independent of Wolpert's Positional Information Theory of genetic control. Goodwin has written elsewhere that his perspective is much the same as the eighteenth century *rational morphologists,* Cuvier and Owen. However, it seems to this writer that Goodwin's position has definite similarities to the *typologist* view of Cuvier's famous adversary, Geoffrey de Saint-Hilaire. One may pursue the history of this controversy in the scientific writings of J. von Goethe. Goethe kept a personal diary of the events in Paris and the Royal French Academy in March of 1830. The translation found in Chapter 10 is from "Principes de Philosophie Zoologique" pages 407-415, in *Goethe's Morphologische Schriften,* edited by W. Troll, Eugen Diederichs Verlag im Jena.

If mention of the name of the great German poet comes as a surprise in this biological context, it is because Goethe made fundamental contributions to plant and animal morphology recognized even today. For example, Goethe, along with de Candole, a distinguished botanist of the early nineteenth century, postulated a *single* leaf type (Blatt) for all the different lateral appendages of the stem such as foliage-leaves, sepals, carpals, bracts, and other flower organs. Goethe's "Blatt," was a concept useful in deriving all appendages from one type, but it carried no phylogenetic implications. In any case, it is not difficult to make analogies between Goethe's "Blatt" and Harper's "modular unit" which serves as the basic concept of all my essays in this book. Indeed, it may serve as a proper philosophical foundation for Harper's ecological theory mentioned earlier. The reader is directed to A. Arber's book, *The Natural Philosophy of Plant Form* (1950) for more details and history.

To Goethe, whom some have named father of morphology, the details were every bit as important as the general forms subscribed to. He was an observer par excellence in human anatomy. It was Goethe who discovered the intermaxillary bone in the human upper jaw, a homologous bone present in the primate jaw, but previously declared absent in humans. Today such observations need statistical proof. F.L. Bookstein (1978) has developed the requisite morphological statistics. In Chapter 8, F.L. Bookstein shows that the growth of the human skull still presents challenges to growth scientists.

The specific essays on development included in this book are focused on the one hand by the work of S. Kauffman et al., who published a model using a difficult diffusion-reaction system to derive the experimentally observed *compartmental boundary lines* in insect imaginal discs. From their perspective, these lines are analogues of patterns achieved in well-known chemical diffusion reaction systems (Kazarinoff, 1981). On the other hand, Goodwin and Goodwin/Trainor use *selection rules* in a *field theory* for the Laplacian operator on a 2-sphere to model pattern formation in metazoans. They have been able recently to use spherical harmonics to establish *anarchic fields* which produce a mouth apparatus or aperture at a hyperbolic point of the fields. In Chapter 7, N. Kazarinoff attempts to bridge the gap between these two approaches in his critical essay, while L.E.H. Trainor examines them more philosophically in Chapter 6 on the emergence of form in physics and biology.

The experimental evidence for the compartmental boundary lines in *Drosophila* is reviewed by developmental biologist M. Russell in Chapter 4. He also explains certain *mirror image symmetry* properties of the *Drosophila* mutant *Engrailed,* in terms of a sphere and three sets of latitude circles in the context of Wolpert's *Positional Information Theory.* Heretofore, French, Bryant, and Bryant had used polar coordinates in a circular disc to describe intercalation (healing) and the appearance of compartment boundary lines. Kazarinoff's article shows that the defect in the model of Kauffman et al. carries over to the diffusion-reaction on the 2-sphere. Thus, dynamical derivation of the compartmental lines is still an unsolved problem.

F.L. Bookstein argues in Chapter 8 that scalar and vector fields may not be enough and that tensor fields of stress and strain should be considered for their power of description of developmental phenomena. He has advanced the work of Goethe and D'Arcy Thompson on human cranical growth in this tensor field context. Yet, his methods based in classical projective and conformal geometry for the most part, yield a profound model of newt limb regeneration. He has developed a theory of cannonical singularities of bilinear maps of measurement quadrilaterals, which has a tensor field description. Heretofore, the intercalation rule in polar coordinates had been used to describe the experiments in which one grafts two limb stumps to the same regeneration site, resulting in emergence of a third limb with reversed polarity between the two. Bookstein's tensor field model does not use positional information. The mechanism is a bifurcation of polarity rather than intercalation.

References

Antonelli, P.L., and B.H. Voorhees. 1983. Nonlinear growth mechanics—I. Volterra-Hamilton systems. *Bull. Math. Biol.*, 45: 103-116.

Arber, A. 1950. *The Natural Philosophy of Plant Form.* Cambridge University Press.

Bookstein, F.L. 1978. *The Measurement of Biological Shape and Shape Change.* Lect. Notes in Biomath. no. 24. Springer-Verlag.

Goodwin, B. 1976. *The Analytical Physiology of Cells and Developing Organisms.* Academic Press.

Graham, R. 1973. Statistical theories of instabilities in stationary nonequilibrium systems with application to lasers and nonlinear optics. In G. Holer, ed., *Quantum Statistics in Optics and Solid State Physics.* Springer-Verlag.

Harper, J. 1977. *The Population Biology of Plants.* Academic Press.

Ikeda, N., and S. Watanabe. 1981. *Stochastic Differential Equations and Diffusion Processes.* North Holland.

Kazarinoff, N, B. Hassard, and Y.-H. Wan. 1981. *Theory and Applications of Hopf Bifurcation.* Cambridge University Press.

Nicolis, G., and I. Prigogine. 1977. *Self-Organization in Nonequilibrium Systems.* John Wiley and Sons.

Thom, R. 1975. *Structural Stability and Morphogenesis.* W.A. Benjamin.

Watanabe, S., and Y. Takahashi. 1980. The probability functionals (Onsager-Machlup functions) of diffusion processes. In D. Williams, ed., *Stochastic Integrals*, pp. 433-463. Proc. LMS, Durham Symposium.

Whittaker, R.H., and P.P. Feeny. 1971. Allelochemics chemical interactions between species. *Science*, 171: 757-771.

Rhoades, D.F. 1979. Evolution of plant chemical defense against herbivores. In Rosenthal and Janzen, eds., *Herbivores, Their Interaction with Secondary Plant Metabolites.* Academic Press.

MECHANICS

1
HAMILTONIAN STATISTICAL MECHANICS AND BIOLOGICAL ORDER: PROBLEMS AND PROGRESS

C.J. LUMSDEN AND L.E.H. TRAINOR

1.1 Introduction

As in the physical sciences, research activity in modern biology is separated into experiment and theory. Experimental biology, through observation in the field and the laboratory, seeks data on living organisms. Theoretical biology sifts this wealth of facts in search of deep conceptual schemes that provide testable explanations of biological phenomena. These general characteristics of experimental and theoretical biology have familiar parallels in the physical sciences.

The practice of theory in biology today enjoys a noticeable division of conceptual orientation. One major area addresses the elaboration of physical models for biological phenomena. In recent times it has been considerably enlivened by the applications of ideas from chemical physics and statistical thermodynamics made possible by remarkable advances in molecular biology. The practitioners of this area of theoretical biology are usually physical scientists with an interest in the physics of life, or life scientists from disciplines such as physiology, biochemistry, and molecular biology, which are closely allied with the physical sciences. By a close alliance to the physical sciences we mean that the theoretical models used in these disciplines characterize biological systems in terms of the observable properties pertinent to physical systems in general. For example, temperatures, pressures, concentrations of chemicals, reaction rates, field gradients, fluid-flow velocities, and molecular weights are important observables in physiology and biochemistry. We shall call this type of theoretical work in biology *theoretical biophysics* to emphasize its physicochemical orientation.

The second major area of theoretical endeavour is concerned with biological disciplines not yet directly connected to the physical sciences. The observable properties considered to be important are not necessarily the familiar physical properties of biological systems. The division of a cell, for example, is an intricate process at present unsuited to complete molecular description. At the same time it involves a succession of forms and functions describable in terms of larger-scale cell constituents, including the chromosomes, cytoskeleton, mitochondria, cell membrane, and the interactions among these objects, which the histologist can readily identify. In a related fashion, the beating heart involves complex molecular interactions among millions of cells, but it can nevertheless be precisely modelled in engineering terms as a physiological pump serving the circulatory system. The development of a multicellular organism from a single cell carries the complexity of cellular and molecular relationships still further. But at the level of gross anatomical description this complexity becomes the passage of the embryo through a crisp sequence of well-described formative stages. Similarly, animal populations interact in the terrestrial environment to form ecological systems in which the basic unit is the intact animal. The major features of demography and population biology have thereby been made susceptible to quantitative theory despite the cellular complexity of the basic units involved. We shall term this style of theoretical work, which stresses the integrated aspects of biological phenomena in terms of uniquely biological observables such as organ shape, gene frequencies, and behavioral predispositions *theoretical biology* in order to distinguish it from the theoretical biophysics orientation.

A great chain of hierarchical organization bridges theoretical biophysics to theoretical biology. It begins in the molecular domain with the synthesis of proteins and nucleic acids by the internal machinery of the cell. The newly synthesized polymers spontaneously adopt elaborate three-dimensional shapes and interact with other molecular components within the cell. New complexes are thereby formed that carry out microphysiological operations crucial to life, including encoding of genetic information, catalysis of specific chemical reactions, and transport of molecules across cell membranes.

From the level of macromolecules the hierarchy rises to the level represented by the intact cell, which constitutes the building block of tissue and organ. As spontaneous natural phenomena cells were apparently born of simple chemical reaction chains in primeval oceans and puddles. The chemical physics of that origin has been the object of theory and experiment for almost a century, but the problem remains unsettled (Eigen and Schuster, 1979; Rowe and Trainor, 1983a, 1983b). Although origins remain shrouded in controversy, structure and function in multicellular systems have been firmly established to be entities obeying thermodynamic laws in which nonequilibrium fluxes of matter, energy, and information maintain long-range macroscopic order. Current evidence suggests that many biochemical reactions occur far from thermodynamic equilibrium (Nicolis and Prigogine, 1977). The physical theory of such nonequilibrium order is now partially understood and has stimulated many new developments in thermodynamics and statistical mechanics (Glansdorff and Prigogine, 1971; Haken, 1977; Prigogine, 1980; Bennett, 1982). The coordinated responses of an organism to its environment and the reactions of organisms to each other in turn form the substance of population systems, the stage on which group behavior and population biology take place. As a result common frontiers linking thermodynamics, physiology, and evolutionary biology meet in the analysis of behavior and the large-scale events to which it gives rise.

Within the hierarchy of biological organization, a significant interrelation between levels can be discerned. From one viewpoint an organic system observed at any level consists of numerous interacting subsystems (for example, the molecular constituents of a biopolymer). From a complementary viewpoint the same organic system is part of a larger biological organization (for example, the cell of which the biopolymer is a subsystem). Each level of the hierarchy has characteristic spatiotemporal scales that are nested with respect to levels above and below (Table 1.1-I). The scales of one level typically exceed in magnitude those of levels below and are exceeded by those above.

Table 1.1-I Spacetime Scales of Biological Order

Organizational Level	*Temporal Scale* $log_{10}(sec)$	*Spatial Scale* $log_{10}(cm)$
Biochemical	-9 to 1	-8 to 1
Cellular	-3 to 7	-4 to 1
Organismic	2 to 7	-1 to 2
Demographic, Ecologic	6 to 9	2 to 6
Microevolutionary	7 to 8 and up	5 to 7
Speciation	10 to 12	6 to 7 and up
Macroevolutionary, Macroecologic	16 up	planetary boundary layers

Progress with understanding the truly macroscopic, integrative phenomena of biology such as morphogenesis and animal behavior in terms of more microscopic physical phenomena such as biochemical reactions has thus far been modest. Indeed, attempts to regard macroscopic biological phenomena as simply a massive problem in the molecular physiology of the component parts persistently founder. This problem is readily traced to the absence of any known set of general rules that specify biological properties directly in terms of the molecular processes which ultimately compose an organic system. It is not even clear as yet to what extent the problem of formulating these rules is solvable (Bremmerman, 1983; see also the chapter in the present volume by Goodwin and Trainor). Some progress with the first steps, from atom to macromolecule and biochemical network, seems to have been made (Perelson and Oster, 1974; Schnakenberg, 1976; Nicolis and Prigogine, 1977; Cantor and Schimmel, 1980; Oster et al., 1980).

Recently we have begun to formulate an alternate approach to the linking of organizational levels of biology (Lumsden and Trainor, 1979a, 1979b; Lumsden and Wilson, 1980, 1981; Lumsden, 1982) and thus of theoretical biophysics to theoretical biology as these terms have been defined above: given the existence of a natural hierarchy of organization, the

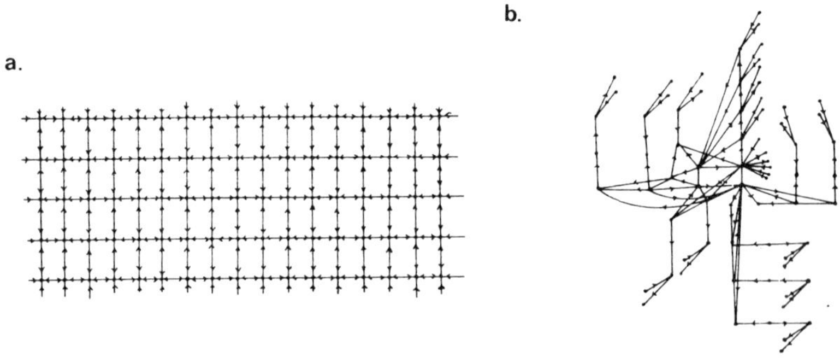

Figure 1.1-I. Subunit interaction patterns for typical systems of interest in physical and biological statistical mechanics.
a.) Physical statistical mechanics. Interactions exhibit an overall uniformity and reciprocity of organization. Interactions of the same general type apply throughout the system.
b.) Biological statistical mechanics. Interactions exhibit an overall nonuniformity and nonreciprocity of organization, depending upon the specific functional properties of the system as a whole. Interactions are highly diverse, with numerous different specific interaction types present in the model. Shown here is an interaction graph of the type used by Rashevsky (1960) to discuss the organization of physiologic and metabolic processes in cellular systems.

level of description is envisaged as being stepped up level by level instead of in a single step. Although it is not yet possible to bridge from atom to organism directly, some of the principal connections between organism and organs, organs and cells, and cells and molecules, have been worked out. These connections may therefore be fruitfully utilized. Ultimately the levels join not only biological description with biological description, but also biological description with microphysical description. The absence of microscopic interpretations of many biological properties is partially counterbalanced by the existence of connections between adjacent levels on organic hierarchy.

At almost any level of description, the subsystem composition of a living organism is complicated (e.g., Totafurno, Lumsden, and Trainor, 1980). There are many identifiable subsystems (each with its own intricate character) coupled together and to the environment in a highly specific fashion (Figure 1.1-I). Neither the properties of the various subsystems nor of their network of interactions are known in anything like complete detail. The environmental couplings are also never known with certainty. Moreover, a less familiar source of uncertainty is created by the natural variation in structure and function observable in evolving biological populations. These ubiquitous variations form the raw material upon which the process of evolution acts (Simpson, 1953; Dobzhansky, 1970).

The problem of synthesizing laws that govern a level of biological organization from the specification of subsystem properties and interactions shares several of the basic properties which motivate the use of statistical mechanics in physical theory: many subsystems are present, along with specific subsystem-subsystem and subsystem-environment couplings; lack of complete information about these couplings and other properties of the system persists; and above all there is a need to recover propositions or laws that express the properties of the assemblage of subsystems as a whole. Physical statistical mechanics is an attempt to reconstruct thermodynamic pictures of physical processes from molecule-level descriptions of the same system. Recent investigations have somewhat clarified the conditions sufficient for such a bridging construction (Zwanzig, 1960; Mori, 1965; Arnold and Avez, 1968; Haake, 1973; Kenkre, 1977; Prigogine, 1980). We have inquired whether similar approaches might now be extended to the problem of order in biological systems. From this viewpoint, one hopes to recover descriptions on a given level of biological organization from suitable operations on variables characterizing subsystems and their interactions. The macroscopic quantities so constructed in turn form microscopic quantities for the synthesis of yet further levels of organization. Resting as it does on the theories relevant to a hierarchy of increasingly macroscopic domains of order, this approach can be termed *layered statistical mechanics.*

In this chapter we summarize a dynamical basis for a layered statistical mechanics. The problem that we believe merits initial study is the structure of the theory as influenced by the dynamical equations used to model the biology. In the simplest cases known to theoretical biology, subsystem behavior is well-modelled by systems of first-order ordinary differential equations giving the time rates of change of the relevant subsystem properties (e.g., Lotka, 1956; Rashevsky, 1960; Rosen, 1972; Hirsch and Smale, 1974; Cooke, 1979). These are usually of the deterministic or Langevin-stochastic type. Stochastic tendencies can also be modelled in terms of Master equations or Fokker-Planck equations (e.g., Haken, 1977), but in this report we deal with the simpler cases represented by deterministic, first-order ordinary differential equations. In practice these equations often form a complicated, highly nonlinear set of dynamical laws that bear little resemblance to the Hamiltonian dynamics used in regular statistical mechanics. Furthermore, many of the dynamical models are not conservative. They are highly dissipative and may show abrupt transitions between regimes of macroscopic order. We shall argue that it is possible to enlarge statistical mechanics to encompass such models of organic processes while preserving and extending much of the formal power of the existing physical theory.

In the conventional theory of statistical thermodynamics the Hamiltonian mechanics achieves a unified description of the motion of the molecular constituents comprising the system. The Hamiltonian picture reveals a remarkable interplay among the generators of the dynamical motion, the symmetries of the system, and the quantities conserved during the system's time development. The unifying effect that the Hamiltonian formulation of dynamics exerts on the central concepts of statistical mechanics can be summarized in the following terms. In the absence of complete information about the system, probability distributions are used to describe its properties. The time development of a probability distribution (statistical ensemble) in the phase space of physical position and momentum variables is determined by the dynamical generator (the Hamiltonian, H) of the system through the Liouville equation. Steady-state or time-independent ensembles are identified via their algebraic commutativity with H, leading to the view of such ensembles as functionals of conserved quantities that are explicitly independent of time.

The conserved quantities commute with H and produce symmetry maps (canonical transforms) that leave the Hamiltonian dynamics invariant. Conversely, the generators of such maps qualify as conserved quantities. Systems in which the conserved quantities are rigorously constant move in regions of phase space consistent with the initial values of the conserved quantities. When these regions define a hypersurface of reduced dimensionality in the overall phase space the system trajectory is fully contained within the region of phase space defined by this phase shell. The space-filling properties of the trajectory on the shell then become the link through which ergodic theory seeks to unite the dynamics with the theory of equilibrium ensembles by equating the time averages incurred during empirical measurement with averages over probability distributions in phase space.

In this report we shall be concerned with systems that are interesting in the sense of statistical mechanics because they consist of many subsystems in interaction. However, modern developments in theory of dynamical systems (e.g., May, 1976; Ford, 1983) show that even simple systems in isolation can exhibit behavior analagous to the filling of phase shells by the trajectories of structurally complex, many-body systems. Such behavior can be likened to a dynamical chaos (Li and Yorke, 1975) in which a lack of predictability about future behavior is inherent in the system's dynamics for any nonzero error in the specification of initial conditions. It occurs, for example, in a system of two harmonic oscillators coupled by cubic terms in the position coordinates in Hamilton's equations if the energy of the system is sufficiently large. To what extent the considerations of statistical mechanics can be usefully applied is a matter of intense research activity at the present time (Ford, 1983). For both large and small physical systems treated with conventional statistical mechanics, the equilibrium ensembles are in turn assessed as attractors of initial ensemble clouds that move on the phase shells in accord with Hamilton's equations (Arnold and Avez, 1968; Sinai, 1973; Ruelle, 1978).

Thus Hamiltonian dynamics interconnects conserved quantities, symmetries, and dynamical generators in a manner that plays an important role in the interpretation of the relationship between molecular and thermodynamic behavior in physical statistical mechanics. If a canonical form could be achieved for biological models, the resultant Hamiltonian pictures would be useful in identifying and interpreting the linkages between levels of organization in biological systems.

1.2 Hamiltonian Dynamics for a Layered Statistical Mechanics

Linear Lagrangians

The question of a statistical mechanics for biological systems has received some consideration in recent years (Kerner, 1957; Goel, Maitra and Montroll, 1971; Conrad, 1972; Liboff, 1972; May, 1974; Iberall, 1975; Kometani and Shimizu, 1975; Lumsden and Trainor, 1979a, 1979b; Lumsden, 1982). The influential early work is that of Kerner

(1957, 1959, 1971, 1972), in which an attempt was made to erect a theory of Gibbs equilibrium ensembles on partially Hamiltonized versions of biologically interesting dynamical models. Applications of Kerner's idea have ranged across genetic and biochemical networks, morphogenesis, behavioral processes, demography, and ecology (Lotka, 1956; Kerner, 1957, 1959; Goodwin, 1963; Cowan, 1970; Rosen, 1970; Kerner, 1971, 1972; May, 1974). While this "bioensemble theory" is largely an *ad hoc* construct, the approach is intuitively appealing and can be said to have enjoyed some early successes (Goodwin, 1963; Leigh, Jr., 1969; Cowan, 1970; Kerner, 1971; Cowan, 1972). In our own work we have tried to develop a more comprehensive approach designed to replace *ad hoc* arguments with firmer foundations and expand the modelling effort into a layered statistical mechanics suited to both equilibrium and nonequilibrium processes.

The formulation of Lagrangian and Hamiltonian equations for the class of models characterized by rate laws of the form

$$\frac{dx_k}{dt} \equiv \dot{x}_k = X_k(\mathbf{x}, t), \quad k = 1,\ldots,N . \tag{1.2-1}$$

has become a subject of continuing research (Lotka, 1956; Kerner, 1957, 1959; Goodwin, 1963; Leigh, Jr., 1969; Rosen, 1970; Kerner, 1971, 1972; Goodwin, 1976; Santilli, 1978; Lumsden and Trainor, 1979a, 1979b; Lumsden, 1982; Paine, 1982). In these ordinary differential equations, the X_k are continuous and suitably smooth functions of the configuration space vector $\mathbf{x} = (x_1,\ldots,x_N)$ and possibly the time, t. Whereas physical systems are coordinatized by generalized coordinates and momenta that usually correspond to particle motions, biological configuration spaces may be coordinatized by such variables as organism population densities, rates of growth, gene frequencies, shape parameters, and so on, depending upon the organizational level embodied in the model.

If the Euler-Lagrange equations

$$\frac{d}{dt}\frac{\partial L}{\partial \dot{x}_k} - \frac{\partial L}{\partial x_k} = 0 \tag{1.2-2}$$

corresponding to the variational principle (fixed end points)

$$\delta \int_{t_1}^{t_2} L\, dt = 0 \tag{1.2-3}$$

for the Lagrangian $L(\mathbf{x}, \dot{\mathbf{x}}, t)$ are to directly reduce to the rate laws (1.2-1), then $\partial L/\partial \dot{x}_k$ must be independent of $\dot{x}_i$ for all i. The Lagrangian itself must therefore be at most linear in the rates:

$$L(\mathbf{x}, \dot{\mathbf{x}}, t) = U_k(\mathbf{x}, t)\dot{x}_k - U_o(\mathbf{x}, t) \quad \text{(summation convention)} \qquad (1.2\text{-}4)$$

The Euler-Lagrange equations (1.2-2) for the Lagrangian (1.2-4) can be written as

$$\Gamma_{km}\dot{x}_m = \Lambda_k \qquad (1.2\text{-}5a)$$

where the quantities Γ_{km} are given by

$$\Gamma_{km} = -\Gamma_{mk} = \frac{\partial U_k}{\partial x_m} - \frac{\partial U_m}{\partial x_k} \qquad (1.2\text{-}5b)$$

and where the quantity

$$\Lambda_k = -\frac{\partial U_k}{\partial t} - \frac{\partial U_o}{\partial x_k} \qquad (1.2\text{-}5c)$$

evidently plays the role of a generalized force for the dynamics. Given invertibility of the matrix $\|\Gamma_{km}\|$, equations (1.2-5a) can be transformed to read

$$\dot{x}_k = \Gamma^{-1}_{km}\Lambda_m \,. \qquad (1.2\text{-}6)$$

Comparison of equations (1.2-6) with the rate laws (1.2-1) then reveals that the existence problem for L can be reformulated as the existence of a set of functions $\{ U_o(\mathbf{x}, t), U_k(\mathbf{x}, t), k = 1, \ldots, N \}$ such that the relations

$$\Lambda_k = \Gamma_{km} X_m \qquad (1.2\text{-}7)$$

are identities over the domain of interest. By calculating the arithmetic difference between the quantities $\partial\Lambda_n/\partial x_k$ and $\partial\Lambda_k/\partial x_n$ we find that the Γ_{km} themselves obey the coupled linear partial differential equations

$$\frac{\partial \Gamma_{kn}}{\partial t} = \frac{\partial}{\partial x_k}(\Gamma_{nm}X_m) - \frac{\partial}{\partial x_n}(\Gamma_{km}X_m) \,. \qquad (1.2\text{-}8)$$

When the dynamics (1.2-1) is autonomous (i.e. the X_k are time independent) it is reasonable to search at first for linear Lagrangians that are also explicitly independent of time. This situation corresponds to equations of the type (1.2-8) with the left hand side set equal to zero. Furthermore, equations (1.2-6) for the $\dot{x}_k$ can then be expressed as

$$\dot{x}_k = -\Gamma^{-1}_{km} \frac{\partial U_o}{\partial x_m} \qquad (1.2\text{-}9)$$

from which we deduce that $U_o(\mathbf{x})$ is a constant of the motion:

$$\frac{dU_o}{dt} = \frac{\partial U_o}{\partial x_k} \dot{x}_k = \frac{\partial U_o}{\partial x_k} (-\Gamma^{-1}_{km}) \frac{\partial U_o}{\partial x_m} \equiv 0 \qquad (1.2\text{-}10)$$

by Γ_{km} antisymmetry. System dynamics is therefore constrained to hypersurfaces of constant U_o.

Kerner (1957, 1971, 1972) used the nonuniqueness of the solution set for (1.2-8) to impose the auxiliary condition

$$U_o = X_k U_k , \qquad (1.2\text{-}11)$$

thus making U_o an additive constant of the motion in analogy with classical mechanics. Given this stipulation equations (1.2-8) can be rewritten as

$$-\frac{\partial U_n}{\partial t} = X_n \frac{\partial U_n}{\partial x_k} + U_k \frac{\partial X_k}{\partial x_n}, \qquad (1.2\text{-}12)$$

a form for which local existence conditions are known (Kerner, 1971). At least in a local sense then, the condition (1.2-11) implies the existence of a linear Lagrangian form (1.2-2) with associated dynamics (1.2-9).

A result of major significance for a statistical mechanics is the additive-integral character of $U_o(\mathbf{x})$. Unfortunately, in the manner of many existence theorems the proof does not take the form of a constructive procedure for generating the U-functions. In many studies, an insight-directed rearrangement of the dynamical system (1.2-1) to the form (1.2-9) is usually attempted, rather than a solution of equations (1.2-12) or (1.2-8) starting from first principles. $U_o(\mathbf{x})$ is thereby identified *a priori* and the $U_k(\mathbf{x})$ can be read off if the condition (1.2-11) is satisfied. Recently, solution procedures based on different conditions than (1.2-11) have been investigated and shown to be a practical means for constructing $U_o(\mathbf{x})$ in low-dimensional systems (Trubatch and Franco, 1974).

The Hamiltonization Problem

The immediate problem in moving from the linear Lagrangian to a canonical form for the rate law models (1.2-1) is at once evident in equation (1.2-4). The usual procedure for constructing a Hamiltonian form,

$$H = p_k \dot{x}_k - L, \quad p_k = \frac{\partial L}{\partial \dot{x}_k}, \tag{1.2-13}$$

leads to the identification $H = U_o(\mathbf{x})$. The canonical equations of motion in the phase space of independent coordinates and momenta $(\mathbf{x}, \mathbf{p})$ are then automatically constrained to

$$\dot{x}_k = \frac{\partial U_o}{\partial p_k} = 0, \quad k = 1, \ldots, N. \tag{1.2-14}$$

The question therefore arises of whether a pure Hamiltonian dynamics that regenerates the full class of solutions to equations (1.2-1) can be constructed from (1.2-2) through (1.2-12) at all.

In the following we show that a dynamics can indeed be constructed by adapting a theory of constrained systems due to Dirac. The Hamiltonian obtained is not, however, the usual physical Hamiltonian for the energy of the system. This generalized Hamiltonian is composed of organic subsystem variables such as chemical concentrations or gene frequencies and, in analogy with the regular physical Hamiltonian, generates the one parameter group of time translations for the dynamics of the biological system.

Singular Lagrangians and Generalized Hamiltonization

Central to the theory of generalized Hamiltonization (Dirac, 1950, 1964; Sudarshan and Mukunda 1974) is the concept of the *singular Lagrangian,* which is a Lagrangian $L(\mathbf{x}, \dot{\mathbf{x}}, t)$ for which the N^2-dimensional matrix of functions

$$\|D_{kn}\| = \|D_{nk}\| = \frac{\partial^2 L}{\partial \dot{x}_k \partial \dot{x}_n} \tag{1.2-15}$$

is singular and therefore noninvertible:

$$\det |D_{kn}| = 0. \tag{1.2-16}$$

If L is singular, then the familiar expression

$$p_j = \frac{\partial L}{\partial \dot{x}_j}, \quad j = 1, \ldots, N \tag{1.2-17}$$

cannot be rearranged to give the $\dot{x}_j$ in terms of the x_j, p_j, and (when applicable) the time, t. The regular route from Lagrangian to Hamiltonian breaks down.

To see this, suppose that $\| D_{kn} \|$ is singular but that equations (1.2-17) *can* be transformed to express the x_k, p_k-dependence of the $\dot{x}_k$. The $\dot{x}_k$ are then the variables in the $(x_k, \dot{x}_k)$-representation that carry the p_k dependence. Differentiation of both sides of equations (1.2-17) with respect to a specific p_k then implies that

$$\frac{\partial p_j}{\partial p_k} = \delta_{jk} = \frac{\partial^2 L}{\partial \dot{x}_n \partial \dot{x}_j} \frac{\partial \dot{x}_n}{\partial p_k} = D_{jn} \frac{\partial \dot{x}_n}{\partial p_k}, \tag{1.2-18}$$

so that the Dirac matrix $\| D_{jn} \|$ *does* have an inverse (given by $\| \partial \dot{x}_n / \partial p_k \|$), contrary to assumption. The Dirac matrix is the Jacobian of the map from the $(x_k, \dot{x}_k)$ into the (x_k, p_k) coordinates, and is singular when the Lagrangian is singular.

Suppose that the Dirac matrix $\| D_{kn} \|$ has some nonmaximal rank $R < N$. As we shall see shortly, for a linear Lagrangian R is in fact zero. Then R of the N velocities $\dot{x}_k$ can be solved for and expressed as functions of the x_k and p_k. For the remainder, the independent unsolved velocities, the definitions (1.2-17) generate a further $(N - R)$ independent relations

$$\phi_\rho(\mathbf{x}, \mathbf{p}) = 0, \quad \rho = R + 1, ..., N \tag{1.2-19}$$

among the p_k and x_k. These $(N - R)$ relations are the *primary constraints* of Dirac's generalized Hamiltonian procedure. We shall be particularly interested in the case where equations (1.2-19) can be rearranged to express the constrained momenta as $p_{R+1}, ..., p_N$ in terms of the configuration coordinates x_k and the independent momenta $p_1, ..., p_R$:

$$p_\rho = F_\rho(x_k, p_j), \tag{1.2-20}$$

where hereafter $k = 1, ..., N; j = 1, ..., R$; and $\rho = R + 1, ..., N$ unless otherwise specified. The primary constraints can then be written as

$$\phi_\rho(\mathbf{x}, \mathbf{p}) = p_\rho - F_\rho(x_k, p_j) = 0 \tag{1.2-21}$$

expressing the fact that in the actual dynamics the phase point is constrained to move on an $N + R$ dimensional surface in 2N dimensional phase space.

If $\dot{x}_\alpha$ refers to the R velocities that can be solved for from equation (1.2-17) and $\dot{x}_\omega$ refers to the rest, then the subscript α must take on R values out of 1, ...,N while ω takes on the remainder. The solved velocities $\dot{x}_\alpha$ will in general be functions of the 2N independent coordinates consisting of the x_k, the p_j and the $\dot{x}_\omega$:

$$\dot{x}_\alpha = G(x_k, p_j, \dot{x}_\omega). \tag{1.2-22}$$

Whereas in the regular Hamiltonization the Legendre transformation

$$p_k \dot{x}_k - L(\mathbf{x}, \dot{\mathbf{x}}) \tag{1.2-23}$$

is rewritten entirely in terms of the 2N independent (x_k, p_k) coordinates, the Dirac theory seeks its representation first in the $(x_k, p_j, \dot{x}_\omega)$ coordinate system:

$$p_k \dot{x}_k - L(\mathbf{x}, \dot{\mathbf{x}}) \equiv H_o(x_k, p_j, x_\omega). \tag{1.2-24}$$

The partial derivatives of this object are

$$\frac{\partial H_o}{\partial x_k} = \dot{x}_\rho \frac{\partial F_\rho}{\partial x_k} - \frac{\partial L}{\partial x_k}, \quad k = 1, \ldots, N \tag{1.2-25a}$$

$$\frac{\partial H_o}{\partial p_j} = \dot{x}_j + \dot{x}_\rho \frac{\partial F_\rho}{\partial p_j}, \quad j = 1, \ldots, R \tag{1.2-25b}$$

$$\frac{\partial H_o}{\partial \dot{x}_\omega} = 0. \tag{1.2-25c}$$

Equation (1.2-25c) asserts the independence of H_o from the unsolved velocities $\dot{x}_\omega$. The two remaining systems of relations can now be rearranged to express the equations of motion for the x_j and p_k :

$$\dot{x}_j = \frac{\partial H_o}{\partial p_j} - \dot{x}_\rho \frac{\partial F_\rho}{\partial p_j}, \quad j = 1, \ldots, R \tag{1.2-26a}$$

$$\dot{p}_k = -\frac{\partial H_o}{\partial x_k} + \dot{x}_\rho \frac{\partial F_\rho}{\partial x_k}, \quad k = 1, \ldots, N. \tag{1.2-26b}$$

Note that the $\dot{x}_j, \dot{p}_k$ are linearly dependent upon the unsolved velocities $\dot{x}_\rho$. Equations (1.2-26) constitute the *generalized Hamilton equations* for the phase coordinates x_j and p_k (Dirac, 1950, 1964).

We can see that in the generalized Hamiltonian theory one begins in a 2N dimensional phase space of independent coordinates x_k, p_k. The actual motion is confined to a hypersurface of lower dimensionality defined by the primary constraints $\phi_\rho(\mathbf{x}, \mathbf{p}) = 0$, $\rho = R + 1, ..., N$ and embedded in the 2N dimensional phase space. Dirac introduced the idea of *weak* and *strong* equations in order to compare the behavior of two or more phase functions on this hypersurface and away from it. Let M be the constraint hypersurface and suppose that $f(\mathbf{x}, \mathbf{p})$ and $g(\mathbf{x}, \mathbf{p})$ are two dynamical variables defined in at least a finite neighborhood of M. The values of $f(\mathbf{x}, \mathbf{p})$ and $g(\mathbf{x}, \mathbf{p})$ on M are obtained by replacing their p_ρ variables by the constraints $F_\rho(x_k, p_j)$, where as before $\rho = R + 1, ..., N$; $k = 1, ..., N$; $j = 1, ..., R$. If $f(\mathbf{x}, \mathbf{p})$ and $g(\mathbf{x}, \mathbf{p})$ become equal after the map $p_\rho \rightarrow F_\rho$, then they are equal *on* M and are said to be *weakly equal.* Following Dirac (1964), this relation is usually written as

$$f \approx g \, . \tag{1.2-27}$$

Equality that automatically holds without attention to the constraints (i.e., not just on M but in a finite neighborhood of the hypersurface) is Dirac's *strong equality* (Mukunda, 1976).

Of course M itself can now be defined in terms of weak equations. In terms of the primary constraint function the structure of M is delineated by

$$\phi_\rho \approx 0 \, , \quad \rho = R + 1, ..., N \, . \tag{1.2-28}$$

We shall see that the weak equality relation plays a significant role in the formal structure of the generalized canonical theory for layered statistical mechanics.

If one introduces a modified "total Hamiltonian" (Dirac, 1964; Mukunda, 1976) with explicit dependence upon the unsolved velocities x_ρ,

$$H(x_k, p_j, \dot{x}_\rho) = H_o(x_k, p_j) + \phi_\rho(\mathbf{x}, \mathbf{p}) \, \dot{x}_\rho \, , \tag{1.2-29}$$

then

$$H \approx H_o \tag{1.2-30}$$

as a consequence of the constraint equations, and the actual dynamics on M cannot distinguish between H and H_o. However, the total Hamiltonian has the advantage that, in terms of a Poisson bracket (PB) operation $[\, \bullet \, , \bullet \,]$ defined in the familiar manner over the entire phase space, all 2N equations of motion can be formally expressed for $k = 1, ..., N$ as

$$\dot{x}_k \approx [x_k, H] \, , \quad \dot{p}_k \approx [p_k, H] \, . \tag{1.2-31}$$

In fact for a suitably extended PB (see below) we obtain the strong conditions (Mukunda, 1976) that for $k = 1, \ldots, N$

$$\dot{x}_k = [x_k, H], \quad \dot{p}_k = [p_k, H] \qquad (1.2\text{-}32)$$

For $k = \rho = R + 1, \ldots, N$ equations (1.2-31) and (1.2-32) are respectively the identities $\dot{x}_\rho \approx \dot{x}_\rho$ and $\dot{x}_\rho = \dot{x}_\rho$. Hence there are again at most $N + R$ independent degrees of freedom.

The construction procedure that leads to the PB dynamics (1.2-31) and (1.2-32) can be summarized in the following way. In terms of phase space functions $f(\mathbf{x}, \mathbf{p})$ and $g(\mathbf{x}, \mathbf{p})$ the PB has the usual formal structure

$$[f, g] = \frac{\partial f}{\partial x_k} \frac{\partial g}{\partial p_k} - \frac{\partial f}{\partial p_k} \frac{\partial g}{\partial x_k}. \qquad (1.2\text{-}33)$$

We now however adopt the convention that *all* differential operations with respect to the x_k and p_k are to be performed before the constraints are imposed. Extended phase functions, such as the total Hamiltonian, which depend explicitly upon the unsolved velocities $\dot{x}_\rho$, have been handled in one of two ways (Dirac, 1950, 1964; Mukunda, 1976).

Dirac accepted the presence of the $\dot{x}_\rho$ but considered its PB with any other quantity as undefined in terms of the differential structure of (1.2-33). The formal PB properties of bilinearity, antisymmetry, product rule, and Jacobi identity were retained. Because $\dot{x}_\rho$ PB's such as $[f, \dot{x}_\rho]$ appear in the Dirac theory in the product form $[f, \dot{x}_\rho]\,\phi_\rho$, they always vanish weakly and therefore have no effect on the M dynamics. Using this convention we obtain the canonical dynamics (1.2-31).

Mukunda (1976) noted that an M-equivalent result obtains if the terms containing $\dot{x}_\rho \phi_\rho$ are treated as untouched by the PB operations in the manner of scalar constants. Then instead of relations such as

$$[f, \phi_\rho \dot{x}_\rho] = [f, \phi_\rho]\,\dot{x}_\rho + [f, \dot{x}_\rho]\,\phi_\rho \approx [f, \phi_\rho]\,\dot{x}_\rho \qquad (1.2\text{-}34a)$$

one obtains

$$[f, \phi_\rho \dot{x}_\rho] = [f, \phi_\rho]\,\dot{x}_\rho. \qquad (1.2\text{-}34b)$$

This convention yields the dynamics (1.2-32). Both conventions lead to identical results relative to the motion on the constraint hypersurface M. Because of its notational convenience, we adopt the Mukunda convention in this chapter. The equivalent expressions in the Dirac convention (1.2-34a) are obtained merely by replacing $=$ with $\approx$.

A dynamical variable $A(\mathbf{x}, \mathbf{p}, t)$ in the constrained system obeys the *Dirac-Poisson* equation of motion

$$\frac{dA}{dt} = [A, H] + \frac{\partial A}{\partial t} . \tag{1.2-35}$$

Thus any quantity $A(\mathbf{x}, \mathbf{p})$ that does not explicitly depend upon the time is a constant of the motion if $[A, H] = 0$. In the space of coordinates x_k, coordinate transformations under which the singular Lagrangian is either invariant or suffers an infinitesimal change that is expressible as a total time derivative correspond to canonical maps in the $(\mathbf{x}, \mathbf{p})$-phase space (Mukunda, 1976). Each such canonical map has a unique generator G that is a conserved quantity in the generalized Hamiltonian dynamics, and has the PB $[G, H] = 0$.

A Hamiltonian Form for the Linear Lagrangian

Let us first consider linear Lagrangians L that lack explicit time dependence. A completely useful statistical mechanics should treat Lagrangian systems in which the time dependence is either implicit or explicit, and in the next section our treatment is extended to include such cases.

From the form of the Lagrangian in (1.2-4) it is at once apparent that

$$D_{kn} \equiv \frac{\partial^2 L}{\partial \dot{x}_k \partial \dot{x}_n} \equiv 0 \tag{1.2-36}$$

so that the Dirac matrix $\|D_{kn}\|$ has rank zero. As a consequence all momenta are of the dependent-type p_ρ, $\rho = 1, ..., N$, while the class of independent momenta p_α is empty. The dependent momenta are given by the *velocity-independent* expressions

$$p_\rho = \frac{\partial L}{\partial \dot{x}_\rho} = U_\rho(\mathbf{x}) \tag{1.2-37}$$

and the primary constraint functions have the form

$$\phi_\rho = p_\rho - F_\rho(\mathbf{x}) = p_\rho - U_\rho(\mathbf{x}) , \quad \rho = 1, ..., N . \tag{1.2-38}$$

The Legendre transform for the Hamiltonian becomes

$$H_o = p_k \dot{x}_k - L = U_o(\mathbf{x}) \,, \tag{1.2-39}$$

while from the generalized Hamiltonian equations (1.2-26b) we have that

$$\dot{p}_k = -\frac{\partial U_o}{\partial x_k} + \dot{x}_\rho \frac{\partial U_\rho}{\partial x_k} \,, \quad k = 1, ..., N \tag{1.2-40}$$

(the set (1.2-26a) is, of course, at most the identities $\dot{x}_\rho = \dot{x}_\rho$, $\rho = 1, ..., N$). But by definition (1.2-17) and equation (1.2-37), $p_k = U_k(\mathbf{x})$, $k = 1, ..., N$ on M, so that

$$\dot{p}_k = \frac{\partial U_k}{\partial x_n} \dot{x}_n \,. \tag{1.2-41}$$

Equating the right hand sides of (1.2-40) and (1.2-41) yields

$$\dot{x}_n = -\Gamma^{-1}_{nk} \frac{\partial U_o}{\partial x_k} \tag{1.2-42}$$

so that equations (1.2-5a) are recovered for the autonomous case where the $U_k(\mathbf{x})$ functions are explicitly independent of time.

The total Hamiltonian is

$$H = H_o + \dot{x}_\rho \phi_\rho = U_o(\mathbf{x}) + (p_\rho - U_\rho(\mathbf{x}))\dot{x}_\rho \,. \tag{1.2-43}$$

Hence, from equation (1.2-32) (or, equivalently (1.2-31))

$$\dot{p}_k = [p_k, H] = -\frac{\partial U_o}{\partial x_k} + \dot{x}_\rho \frac{\partial U_\rho}{\partial x_k} \tag{1.2-44}$$

Equating the right hand side with (1.2-41) then yields the equations (1.2-42) as before.

From the Dirac-Poisson dynamics (1.2-35) with $A = H$ we have immediately

$$\dot{H} = 0 \tag{1.2-45}$$

The Dirac total Hamiltonian is a constant of the constrained motion. If $A(\mathbf{x})$ is any function of the biological coordinates x_k, then for H corresponding to the linear Lagrangian the equation of motion for the

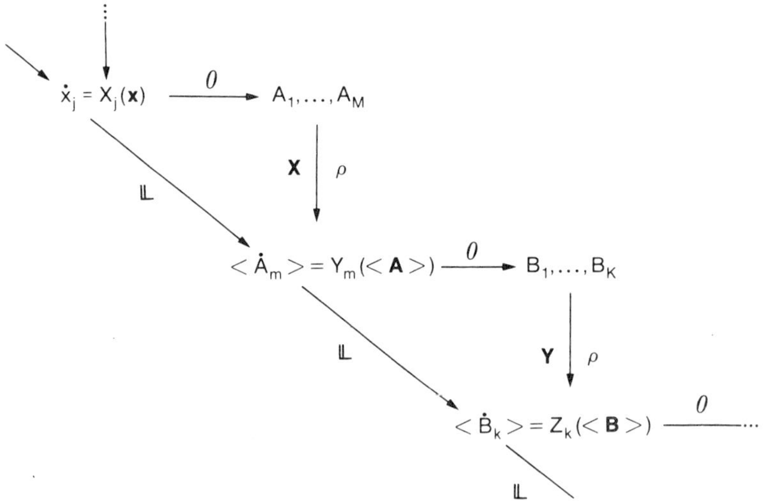

Figure 1.3-II. Layered statistical mechanics is a procedure for synthesizing hierarchically organized levels of description. Let the variables appropriate to one particular level be x_j, $j = 1, \ldots, N$, such that $\dot{\mathbf{x}} = X(\mathbf{x})$. The system variables for the behavior of the set $\{x_i\}$ as a whole are $A_m(x)$, $m = 1, \ldots, M$. $\mathcal{O}$ represents an algorithm for producing whole-system variables A_m. As explained in the text, the flow field **X** and the statistical ensemble ρ **(x, t)** induce a mapping 𝕃 to the more macroscopic level of description given by the expectation values $\langle A_m \rangle(t)$. The transition to $\langle A_m \rangle(t)$ is interesting if these macrovariables in turn obey a closed or almost-closed set of equations of motion. From the $\langle A_m \rangle$-level, the procedure can be repeated to seek a description in which the $\langle A_m \rangle$ comprise the microscopic variables sustaining a still more macroscopic level of organization $\langle B_k \rangle$, and so on. The interlevel transitions are best understood in the case of Hamiltonian dynamics, stimulating the search for Hamiltonian formulations of the rate laws on each level.

observable is

$$\dot{A} = [A, U_0(x) + (p_\rho - U_\rho(x))\dot{x}_\rho] . \qquad (1.2\text{-}46)$$

Equation (1.2-46) will be instrumental in our development of a canonical basis for the statistical mechanics.

1.3 Generalized Liouville Equations for Layered Statistical Mechanics

Ensemble Equations of Motion

This section introduces the problem of constructing a layered statistical and its extension to a generalized Poisson bracket place statistical ensembles for different levels of description in one unified framework. The Hamiltonian structure of the dynamics clarifies several important points. First, the relation is obtained between the generators of the ensemble motion and the Hamiltonian derived in the previous section. Second, the algebraic structure of this relation provides a concise geometric interpretation of the ensemble dynamics. Third, the stationary solutions of our generalized Liouville equations express the conditions for equilibrium ensembles in a way that naturally emphasizes the significance of conserved quantities and dissipations. A new representation of these equilibrium ensembles is obtained in terms of their information-entropy maximizing properties. Finally, the Liouville equations suggest useful parallels between hierarchical description in biology and macroscopic descriptions of physical many-body systems.

In a statistical mechanics, nonequilibrium phenomena arise in circumstances best represented by the free development of initial conditions forward in time (perhaps toward asymptotic states), from explicit time dependence in the microscopic equations of motion, or from a combination of these effects. External driving from the environment is the most familiar source of explicit time dependence. However, it is important to emphasize that completely autonomous but dissipative rate laws of the form (1.2-1) can also generate Lagrangians and Hamiltonians which depend explicitly upon the time. For a two-dimensional dynamical system one can in fact demonstrate a close connection between the dissipation expressed by the compressibility of the flow $\dot{\mathbf{x}} = \mathbf{X}$ and the explicit time dependence of the linear Lagrangian.

The fundamental step in a layered statistical mechanics is to connect two levels of organization. If the linkage procedure is sufficiently general, repeated application will chain together successive levels of organization (Figure 1.3-II). Given a dynamics in the form of well-behaved ordinary differential equations

$$\dot{x}_k = X_k(\mathbf{x}, t), \quad k = 1, ..., N \tag{1.3-1}$$

describing subsystems, their interactions, and couplings to the environment, one seeks a related set of equations expressing the behavior of the subsystem assemblage as a whole. If $A_m(\mathbf{x})$, $m = 1, ..., M \ll N$ are system properties expressed in terms of the subsystem (= microscopic) variables and $\rho(\mathbf{x}, t)$ is the statistical ensemble density expressing the likelihood that the system as a whole is in microstate $\mathbf{x}$ at time t, then the theoretical constructs representing the system-level of organization are the expectation values

$$\langle A_m \rangle(t) = \int d\mathbf{x}\, A_m(\mathbf{x}) \rho(\mathbf{x}, t) . \tag{1.3-2}$$

If, for example, it can be shown that

$$\frac{d\langle A_m \rangle}{dt}(t) = F_m(\langle A_1 \rangle, ..., \langle A_m \rangle) , \tag{1.3-3}$$

where we have in general that

$$\frac{d\langle A_m \rangle}{dt}(t) = \int d\mathbf{x}\, A_m(\mathbf{x})\, \partial_t \rho(\mathbf{x}, t), \quad \partial_t \equiv \frac{\partial}{\partial t} \tag{1.3-4}$$

then the subsystem interactions sustain a domain of whole-system behavior described by its own closed set of autonomous dynamical laws. The identification of dynamics (1.3-1) that give rise to such remarkable forms of collective behavior as (1.3-3) is one of the central problems to be addressed by a layered statistical mechanics.

Equation (1.3-4) indicates that the rate of change $\partial_t \rho(\mathbf{x}, t)$ is the link between dynamics on the $\mathbf{x}$-level (1.3-1) and organization on the A-level (1.3-3). The continuity equation for the motion of the ensemble ρ relates $\partial_t \rho(\mathbf{x}, t)$ to equations (1.3-1):

$$\partial_t \rho + \frac{\partial}{\partial x_k}(\rho \dot{x}_k) = 0 , \tag{1.3-5}$$

where we have taken $\rho = \rho(\mathbf{x}, t)$ to be a normalized ensemble density (Kurth, 1960; Landau and Lifshitz, 1969). We are therefore considering cases in which all relevant interactions within the system and between the system and the environment are included in the smooth differential changes of the dynamics (1.3-1). Under these circumstances an ensemble treatment is necessary when there is missing information or uncertainty about initial conditions, limiting predictions to likelihoods that the system will be found in given regions of its phase space. When complete information is available the ensemble density $\rho(\mathbf{x}, t)$ is equivalent to the Dirac delta function $\boldsymbol{\delta}(\mathbf{x}, t)$, and the motion of the ensemble cloud collapses to the path of a single trajectory $\mathbf{x}(t)$.

We should emphasize that the standard approach to constructing ensembles for complex physical systems has a very different viewpoint in equilibrium or near-equilibrium theory. Ensemble theories for equilibrium conditions, whether of the Gibbs canonical or grand canonical variety, consider macroscopic systems embedded in a very large external heat bath (canonical systems) or heat and particle baths (grand canonical systems). In such cases the dynamical behavior of the system in question is repeatedly interrupted in a nonpredictable fashion by interactions between the system and the surrounding bath, which distribute the system in a probabilistic manner over its accessible microscopic states. The connection between ensembles expressing missing information about initial conditions and the canonical ensembles is established by the approach of the nonequilibrium ensembles $\rho(\mathbf{x},t)$ to the canonical forms at $t \to \infty$ in systems that are relaxing toward thermodynamic equilibrium.

Unfortunately, comparatively little is known about maps of the form

$$\int d\mathbf{x} A_m(\mathbf{x})\, \partial_t \rho(\mathbf{x},t) \rightarrow F_m(\langle \mathbf{A} \rangle) \tag{1.3-6}$$

for nonequilibrium ensembles $\rho(\mathbf{x}, t)$ given only that $\mathbf{x}$ follows rate laws of the form (1.3-1). In contrast the transition is relatively well understood when $\mathbf{x}$ obeys a Hamiltonian mechanics and is already the subject of extensive literature (e.g., Zwanzig, 1960; Mori, 1965; Ballescu and Wallenborn, 1971; Haake, 1973; Kenkre, 1977; Prigogine, 1980; Grabert, 1982). This circumstance greatly increases the interest and relevance of Hamiltonian formulations for a layered statistical mechanics.

Limitations of Canonical Ensemble Theories

Given the impressive efficacy of equilibrium statistical mechanics in physical theory (e.g., Landau and Lifshitz, 1969) one is lead to inquire why a direct transcription of canonical ensembles into the appropriate biological terms is not the answer to a layered statistical mechanics. Although applications of canonical or Gibbs ensembles in theoretical biology have produced useful insights (e.g., the reference cited in section 1.2), the methodology does not appear to us to constitute a satisfactory means for explaining multilevel order in organic systems. To clarify our argument let us briefly summarize the elements of the standard equilibrium theory as it is developed in physics:

1. Since Lagrange's equations are second order in time, integration involves two constants for each degree of freedom, fixed by initial conditions or their equivalent. Thus, a system of N particles possesses 6N independent functions of the coordinates and momenta that remain independent of the time.
2. For a small but macroscopic system (e.g., a typical experimental sample) connected to a large heat bath at equilibrium, the state of the system fluctuates as a result of heat exchanges with the bath. One presumes the existence of distribution function ρ that gives the relative probability for finding the system in any particular microscopic state.
3. Actual measurements on the system involve macroscopic variables. Such measurements cannot be taken at an instant of time but require microscopically long time averages. The system meanwhile has fluctuated through an enormous number of microscopic states, presumably distributed according to ρ. It is as though one made instantaneous measurements on a large ensemble of systems distributed instantaneously in their microscopic states according to ρ. Thus time averages are replaced by ensemble averages weighted according to ρ.
4. Predictions about the system thus depend entirely on a knowledge of ρ and the problem reduces (statistically) to a determination of ρ.
5. Crucial to this determination is the assumption of equal *a priori* probabilities for equal volumes of phase space (the space of all the coordinates and momenta). In quantum statistical mechanics this translates into equal *a priori* probabilities of quantum states.
6. For equilibrium ensembles, the distribution function is stationary so that

$$\frac{d\rho}{dt} = 0$$

   Hence, ρ must be expressible in terms of constants of the motion.
7. From the nature of independent probabilities, if the sample size is increased by putting samples I and II together, the overall distribution function ρ is given by

$$\rho = \rho_I \cdot \rho_{II}$$

8. Since ρ is multiplicative, $\ln \rho$ is additive ("extensive") and must be

expressible linearly in terms of additive constants of the motion only. These are determined by space-time invariances of the system. Under the usual experimental conditions only the energy (Hamiltonian) of the system survives.

9. Thus $\ell n\ \rho \propto H$ and $\rho \propto e^{\lambda H}$, the classic form of the Gibbs canonical equilibrium ensemble. The constant λ is easily identified as $-\beta = 1/k_B T$ (for example, by using an ideal gas sample) where k_B is Boltzmann's constant and T is the temperature and depends, of course, only on properties of the heat bath.

To typify the problems posed by biological implementations of this protocol consider a network of interacting species and the analog of a Gibbs ensemble for this ecological system. For convenience, we make reference to points 1 to 9. Point 1 involves the Hamiltonization of the equations of motion. We shall present evidence in the remainder of this report that this part of the problem can be considered resolved. Point 2, however, causes great difficulties. What constitutes the equivalent of the particle bath with a fixed temperature, which causes the population system to fluctuate through a very large number of states (population states) in any "macroscopic" time interval? To give benefit-of-the-doubt, let us suppose that the environment in which the species is located has some dominating quality (temperature analog) which forces the population system through its various states in some appropriate time scale so that a distribution function can be established. Points 4 and 5 then present fundamental problems. In ordinary mechanics, the equal *a priori* probability of equal volumes of phase space is not so much a matter of proof as it is of empirical experience. Much more familiarity would be necessary in the biological examples to assert a similar principle holds there. As we have pointed out previously (Lumsden and Trainor, 1979a, 1979b), time-dependent Lagrangians that arise naturally in dissipative systems lead to ensemble compressibility and to preferred states in the ensemble.

Points 6 and 7 would seem to present no essential problems in the analogy. If the analog of a heat bath is valid (in some as yet unspecified context) then it is reasonable to talk of equilibrium distributions and statistical independence of macroscopic samples. Point 8, however, presents a troubling difficulty. For the nonlinear dissipative dynamics suited to modeling biological systems the whole question of constants of the motion is largely unexplored. No argument available to us exists to state that the additive constants are derivable from symmetries of the system, nor is there a substantial argument to show that of all additive constants only a Hamiltonian analog survives.

Naturally, the extension to nonequilibrium systems depends crucially on a solid foundation for equilibrium systems, if one is to learn from the experience of equilibrium statistical mechanics in the physical context.

We feel that in the biological case, where the meaning of ensemble equilibrium has yet to be properly understood, some other starting point, such as the fully time-dependent problem posed by the continuity equation (1.3-5), will be necessary to resolve all of the outstanding difficulties. Moreover, a Hamiltonian formulation of the ensemble dynamics (1.3-5) is the natural basis from which to search for higher-level kinetic descriptions of the form (1.3-3). With Gibbs ensembles one is automatically restricted to static, unchanging properties that are the analogues of thermodynamic equilibrium quantities in physical systems.

Dirac Picture for the Ensemble Motion

Our first step is to construct a generalized Hamiltonian dynamics for $\rho(t)$ in a form that returns the fundamental equation of motion (1.3-5) via the Euler-Lagrange equations. In this subsection and the next we discuss the situation associated with linear Lagrangians that possess strictly implicit time dependence. The consequences of explicit time dependence are summarized later.

The time rate of change of a dynamical variable $A(\mathbf{x}, \mathbf{p}, t)$ is related to the Dirac total Hamiltonian H by the expression $\dot{A}(\mathbf{x}, \mathbf{p}, t) = [A, H] + (\partial A)/(\partial t)$ from which we obtained the basic identities $\dot{x}_k = [x_k, H]$, $k = 1, \ldots, N$. We can then express the $\rho(t)$-dynamics (1.3-5) in terms of the commutator between ρ and H :

$$\partial_t \rho = -[\rho, H] + \rho D(\mathbf{x}) \qquad (1.3\text{-}7)$$

where $-D(\mathbf{x})$ is formally the compressibility

$$\partial[x_k, H] / \partial x_k \qquad (1.3\text{-}8)$$

of the rate field $X_k(\mathbf{x})$, $k = 1, \ldots, N$. If the flow is everywhere-incompressible, then $D(\mathbf{x}) \equiv 0$ and the temporal behavior of the ensemble density is given by

$$\partial_t \rho = -[\rho, H], \qquad (1.3\text{-}9)$$

which is the Dirac-Poisson extension of the classical Liouville equation.

Practical analysis of $\rho(\mathbf{x}, t)$ using the Dirac theory is of course somewhat complicated by the constraints ϕ. Their presence imposes additional formal structure which is absent in the Hamiltonization of nonsingular Lagrangian systems. We have been able to reduce this formal complexity by imposing the constraint conditions and considering the statistical mechanics as embedded in the Dirac constraint hypersurface. Equivalently, for the dynamics of the **p**-independent density $\rho(\mathbf{x},t)$, the ensemble dynamics is restricted to the projection of the constraint hypersurface onto **x**-space. We find that much of the canonical structure of the present dynamical theory is preserved during the projection process. Furthermore, the auxiliary constraint conditions are satisfied and need not be carried along separately from the dynamical equations. One cannot escape all formal complexity, however. In this case it arises directly from the constrained dynamics, which is canonical in a generalized sense: the Lie bracket of the equations of motion is not the Poisson bracket but a more general object. Nevertheless, the description has many properties analogous to the Poisson canonical dynamics, and it turns out to be directly related to a *bona fide* Poisson canonical dynamics in **x**-space, as well as to the Dirac canonical dynamics in $(\mathbf{x}, \mathbf{p})$-space (Lumsden and Trainor, 1980).

Statistical Mechanics in the Dirac Hypersurface

In the previous section we saw that imposition of the conditions

$$\phi_\rho \approx 0\,, \quad \rho = 1, \ldots, N$$

on the constraints ϕ_ρ led to the determining equations

$$\dot{x}_\rho \approx -\Gamma^{-1}_{\rho\nu} \frac{\partial U_0}{\partial x_\nu}(\mathbf{x}) \qquad (1.3\text{-}10)$$

for the unsolved velocities $\dot{x}_\rho$, $\rho = 1, \ldots, N$. The form of equations (1.3-10) is intriguingly reminiscent of the ordinary Hamiltonian dynamics

$$\dot{x}_k = [x_k, H] = \boldsymbol{\delta}_{kn} \frac{\partial H}{\partial p_n}$$

$$\dot{p}_k = [p_k, H] = -\boldsymbol{\delta}_{kn} \frac{\partial H}{\partial x_n}$$

This observation suggests that the action of $\|\boldsymbol{\Gamma}_{\rho\nu}\|$ might induce some type of generalized bracket action on the constraint hypersurface M. Equivalently, since $\rho(\mathbf{x}, t)$ is independent of the momenta p_j, $\|\boldsymbol{\Gamma}_{\rho\nu}\|$

might create a natural but generalized bracket action on ρ in the **x**-space of biological coordinates itself. We have found that this is indeed the case, and the the time evolution of the system in these subspaces can be succinctly treated by means of the theory of generalized Poisson brackets (GPB) (Lee, 1943; Martin, 1959; Mukunda and Sundarshan, 1968).

In GPB theory one defines a skew-symmetric metric tensor

$$\eta^{jk}(\mathbf{x}) = -\eta^{kj}(\mathbf{x}) \tag{1.3-11}$$

on the **x**-space of the Dirac hypersurface such that $\eta^{jk}(\mathbf{x})$ satisfies the partial differential equations

$$\eta^{jk}\frac{\partial\eta^{\ell n}}{\partial x_k} + \eta^{\ell k}\frac{\partial\eta^{nj}}{\partial x_k} + \eta^{nk}\frac{\partial\eta^{j\ell}}{\partial x_k} \equiv 0. \tag{1.3-12}$$

The GPB of any two functions $f(\mathbf{x})$, $g(\mathbf{x})$ is then defined to be

$$[\![f, g]\!] \equiv \eta^{jk}\frac{\partial f}{\partial x_j}\frac{\partial g}{\partial x_k} = -[\![g, f]\!]. \tag{1.3-13}$$

Equations (1.3-12) guarantee the Jacobi identity

$$\sum_{\substack{f,g,h \\ \text{cyclic}}} [\![[\![f, g]\!], h]\!] = 0 \tag{1.3-14}$$

for the GPB.

A transformation $\phi : \mathbf{x} \rightarrow \mathbf{x}'$ is called a generalized canonical map if it preserves the GPB:

$$\phi : \{[\![f, g]\!] = h\} \rightarrow [\![f', g']\!] = h' \tag{1.3-15a}$$

$$f'(\mathbf{x}') = f(\mathbf{x}), \quad g'(\mathbf{x}') = g(\mathbf{x}) \tag{1.3-15b}$$

$$\begin{aligned} h'(x') &= \eta^{jk}(\mathbf{x}')\frac{\partial f'}{\partial x_j'}(\mathbf{x}')\frac{\partial g'}{\partial x_k'}(\mathbf{x}') = h(\mathbf{x}) \\ &= \eta^{jk}(\mathbf{x})\frac{\partial f}{\partial x_j}(\mathbf{x})\frac{\partial g}{\partial x_k}(\mathbf{x}) \end{aligned} \tag{1.3-15c}$$

$$\eta^{jk}(\mathbf{x})\frac{\partial x_n'}{\partial x_j}\frac{\partial x_m'}{\partial x_k} = \eta^{nm}(\mathbf{x}'). \tag{1.3-15d}$$

This map is said to be regular if the new variables map into the same ranges as the old variables, say $0 \leqslant x_k \leqslant m_k$, $k = 1, \ldots, N$ where m_k is the maximum value assumed by x_k.

Let $K(\mathbf{x}, \lambda)$ be a one-parameter function on **x**-space. Then in analogy to the standard canonical dynamics the solutions of differential equations of the form

$$\frac{dx_k}{d\lambda} = [\![x_k, K(\mathbf{x},\lambda)]\!] \qquad (1.3\text{-}16)$$

are in fact generalized canonical maps of **x**-space into itself (Sudarshan and Mukunda, 1974). $K(\mathbf{x}, \lambda)$ can therefore be legitimately referred to as the generator of the maps, where the λ-dependence can be implicit or explicit.

Now it is clear from its PB structure that the Dirac-Poisson bracket using the Dirac total Hamiltonian is a particular case of GPB. Under its action the generalized canonical map moves the nonequilibrium ensemble in $(\mathbf{x}, \mathbf{p})$- phase space and maps the constraint hypersurface into itself. However, a remarkable GPB structure is also latent in the intrinsic geometry of the Dirac hypersurface. The crucial step towards a constraint-free theory in this context is the recognition that the Γ_{km} functions (or equivalently the primary constraint PB's $[\phi_k, \phi_m])$ directly determine *bona fide* metric functions of the GPB η-type.

Let $A_j(\mathbf{x})$, $j = 1, \ldots, N$ be a set of suitably smooth functions defining a vector field on **x**-space. Then the antisymmetric tensor

$$\sigma_{jk} = \frac{\partial A_k}{\partial x_j} - \frac{\partial A_j}{\partial x_k} = -\sigma_{kj}$$

$$\equiv \partial_j A_k - \partial_k A_j$$

corresponding to the curl of this vector field obeys the equations

$$\partial_\ell \sigma_{jk} + \partial_j \sigma_{k\ell} + \partial_k \sigma_{\ell j} = 0 ,$$

which is tantamount to satisfying the Jacobi identity (1.3-14) for the inverse tensor $\sigma^{j\ell}$.

The implication for linear Lagrangians is immediate. The U_k-functions lay down a smooth vector field on **x**-space, whereby the elements of $\| -\Gamma_{jk} \|^{-1} \equiv \| \Gamma^{jk} \|$ define a natural GPB structure

$$[\![f, g]\!] = \Gamma^{jk} \frac{\partial f}{\partial x_j} \frac{\partial f}{\partial x_k} . \tag{1.3-17}$$

With the choice $f = x_m$, $g = U_o(\mathbf{x})$ corresponding to a linear Lagrangian explicitly independent of time we have

$$[\![f, g]\!] = \Gamma^{jk} \frac{\partial x_m}{\partial x_j} \frac{\partial U_o}{\partial x_k}(\mathbf{x}) = \Gamma^{mk} \frac{\partial U_o}{\partial x_k}(\mathbf{x}) .$$

Hence, the dynamics (1.3-1) assumes the form

$$\dot{x}_m = [\![x_m, U_o]\!] , \quad m = 1, \ldots, N . \tag{1.3-18}$$

Equations (1.3-18) imply that $U_o(\mathbf{x})$ is the generator of generalized canonical maps which take the biological coordinate space, regarded as a subspace of the Dirac hypersurface, into itself.

Passing through the Dirac to the GPB dynamical description, we replace the configuration velocity terms $\dot{x}_k$ in the ensemble dynamics

$$\partial_t \rho + \dot{x}_k \frac{\partial \rho}{\partial x_k} = 0$$

with the GPB expression $[\![x_k, U_o]\!]$, giving

$$\partial_t \rho = - [\![\rho, U_o]\!] \tag{1.3-19}$$

for incompressible flow in **x**-space. For compressible flow, equal *a priori* probabilities are no longer conserved and equation (1.3-19) must be modified to

$$\partial_t \rho = - [\![\rho , U_o]\!] + \rho D(\mathbf{x}) \tag{1.3-20}$$

where

$$D(\mathbf{x}) = - \partial [\![x_k, U_o]\!] / \partial x_k .$$

Stationary Ensembles

In a theory of ensemble dynamics, one of the most interesting questions concerns the asymptotic properties of the ensembles as $t \to \infty$.

Attention generally focuses on the existence of stationary ensembles and the extent to which time-dependent ensembles approach the stationary solutions as $t \to \infty$. If a stationary attractor $\rho_0(\mathbf{x})$ exists for $\rho(\mathbf{x}, t)$ then the macroscopic properties $\langle A_m \rangle(t)$ approach steady state values $\langle A_m \rangle_0$ and the system can be said to possess self-regulating or homeostatic tendencies on the **A**-level of macroscopic order. Provided that questions regarding the structure of initial ensembles can be settled, such analyses complement arguments from ergodic theory in justifying specific forms for the equilibrium ensembles. As a first application of the Hamiltonian theory we examine the question of stationary ensembles.

In his early papers, Kerner (1957, 1971) suggested an approach based on Gibbs ensembles. If the dynamics (1.3-1) is conservative and admits an isolating first integral, then the corresponding conserved quantity $U_0(\mathbf{x})$ is a good macrovariable A for the system. Ensembles on the **x**-space shells corresponding to U_0 = constant give rise to statistical predictions about the system's behavior. A Gibbs ensemble $\rho \propto e^{-\lambda U_0}$ was posited as an intuitively attractive form for this ensemble, but was suggested on purely an *ad hoc* basis and has retained this status during extensive applications. It is therefore of considerable importance to inquire whether the Gibbs form is a stationary solution of the appropriate Dirac-Liouville equation.

From the Liouville equation (1.3-7) it is apparent that for a general biodynamics the stationary ensemble condition is

$$-[\rho, H] + \rho D(\mathbf{x}) \approx 0 \qquad (1.3\text{-}21)$$

The compressibility term, when nonzero, can produce a condensation of $\rho(\mathbf{x}, t)$ into selected x-space regions. The problems that lead to the Gibbs ensemble discussed by Kerner and later workers correspond to the case of a conservative, incompressible dynamics. The condition (1.3-21) then specializes to

$$-[\rho, H] \approx 0 \qquad (1.3\text{-}22)$$

with $[H, H] \approx 0$. The constant of the motion $H \approx U_0$ appears to be isolating in the conservative models studied to date (Kerner, 1957; Goodwin, 1963; Cowan, 1970; Kerner, 1971).

Equal initial *a priori* probabilities are motivated in the symplectic geometries of the Dirac and the GPB dynamics (Hobson, 1971). The ensemble of optimal information entropy is then (Jaynes, 1963; Hobson, 1971) proportional to $e^{-\lambda H}$ where λ is a constant analogous to the inverse temperature. From equation (1.3-22) we see that such an ensemble

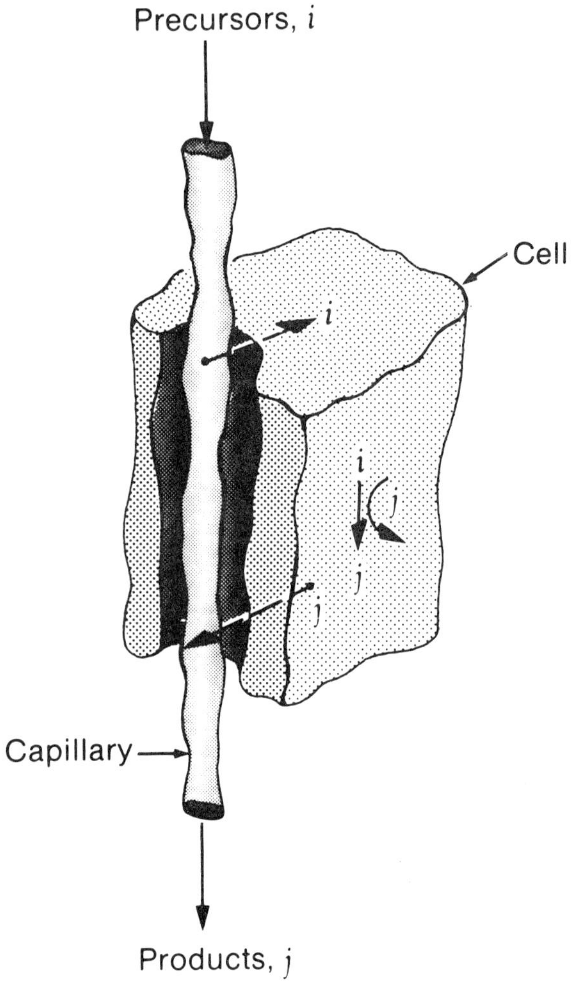

Figure 1.3-III. Physical arrangement of the capillary and cell for the transport/metabolism example discussed in the text. The microcirculatory flow delivered in the capillary supplies precursor molecules i which are autocatalytically converted within the cell into products j. Exchange between the vascular compartment and the intracellular space is mediated by membrance transport mechanisms. Both precursors i and products j act as intracellular messengers modulating steps in membrane transport and in the biosynthesis of the product molecules. It can be shown that this kinetics of exchange and interconversion sustains a concisely describable level of whole-cell behavior.

is stationary, and from the properties of the Dirac weak equalities that (up to normalization factors)

$$\rho \approx e^{-\lambda H} \approx e^{-\lambda U_0} . \qquad (1.3\text{-}23)$$

The form $e^{-\lambda U_0}$ is the form conjectured by Kerner. We are therefore able to report that the "Gibbsian bioensemble" is an information-optimal, stationary solution of the Dirac-Liouville equation in a one-integral, conservative biodynamics. The same conclusion follows from an analysis based on the GPB Liouville equation (1.3-18). In future work it will therefore be of interest to study the Gibbsian bioensemble as an asymptotic attractor of initial ensembles. Some progress in this direction has been achieved, and is the topic of a separate report (Lumsden and Trainor, 1979b).

An Example: Transport-Metabolism Behavior in a Cell

In order to view these general developments in more concrete terms, let us consider a simple model of the process in which a cell utilizes precursor materials to synthesize product molecules such as hormones or metabolites, which are then passed into the microcirculation for transport to target sites within the organism. The anatomical arrangement of the system is illustrated in Figure 1.3-III. Blood flow carries the precursors to the vicinity of cell, where they move from a nearby capillary into the extravascular space and are taken across the plasma membrane of the cell by a transport process. The precursors act as intracellular messengers in the transport mechanism, stimulating their own uptake: if $N_i(t)$ is the number of precursor molecules of type i within the cell at time t, then during the interval [t, t+dt] a total of

$$dN_i = \varepsilon_i N_i(t)dt, \quad \varepsilon_i > 0 \qquad (1.3\text{-}24)$$

additional molecules of this type are extracted from the extracellular pool by the membrane transport machinery and injected into the cytoplasm. Conversely, if $N_j(t)$ is the number of product molecules of type j within the cell at time t then during the same time interval a total of

$$dN_j = -\varepsilon_j N_j(t)dt, \quad \varepsilon_j < 0 \qquad (1.3\text{-}25)$$

diffuse out of the cell and into the capillary, where they are convected away from the synthesis site by the flowing blood.

Within the cell, biochemical reactions transform the precursors into product molecules. For this example we consider steps in biochemical synthesis that are autocatalytic in the sense that the conversion of precursors i to products j will not occur unless a product molecule is also active in the process. In this way we model the action of molecules that function as messengers to switch on their own synthesis. In the simplest case one product molecule and one percursor must be present in order for the transformation of the percursor to occur, and the supply of catalysts is not limiting. If σ_{ji} is the total cross section for binary encounters between percursor i and product j within the cell and ω_{ji} is the fraction that occur in the catalytic sites of the $i \rightarrow j$ enzyme, then during $[t, t+dt]$ N_i decreases by the amount $\omega_{ji}\,\sigma_{ji}\,N_i N_j dt$ while N_j increases by $(\beta_i/\beta_j)\,[\,\omega_{ji}\,\sigma_{ji}\,N_i N_j dt\,]$, where β_i/β_j models the increase in j per precursor i consumed. Factors $\beta_i/\beta_j < 1$ allow for wastage effects due to errors in the transformation process, which result in molecules i being converted to some other combination of products than j. If α_{ij} is defined to be the product $\beta_i\,\sigma_{ji}\,\omega_{ji}$, then the intracellular changes in i and j are

$$dN_j = \frac{\alpha_{ij}}{\beta_j} N_j N_i dt \tag{1.3-26a}$$

$$dN_i = -\frac{\alpha_{ij}}{\beta_i} N_j N_i dt = \frac{\alpha_{ji}}{\beta_j} N_i N_j dt \tag{1.3-26b}$$

where $\alpha_{ij} = -\alpha_{ji}$ and $\alpha_{ii} = 0$. Combining the membrane transport terms (1.3-24, 1.3-25) with the cytoplasmic reaction terms (1.3-26a, 1.3-26b) and summing over the possible end products of each precursor we see that each molecular species N_k follows a kinetic equation of the Lotka-Volterra form

$$\dot{N}_k = \varepsilon_k N_k + \beta_k^{-1}\,\alpha_{nk}\,N_k N_n\,, \quad \text{(summation convention)} \tag{1.3-27}$$

where $\varepsilon_k > 0$ if k is a precursor and $\varepsilon_k < 0$ if k is a product molecule.

Let N_k^o denote the steady-state solutions of equation (1.3-27), which require that $\|\,\alpha_{nk}\,\|$ be nonsingular, i.e., that the total number of interacting molecular species is even. Then in the scaled coordinates

$$v_k \equiv \ln\,(N_k / N_k^o) \tag{1.3-28}$$

the kinetics (1.3-27) can be written as the incompressible flow (div $\dot{\mathbf{v}} = 0$)

$$\dot{v}_k = \frac{\alpha_{nk}}{\beta_k} q_n(e^{v_n} - 1) = [v_k, H] = [\![v_k, U_o]\!] . \qquad (1.3\text{-}29)$$

Following the procedure outlined in section 1.2, we readily find that H is the Dirac total Hamiltonian

$$H = U_o(\mathbf{v}) + \dot{v}_k \phi_k \qquad (1.3\text{-}30)$$

with the primary constraints $\phi_k = p_k - U_k(\mathbf{v})$ vanishing for the allowed motion. The $U_k(\mathbf{v})$ functions for the Lotka-Volterra dynamics (1.3-27) are linear combinations (Kerner, 1971)

$$U_k(\mathbf{v}) = ½ \Lambda_{ik} v_i , \qquad (1.3\text{-}31)$$

where the matrix $\| \Lambda_{ij} \|$ is the inverse $\| \alpha_{ij} / \beta_i \beta_j \|^{-1}$. The generalized Poisson bracket $[\![v_k, U_o]\!]$ is

$$\dot{v}_k = \Gamma^{kn} \frac{\partial U_o}{\partial v_n} , \quad \Gamma^{kn} \equiv \frac{\alpha_{nk}}{\beta_k \beta_n} \qquad (1.3\text{-}32)$$

and $U_o(\mathbf{v})$ is the conserved additive function

$$N^o_\ell \beta_\ell (\exp(v_\ell) - v_\ell) , \quad \dot{U}_o \equiv 0 . \qquad (1.3\text{-}33)$$

The linear Lagrangian for the transport-metabolism network is the sum of the terms involving the $U_k(\mathbf{v})$ and $U_o(\mathbf{v})$:

$$L = U_k(\mathbf{v}) \dot{v}_k - U_o(\mathbf{v}) , \qquad (1.3\text{-}34)$$

where the $U_k(\mathbf{v})$ are given by equation (1.3-31) and U_o by (1.3-33). Since the flow $\dot{\mathbf{v}}$ is incompressible, the Liouville equations for the time development of the statistical ensemble $\rho(\mathbf{v}, t)$ have the form

$$\partial_t \rho = - [\rho , H] \qquad (1.3\text{-}35)$$

and

$$\partial_t \rho = - [\![\rho, U_o]\!] \tag{1.3-36}$$

in the $(\mathbf{v}, \mathbf{p})$-phase space and in the v-subspace of the Dirac hypersurface respectively.

Applying the results of the preceding subsection we observe that the Gibbs ensemble

$$\rho \propto e^{\lambda U_o} \tag{1.3-37}$$

is the stationary, information-entropy optimal solution to the Liouville equations. Since $U_o(\mathbf{v})$ is simple-additive inthe molecular species ℓ, ρ can be expressed as a product of one-species densities ρ_ℓ:

$$\rho \propto \prod_\ell \rho_\ell \tag{1.3-38}$$

where up to normalization constants the ρ_ℓ are given by

$$\rho_\ell \equiv \rho(v_\ell) = \exp[\lambda N_\ell^o \beta_\ell (e^{v_\ell} - v_\ell)] \tag{1.3-39}$$

in the $\mathbf{v}$-coordinates. In the molecular abundance coordinates $n_\ell = N_\ell / N_\ell^o$ the $\rho(v_\ell)$ are gamma distributions of the form

$$\rho(n_\ell) = n_\ell^{\gamma - 1} e^{-\gamma n_\ell}, \quad \gamma = N_\ell^o \beta_\ell \lambda. \tag{1.3-40}$$

The exponential factor λ in equation (1.3-37) is analagous to the inverse temperature factor $(k_B T)^{-1}$, k_B = Boltzmann's constant, in a physical system. It is therefore a macroscopic parameter characteristic of the behavior of the transport-metabolism network as a whole. Just as the cytoplasm has a physical temperature, T, the network (1.3-27) has a "transport/metabolism temperature" $T^* \propto \lambda^{-1}$ that expresses the mean value of U_o driving the reaction system (eqn. (1.3-32)). Temperature-analogues for kinetics of the Lotka-Volterra type have previously been considered in some detail, and a number of important relationships between them and the fluctuation behavior and steady-state properties of the kinetics (1.3-27) have been reported (e.g., Kerner, 1972).

Extensions to Time-Dependent Lagrangians

Both the Dirac and the GPB methods can be extended to cover situations involving Lagrangians which depend explicitly on time (Lumsden and Trainor, 1979a):

$$L = L(\mathbf{x}, \dot{\mathbf{x}}, t) = U_j(\mathbf{x}, t)\dot{x}_j - U_o(\mathbf{x}, t) \tag{1.3-41}$$

In applications to problems of biological order, time-dependent Lagrangians are encountered in models subjected to external driving forces and in models with internal dissipation (self-regulation, homeostasis).

The GPB can be generalized through the time-dependent functions

$$\Gamma_{km}(\mathbf{x}, t) = \frac{\partial U_k}{\partial x_m}(t) - \frac{\partial U_m}{\partial x_k}(t) \tag{1.3-42}$$

leading to an extended bracket

$$⦃ f, g ⦄ = \Gamma^{km}(\mathbf{x}, t) \frac{\partial f}{\partial x_k} \frac{\partial g}{\partial x_m} \tag{1.3-43}$$

where f and g can have an explicit time dependence and the Jacobi identity is satisfied at every instant t. Since the Euler-Lagrange equations now contain terms due to the explicit time dependence, the complexity of the $\dot{x}_k$ equations is somewhat increased:

$$\dot{x}_k = ⦃ x_k, U_o ⦄ + \Gamma^{km} \partial_t U_m \tag{1.3-44}$$

and the generalized Liouville equation contains these new terms as well:

$$\partial_t \rho = - ⦃ \rho, U_o ⦄ + \Gamma^{m\ell} \frac{\partial \rho}{\partial x_\ell} \partial_t U_m + \rho D(\mathbf{x}, t) \tag{1.3-45}$$

The Dirac-Poisson bracket is, in contrast, undisturbed by an explicit time dependence, and for a time-dependent Dirac total Hamiltonian the Liouville equation again has the simple canonical structure

$$\partial_t \rho \approx - [\rho, H(t)] + \rho D(\mathbf{x}, t) . \tag{1.3-46}$$

$D(\mathbf{x}, t)$ vanishes for incompressible flow. The constraints $\phi_\rho(\mathbf{x}, \mathbf{p}, t)$ are now time dependent, so that as the dynamics unfolds on the constraint hypersurface, the constraint hypersurface is itself unfolding within the $(\mathbf{x}, \mathbf{p})$-phase space.

1.4 Discussion

Our work has shown that systems of rate laws $\dot{x}_k = X_k(\mathbf{x}, t)$ can often be cast into a Hamiltonian form useful for developing a layered statistical mechanics. For example, let us suppose the system is an ecology in which the x_j are population abundances depending on time variable t. Conceptually, one can think of the Hamiltonian phase space as a space of N population variables and N canonical momenta involving the rates of change of the population variables. However, the system is then described by a point in phase space whose motion is restricted to a well-defined N-dimensional hypersurface in a manner that reduces the second order dynamics to a first order dynamics. From this basis we have been able to develop equations of motion for the ensembles $\rho(\mathbf{x}, t)$ and begin to study the dynamics of macrovariables $\langle A_m \rangle(t)$ for various levels of biological order (c.f. section 1.3; also Lumsden and Trainor, 1979a, 1979b; Lumsden and Wilson, 1980, 1981; Lumsden, 1982).

Notwithstanding the difficulties with applications in theoretical biology of the ordinary Gibbs ensemble, there remain important and challenging problems in the layered statistical mechanics of biological systems. To continue our example from population biology, it is at least suggestive that a large number of interacting species embedded in some larger and more substantial ecosystem (heat bath) could have an average viability level ("temperature") about which fluctuations take place, or a temperature-like property associated with the informational properties of the underlying ensemble. Is it possible for such a system to be described by an innovative statistical mechanics, not necessarily the analog of Gibbs theory, but one that would provide deep insights into the properties of hierarchical organization and emergence? If so what is this theory and how can one proceed to its construction? This is the primary challenge. Substantial progress to meet it has been made, particularly with regard to casting dynamics in Hamiltonian form and to formulating the generalized statistical development in Liouvillean terms; but many central problems remain and the subject provides an exciting field for exploration and discovery.

Acknowledgements

We would like to thank Elizabeth Coote for her careful typing of the manuscript. The work discussed here was supported by the Natural Sciences and Engineering Research Council of Canada.

References

Altenberger, A.R. 1972. On the theory of generalized master and kinetic equations. *Physica,* 62: 379-392.

Arnold, V.I., and A. Avez 1968. *Ergodic Problems of Classical Mechanics.* Benjamin, New York.

Balescu, R., and J. Wallenborn. 1971. On the structure of the time-evolution process in many-body systems. *Physica,* 54: 477-503.

Bennett, C.H. 1982. The thermodynamics of computation—a review. *International Journal of Theoretical Physics,* 21: 905-940.

Bremmerman, H. 1983. Parasites at the origin of life. *Journal of Mathematical Biology,* 16: 165-180.

Cantor, C.R., and P.R. Schimmel. 1980. *Biophysical Chemistry: Part I, the Conformation of Biological Macromolecules.* W.H. Freeman and Company, San Francisco.

Conrad, M. 1972. Statistical and hierarchical aspects of biological organization. In C.H. Waddington, ed., *Towards a Theoretical Biology,* Vol. 1, pp. 189-228. Aldine Publishing, Chicago.

Cooke, K.L. 1979. Mathematical approaches to culture change. In C. Renfrew and K.L. Cooke, eds., *Transformations: Mathematical Approaches to Culture Change,* pp. 45-81. Academic Press, New York.

Cowan, J.D. 1970. A statistical mechanics of nervous activity. In *Some Mathematical Questions in Biology: Lectures on Mathematics in the Life Sciences,* Vol. 2, pp. 3-57. American Mathematical Society, Providence, RI.

Cowan, J.D. 1972. Stochastic models of neuroelectric activity. In S.A. Rice, K.F. Freed, and J.C. Light, eds., *Statistical Mechanics: New Concepts, New Problems, New Applications,* pp. 109-121. University of Chicago Press, Chicago.

Dirac, P.A.M. 1950. Generalized Hamiltonian dynamics. *Canadian Journal of Mathematics,* 2: 129-148.

Dirac, P.A.M. 1964. *Lectures in Quantum Mechanics.* Yeshiva University, New York.

Dobzhansky, T. 1970. *Genetics of the Evolutionary Process.* Columbia University Press, New York.

Eigen, M., and P. Schuster. 1979. *The Hypercycle: A Principle of Natural Self-organization.* Springer-Verlag, New York.

Ford, J. 1983. How random is a coin toss? *Physics Today,* 36: 40-47.

Glansdorff, P., and I. Prigogine. 1971. *Thermodynamic Theory of Structure, Stability and Fluctuations.* John Wiley and Sons, Toronto.

Goel, N.S., S.C. Maitra, and E.W. Montroll. 1971. On the Volterra and other nonlinear models of interacting populations. *Reviews of Modern Physics,* 43: 231-276.

Goodwin, B.C. 1963. *Temporal Organization in Cells.* Academic Press, New York.

Goodwin, B.C. 1976. *Analytical Physiology of Cells and Developing Organisms.* Academic Press, New York.

Grabert, H. 1982. *Projection Operator Techniques in Nonequilibrium Statistical Mechanics.* Springer-Verlag, New York.

Haake, F. 1973. Statistical treatment of open systems by generalized master equations. *Springer Tracts in Modern Physics,* 66: 98-168.

Haken, H. 1977. *Synergetics: an Introduction. Nonequilibrium Phase Transitions and Self-Organization in Physics, Chemistry and Biology.* Springer-Verlag, New York.

Hirsch, M.W., and S. Smale. 1974. *Differential Equations, Dynamical Systems, and Linear Algebra.* Academic Press, New York.

Hobson, A. 1971. *Concepts in Statistical Mechanics.* Gordon and Breach, New York.

Iberall, A.S. 1975. On nature, man and society: a basis for scientific modelling. *Annals of Biomedical Engineering,* 3: 344-385.

Jaynes, E.T. 1963. Information theory and statistical mechanics. In G. Uhlenbeck et al., eds., *Statistical Physics 3: Brandeis 1962,* pp. 181-218. Benjamin, New York.

Kenkre, V.M. 1977. The generalized master equation and its applications. In U. Landman, ed., *Statistical Mechanics and Statistical Methods in Theory and Applications,* pp. 441-461. Plenum Press, New York.

Kerner, E.H. 1957. A statistical mechanics of interacting biological species. *Bulletin of Mathematical Biophysics,* 19: 121-146.

Kerner, E.H. 1959. Further considerations on the statistical mechanics of biological associations. *Bulletin of Mathematical Biophysics,* 21: 217-255.

Kerner, E.H. 1971. Statistical-mechanical theories in biology. In I. Prigogine and S.A. Rice, eds., *Advances in Chemical Physics,* Vol. 19, pp. 325-352. John Wiley and Sons, New York.

Kerner, E.H. 1972. *Gibbs Ensemble: Biological Ensemble.* Gordon and Breach, New York.

Kurth, B. 1960. *Axiomatics of Classical Statistical Mechanics.* Pergamon Press, New York.

Landau, L.D., and E.M. Lifshitz. 1969. *Statistical Physics,* Second Edition. Addison-Wesley, Toronto.

Lee, H.-C. 1943. A kind of even-dimensional differential geometry and its application to exterior calculus. *Journal of Mathematics,* 65: 433-438.

Leigh, Jr., E.G. 1969. The ecological role of Volterra's equations. In *Some Mathematical Problems in Biology: Lectures on Mathematics in the Life Sciences,* Vol. 1, pp. 1-61. American Mathematical Society, Providence, R.I.

Li, T.-Y., and J.A. Yorke. 1975. Period three implies chaos. *American Mathematical Monthly,* 82: 985-992.

Liboff, R.L. 1972. Physical laws and the structure of society. *The Cornell Engineer,* 36: 3-15.

Lotka, A.J. 1956. *Elements of Mathematical Biology.* Dover Publications, Toronto.

Lumsden, C.J. 1982. The social regulation of physical caste: the superorganism revived. *Journal of Theoretical Biology,* 59: 749-781.

Lumsden, C.J., and L.E.H. Trainor. 1979a. On the statistical mechanics of constrained biophysical systems. *Journal of Statistical Physics,* 20: 657-669.

Lumsden, C.J., and L.E.H. Trainor. 1979b. Nonequilibrium ensembles of self-organizing systems. *Canadian Journal of Physics,* 57: 23-38.

Lumsden, C.J., and L.E.H. Trainor. 1980. On the Hamiltonian content of biodynamics. *Bulletin of Mathematical Biology,* 42: 327-337.

Lumsden, C.J., and E.O. Wilson. 1980. Translation of epigenetic rules of individual behavior into ethnographic patterns. *Proceedings of the National Academy of Sciences (U.S.A.),* 77: 4382-4386.

Lumsden, C.J., and E.O. Wilson. 1981. *Genes, Mind, and Culture: the Coevolutionary Process.* Harvard University Press, Cambridge, MA.

Martin, J.L. 1959. Generalized classical dynamics, and the 'classical analogue' of a Fermi oscillator. *Proceedings of the Royal Society (London),* Series A, 251: 536-542.

May, R.M. 1974. *Stability and Complexity in Model Ecosystems,* Second Edition. Princeton University Press, Princeton, NJ.

May, R.M. 1976. Simple mathematical models with very complicated dynamics. *Nature,* 261: 459-467.

Mori, H. 1965. Transport, collective motion, and Brownian motion. *Progress of Theoretical Physics,* 33: 423-455.

Mukunda, N. 1976. Symmetries and constraints in generalized Hamiltonian dynamics. *Annals of Physics,* 99: 408-433.

Mukunda, N. and E.C.G. Sudarshan. 1968. Structure of the Dirac bracket in classical mechanics. *Journal of Mathematical Physics,* 9: 411-417.

Oster, G.F., G. Odell, and P. Alberch. 1980. Mechanics, morphogenesis and evolution. In G. Oster, ed., *Lectures on Mathematics in the Life Sciences,* pp. 165-255. American Mathematical Society, Providence, RI.

Paine, G.H. 1982. The development of Lagrangians for biological models. *Bulletin of Mathematical Biology,* 44: 749-760.

Perelson, A.S., and G.F. Oster. 1974. Chemical reaction dynamics: Part II: reaction networks. *Archive for Rational Mechanics and Analysis,* 57: 31-98.

Prigogine, I. 1980. *From Being to Becoming: Time and Complexity in the Physical Sciences.* W.H. Freeman and Company, San Francisco.

Rashevsky, N. 1960. *Physico-Mathematical Foundations of Biology,* Volumes I and II, Third Edition. Dover Publications, New York.

Rosen, R. 1970. *Dynamical Systems Theory in Biology.* John Wiley and Sons, New York.

Rosen, R. 1972. *Foundations of Mathematical Biology,* Volumes I, II, III. Academic Press, New York.

Rowe, G.A., and L.E.H. Trainor. 1983a. On the informational content of viral DNA. *Journal of Theoretical Biology,* 101: 151-170.

Rowe, G.A., and L.E.H. Trainor. 1983b. A thermodynamic theory of codon bias in viral genes. *Journal of Theoretical Biology,* 101: 171-203.

Ruelle, D. 1969. *Statistical Mechanics: Rigorous Results.* Benjamin, New York.

Ruelle, D. 1978. *Thermodynamic Formalism: the Mathematical Structures of Classical Equilibrium Statistical Mechanics.* Addison-Wesley, Reading, MA.

Santilli, R.M. 1978. *Foundations of Theoretical Mechanics, I: the Inverse Problem in Newtonian Mechanics.* Springer-Verlag, New York.

Schieve, W.C. 1974. Aspects of non-equilibrium quantum statistical mechanics: an introduction. In W.C. Schieve and J.S. Turner, eds., *Lectures in Statistical Physics,* pp. 1-52. Springer-Verlag, New York.

Schnakenberg, J. 1976. Network theory of microscopic and macroscopic behavior of master equation systems. *Reviews of Modern Physics,* 48: 571-585.

Simpson, G.G. 1953. *The Major Features of Evolution.* Columbia University Press, New York.

Sinai, Ya. G., 1973. Ergodic theory. *Acta Physica Austrica Supplement X:* 575-608.

Sudarshan, E.C.G., and N. Mukunda. 1974. *Classical Dynamics: A Modern Perspective.* John Wiley and Sons, New York.

Totafurno, J., C.J. Lumsden, and L.E.H. Trainor. 1980. Structure and function in biological hierarchies: an Ising model approach. *Journal of Theoretical Biology,* 85: 171-198.

Trubatch, S.L., and A. Franco. 1974. Canonical procedures for population dynamics. *Journal of Theoretical Biology,* 48: 299-324.

Vstovsky, V.P. 1976. On the dynamical basis of macroscopic theory. *Journal of Statistical Physics,* 15: 105-121.

Zwanzig, R. 1960. Ensemble method in the theory of irreversibility. *The Journal of Chemical Physics,* 33: 1338-1341.

Zwanzig, R., 1961a. Statistical mechanics of irreversibility. In W.E. Brittin et al., eds., *Lectures in Theoretical Physics,* Vol. III, pp. 106-141. Wiley-Interscience, New York.

Zwanzig, R., 1961b. Memory effects in irreversible thermodynamics. *Physical Review,* 124: 983-992.

2

PRIMARY PRODUCTIVITY AND THE ECOLOGY OF SESSILE ORGANISMS

2A

Non-Euclidean Allometry and the Growth of Forests and Corals

P.L. Antonelli

2A.1 Huxley's Allometric Equation

We are all familiar with "growth" and "decay" as described by

$$x(t) = C \cdot e^{\alpha t} \qquad (2A.1\text{-}1)$$

where C and α are constants. In the case $\alpha > 0$, $C > 0$, *exponential growth* is described and $x(t)$ represents the amount of material, living or dead, that has accumulated up to time t in some observable natural process. In the case $\alpha < 0$; $C > 0$, $x(t)$ represents the amount remaining at time t in an exponential decay process. Huxley's law is usually associated to growth processes so in the ensuing discussion I shall restrict myself to $\alpha > 0$, $C > 0$.

The system I wish to use to help illustrate Huxley's idea may be taken to be a single growing plant (e.g. a tree) or animal (e.g. a horse). In fact, I prefer to consider plants over animals. Therefore, the dependent variable $x(t)$ will be interpreted as, say, the weight of the total number of leaves or roots at time t during the life of the plant. It may help the reader to visualize in his mind's eye a sort of time-lapse movie of a simple growing plant. Further suppose that proper soil conditions requisite for optimal growth are satisfied. These amount to supplying all the space, light, water, and various other nutrients required for the best possible (i.e. optimal) growing. We may summarize by saying that *growth is unlimited by resources.* It is as if *resources are infinite* in the situation depicted by (2A.1-1). One is reminded of *Jack and the Beanstalk!*

Suppose that the root weight of our beanstalk is given by (2A.1-1), while leaf weight is given by

$$y(t) = De^{\beta t} \tag{2A.1-2}$$

with D, β positive constant. It is easy to derive the expression

$$y = Mx^{\beta/\alpha} \quad \text{(Huxley Allometric Law)} \tag{2A.1-3}$$

where

$$M = \frac{D}{C^{\beta/\alpha}}, \text{ follows} \tag{2A.1-4}$$

from (2A.l-1) and (2A.l-2) by elimination of the time variable t. The equation (2A.1-3) is called *Huxley's Allometric Law* (Huxley, 1932). The importance of (2A.1-3) is that β/α is characteristic of the species (i.e. different for different species) and knowing either the root weight or leaf weight determines the other. This may not seem important but in practise $x(t), y(t)$ are average values taken over a sample of plants from, say, an entire crop. Thus it becomes possible to estimate one of the parameters x, y from knowing the other, and this from just a small randomly selected sample! This is exactly what the forester J. Kittredge (1944) had in mind in 1944 when he estimated amount of foliage of trees by merely knowing their average trunk diameter. Kittredge found that Huxley's equation held for many species of spruce and pine trees. The dependent variable $y(t)$ was *average needle weight* of a pine species while $x(t)$ was the *average diameter of the trunk at breast height.* A few of his allometric equations are given here in logarithmic form.

canyon live oak	$\ell n\, y = 2.66\, \ell n\, x - 1.36$
douglas fir	$\ell n\, y = 1.96\, \ell n\, x - 0.91$
red pine	$\ell n\, y = 2.56\, \ell n\, x - 0.94$

Figure 2A.1-I

These are typical results in the Kittredge study. His results hold for *all ages of trees up to the time of full growth.*

Other scientists have found Huxley's equation holds for many plants (e.g. cotton, peas, carrots, turnips) and animals as diverse as fish and rodents. In the animal studies, $x(t)$ and $y(t)$ were typically calcium, phosphorous, glycogen, water, or some other simple chemical compound found in animal embryos. The reader should consult J. Needham (1934) for more discussion of these animal matters. In 1971 Patefield and Austin found that Huxley's law held for the beet plant β *Vulgaris L. as long as the plants were fairly young.* When they plotted average leaf weight against time, statistical procedures indicated a *logistic curve* instead of exponential growth. Whereas exponentials come from the equation

$$\frac{dx}{dt} = \alpha x \,, \tag{2A.1-5}$$

this *logistic* comes from

$$\frac{dx}{dt} = \alpha x \left(1 - \frac{x}{K}\right), \tag{2A.1-6}$$

where $0 < x < K$. The method of partial fractions can be used to derive

$$x(t) = \frac{K}{1 + be^{-\alpha t}} \quad \text{(Logistic Equation)} \tag{2A.1-7}$$

from integration of (2A.1-6). Here, $b > 0$ and $K > 0$ are positive constants. We say they determine *initial conditions* for (2A.1-6) because

$$x(0) = \frac{K}{1 + b} \,. \tag{2A.1-8}$$

In a similar fashion, one derives

$$y(t) = \frac{L}{1 + ae^{-\beta t}} \tag{2A.1-9}$$

from

$$\frac{dy}{dt} = \beta y \left(1 - \frac{L}{y}\right). \tag{2A.1-10}$$

The logistic (2A.1-7) is plotted in Figure 2A.1-II below

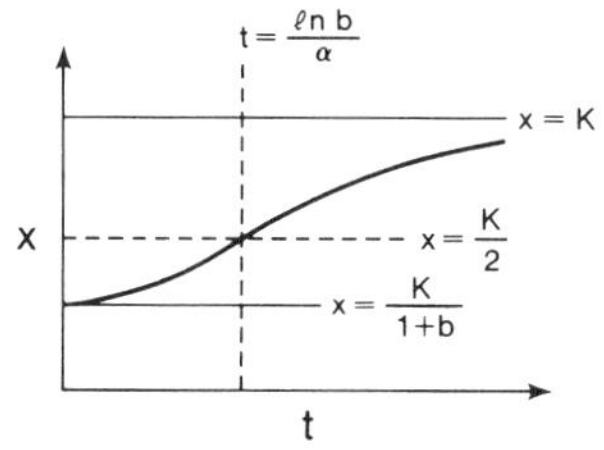

Figure 2A.1-II. Graph of Logistic (2A.1-7)

Application of differential calculus allows us to conclude that $dx/dt > 0$ for all t and that there is exactly one point of inflection at $t = (\ell n\, b)/\alpha$, $x = K/2$ and that $\lim_{t\to\infty} x(t) = K$ so that $x(t)$ is asymptotic to the line $x = K$.

2A.2 The Logistic Form of Huxley's Law

The logistic arose first in the work of the mathematician P. Verhulst in 1838 and later in the work of R. Pearl and L. Reed in 1920. These two demographers found that the logistic described human population growth in the U.S.A. from 1790 to 1920.

The logistic has found wide application in epidemic theory. Indeed, one of the very first applications was by Sir Ronald Ross who studied malarial infections in human beings. Other examples may be found in Lotka's *Elements of Mathematical Biology* (1956).

We can see why logistics are different than exponentials if we examine the differential equation (2A.1-6). The difference lies with the additional term $-\alpha x^2/K$. It is this negative term which causes the concavity to switch from concave up to concave down as t passes through the time $(\ell n\, b)/\alpha$. It causes the derivative dx/dt to slow down its rate of increase after $x = k/2$ has been reached.

It is easy to see that if x is close to zero, $-\alpha x^2/k$ may be neglected by comparison to αx so that (2A.1-5) results. This leads us to *suspect that the logistic is exponential for small* x *and for small* t. Another thing to notice about logistics is that the dependent variable $x(t)$ is never larger than K (if $b > 0$). One imagines that as the system grows, more and more resources are needed to support the living bulk already accumulated, so that eventually the system must come to equilibrium at (or near) $x = K$. At this point all growth effectively ceases. If we begin with $x(t)$ and $y(t)$ as in (2A.1-7) and (2A.1-9) elimination of the variable t results in

$$y = \frac{L}{a\left(\frac{K-x}{bx}\right)^{\beta/\alpha} + 1} \quad \text{(Logistic Huxley Law).} \tag{2A.2-1}$$

I shall call (2A.1-1) the *allometric law for resource limited growth.* If we assume x is small compared to K, then (2A.1-1) reduces to

$$y = M' \cdot x^{\beta/\alpha} \tag{2A.2-2}$$

where

$$M' = \frac{(b/K)^{\beta/\alpha}}{a/L} . \tag{2A.2-3}$$

Since initial conditions $(t = 0)$ for exponential growth must agree with those of logistic growth if our suspicions are correct, we must have

$$x(0) = C = \frac{K}{1+b} , \quad D = \frac{L}{1+a} = y(0) . \tag{2A.2-4}$$

Then the constant M in Huxley's Law must be

$$M = \frac{D}{C^{\beta/\alpha}} = \frac{((1+b)/K)^{\beta/\alpha}}{(1+a)/L} . \tag{2A.2-5}$$

But since x, y must both be close to zero, we must take a, b large in (2A.2-4). Then $1 + b \sim b$ and $1 + a \sim a$ so that $M' \sim M$. Similarly,

$$\left.\frac{dx}{dt}\right]_{t=0} = \frac{Kb}{(1+b)^2} \tag{2A.2-6}$$

must hold for the x-logistic; while

$$\left.\frac{dx}{dt}\right]_{t=0} = \frac{K}{1+b} \tag{2A.2-7}$$

must hold for the x-exponential. We see that these two slopes at $t = 0$ agree for large b, as they should if the x-exponential is to agree with the x-logistic for small enough x and t.

Thus we have seen that the reduction of the resource limited allometric law to Huxley's requires the parameters a and b to be large and t to be small. *Since growth is more realistically resource limited than resources unlimited it seems clear why one should not expect Huxley's original law* (2A.1-3) *to work for more than early stages of the growth process.* This agrees with the experimental findings of Patefield and Austin and Kittredge.

2A.3 Huxley's Law for Relative Proportions

In this section we wish to examine the approach to equilibrium $x = K$ in the logistic (2A.1-7). We will use the differential equation (2A.1-6) and introduce a *new variable:*

$$x^* = \frac{K - x}{x} \tag{2A.3-1}$$

which we shall call the *relative* x-*proportion.* Use of the chain rule yields

$$\frac{dx^*}{dt} = \frac{-\alpha K}{x}\left(1 - \frac{x}{K}\right) = -\alpha x^* . \tag{2A.3-2}$$

The solution of (2A.3-2) is obviously

$$x^*(t) = b \cdot e^{-\alpha t} . \tag{2A.3-3}$$

I know the constant is b in (2A.3-3) because $x^*(0) = b$. Now the same procedure can be applied to the y-logistic (2A.1-10). One obtains

$$y^*(t) = ae^{-\beta t} . \tag{2A.3-4}$$

If we apply the derivation of Huxley's original equation to (2A.3-3) and (2A.3-4), the result is

$$y^* = M^* \cdot (x^*)^{\beta/\alpha} \quad \text{(Relative Proportion Huxley Law)} \tag{2A.3-5}$$

where

$$M^* = \frac{a}{b^{\beta/\alpha}} . \tag{2A.3-6}$$

The equations (2A.3-5) and (2A.3-6) are equivalent to (2A.2-1).

If I am interested in x-values near K , then I should look at x^*-values close to zero. We see that the approach to equilibrium is increasingly slow and in the limit $t \to \infty$ the slope vanishes altogether. These results are typical of any exponential decay. It is important to realize that the *equation* (2A.3-5) *holds for all time* t . It may be called *Huxley's Allometric Law for Relative Proportions.* This form of the Allometric Law for resource limited growth is preferable over (2A.2-1) because it has the classical form of Huxley's allometric law. Thus we see that in spite of resource limitations the allometric law obtains and because of our work in section 2A.2 of this essay we know that for x close to zero (2A.3-5) the relative proportion Huxley Law, reduces to (2A.1-3) the original (classical) Huxley Law.

2A.4 The Allometric Differential Equation

If we suppose the allometric relation

$$x = b \cdot r^{\alpha}, \tag{2A.4-1}$$

holds, then the equation

$$\frac{d^2x}{dr^2} - \frac{1}{x}\left(\frac{dx}{dr}\right)^2 + \left(\frac{b}{x}\right)^{1/\alpha}\frac{dx}{dr} = 0, \tag{2A.4-2}$$

also is satisfied. This is the *allometric differential equation.* It is worth noting that the mapping $\ell n^* : \mathbf{R}^1_+ \times \mathbf{R}^1_+ \to \mathbf{R}^1_+ \times \mathbf{R}^1_+$ defined by

$$\ell n^*(x, r) = (\ell n\, x, \ell n\, r) = (\phi(x), s), \tag{2A.4-3}$$

converts (2A.4-2) into,

$$\frac{d^2\phi}{ds^2} = 0. \tag{2A.4-4}$$

This is the *standard form* of the allometric equation. A similar log transformation will be useful later on when we discuss *Gompertz growth* (see essay 2F), and the *Gompertz community differential equations.*

2A.5 Volterra's Record Variable

In 1936, Vito Volterra wrote a paper in which he described a method of obtaining his ecological equations (based on a mass action law) from the principle of Maupertuis (see the translation of Volterra's 1936 paper in Chapter 9). His idea was to introduce an auxillary variable which he called "La quantite de vie." If N denotes the (large) number of animals or plants of a certain species, the new variable x is defined by

$$x = \int_{t_0}^{t_1} N\, dt. \tag{2A.5-1}$$

Volterra did not carry this idea very far and subsequent ecological theory has not as yet found it useful. Our point of view is that there are several large classes of biological systems exhibiting both ecological and allometric properties, and for these, Volterra's idea becomes centrally important. These systems are "closed" in the sense that each species deposits into the environment a substance which accumulates in the system and whose presence affects the ecological and/or developmental interactions occuring there.

On an ideal coral reef each polyp species produces aragonite calcium carbonate in characteristic physical shapes during the reef building process (Maragos, 1934). This is an example of what we term a *closed eco-developmental system*. Biological systems exhibiting *allelopathic interactions* in which each species deposits a toxin inhibiting growth and development of the other species in the system (while perhaps affecting his own progress) are also examples. The most familiar examples of allelopathy are among plants, where it is virtually universal (Jackson and Buss, 1975). However, it is known that some sponge species living on reefs also exhibit allelopathy for some corals (Jackson and Buss, 1975).

The nonlinear coupling of the species numbers will be found to strongly affect the qualitative behavior of the allometric growth process. This can be seen directly in the behavior of the growth curves in the general theory. These are geodesics in a negatively curved space. Consequently, these curves are unstable relative to initial conditions and thereby exhibit indeterminism during *initial growth stages* (see section 2F.3). Another important result is that in *later stages* the system is strongly resistant to changes caused by environmental noise. We term this special property *stochastic canalization*. Furthermore, both these properties are generic, so that they hold true for systems whose interaction coefficients are close to the given mass action coefficients. We discuss this briefly in section 2A.7.

The consequences for reefs are as follows. During initial stages of growth nonpredictability of the Aragonite abundances on a reef are to be expected. While later on, this is replaced by a strong resistance to environmental noise such as reef boring sponges, parrot fish grazing, and fluctuations in the light intensity due to cloud cover variations (Porter, 1976; Stearn and Scoffin, 1977). Using this basic model, starfish predation of a coral reef has been studied. Stable limit cycle behavior has been proved to hold (see essay 2B). The basic mathematical model presented here has also been used to model allelopathic plant communities (see essays 2C and 2F), and plant/herbivore interactions (see essays 2D and 2F).

2A.6 Volterra-Hamilton Systems

By a *closed eco-developmental system* $E(P_1, \ldots, P_n \mid p_1, \ldots, p_n)$, we shall mean a set of n kinds of "Producers" $P_1, \ldots, P_n$ with each set P_i

consisting of N^i individuals, each depositing a product p_i into the environment E. The *total* amount of product of kind i due to *all* N^i individuals in p_i is denoted x^i. These variables are clearly monotone increasing and are used to span coordinate space H^n, which we call *Allometric Space.* Producers can interact *passively,* as in the case of the classical mass action law of Volterra mechanics or *actively,* in which case the interaction coefficient depends explicitly on total amounts of the products. In the present paper, we present a passive model and give applications towards a theory of the development of a coral reef.

By a n-dimensional *Volterra-Hamilton system* we mean a system of ordinary differential equations

$$\Gamma: \begin{cases} \dfrac{dx^i}{dt} = k_{(i)} N^i \\ \dfrac{dN^i}{dt} = -\Gamma^i_{jk} N^j N^k + \gamma^i_j N^j \end{cases} \qquad i,j,k = 1,2,\ldots,n. \tag{2A.6-1}$$

Here, all coefficients are constants and the Einstein summation convention on upper and lower indices is used except for $k_{(i)}$. The second equation is termed a *(passive) mass action law* because of the constancy of coefficients. Substitution of the first equation into the second yields the system

$$\frac{d^2x^i}{dt^2} = -\tilde{\Gamma}^i_{jk} \frac{dx^j}{dt}\frac{dx^k}{dt} + \gamma^i_j \left(\frac{1}{k_{(j)}} \frac{dx^j}{dt} \right) \tag{2A.6-2}$$

where $i = 1,2,\ldots,n$ and,

$$\tilde{\Gamma}^i_{jk} = \frac{k_{(i)}}{k_{(j)}k_{(k)}} \Gamma^i_{jk}\,. \tag{2A.6-3}$$

From (2A.6-1) we can see that the constants $k_{(i)}$ are rates of deposit of the ith product per individual in P_i. There is good reason to believe that the aragonite deposit rate is of the order of one micron thickness per hour in clear sunlight (Alexander, 1979). This would be the case for the class of coral that have specialized in small polyp (diameter 1 mm) "photon capture" as opposed to the large polyp (diameter 2 cm) "zooplankton capture." The former largely live off the sun. We shall suppose

$$k_{(i)} = k_{(j)} = 1 \text{ for all } i,j, \tag{2A.6-4}$$

in the present paper. This is no loss of generality. If we were to suppose the $k_{(i)}$ negative, the processes would be of *depletion type.* For example, bioerosion on a coral reef constitute processes of this kind. The systematic action of reef boring sponges and sea urchins are good examples. It is important to realize that starfish predation is *not* a depletion process in our view, rather, it is predator-prey. Proof of existence of a stable limit cycle for this starfish-coral system is given in essay 2B of this chapter.

The classical role of the constants γ^i_j in Volterra theory is that of intrinsic rates of growth, or just the rate of proliferation of a species on its own without outside influence. We shall suppose throughout that these rates are equal for all species of coral considered. (This can be arranged by the addition of phosphate to cultures of plants (see essay 2D).) Once this is assumed an *affine parameter change* will bring the system (2A.6-2) into the more familiar *geodesic form*

$$\frac{d^2x^i}{ds^2} = -\Gamma^i_{jk}\frac{dx^j}{ds}\frac{dx^k}{ds}, \quad \text{where } s = e^{\lambda t}. \tag{2A.6-5}$$

Solution curves of (2A.6-5) need not be geodesics in any Riemannian geometry. In order for such to be the case one must establish the existence of a positive definite 2nd order covariant tensor field g_{ij} on H^n which solves the equation of Levi-Civita′

$$\Gamma^i_{jk} = \tfrac{1}{2} g^{\lambda i}\left(\frac{\partial g_{j\lambda}}{\partial x^k} + \frac{\partial g_{k\lambda}}{\partial x^j} - \frac{\partial g_{jk}}{\partial \lambda^\lambda}\right). \tag{2A.6-6}$$

In 1974, B.H. Voorhees and this author proved that

$$g_{ij} = e^{+2\alpha_i x^i}\delta_{ij} \tag{2A.6-7}$$

where $\alpha_1, \ldots, \alpha_n$ are positive constants, satisfied (2A.6-6) and constitutes a Riemannian metric on Allometric Space H^n. The coefficients are then

$$\Gamma^i_{ii} = +\alpha_i$$

$$\Gamma^i_{ij} = \Gamma^i_{ji} = +\alpha_j, \quad i \neq j$$

$$\Gamma^i_{jj} = -\alpha_i, \quad i \neq j$$

$$\Gamma^i_{jk} = 0, \quad i \neq j \neq k. \tag{2A.6-8}$$

In 1977, E. Rhunau proved that if

$$\delta_{is}\Gamma^s_{k\ell} + \delta_{\ell s}\Gamma^s_{ki} = \gamma_k \delta_{i\ell}, \tag{2A.6-9}$$

holds for some set of constants $\gamma_1, \ldots, \gamma_n$, then the metric (2A.6-7) is the only solution to (2A.6-6) (Rhunau, 1977). In this case, $\gamma_n = -(2/n)\,\alpha_k$. The equations (2A.6-9) are just the integrability conditions for the system of partial differential equations

$$g_{ik,\ell} = g_{is}\Gamma^s_{k\ell} + g_{ks}\Gamma^s_{i\ell}, \tag{2A.6-10}$$

where the comma denotes covariant differentiation relative to g_{ij}. E. Rhunau conjectured that (2A.6-9) is necessary as well as sufficient for integrability of (2A.6-10). The present author has indeed verified this conjecture. A proof is given in essay 2E of this chapter. Thus, the Antonelli-Voorhees *metric is the only one possible for a Volterra-Hamilton system with constant coefficients* (i.e. passive)!

2A.7 Stochastic Canalization of Allometric Brownian Motion

The Allometric Brownian motion equation is

$$\frac{\partial P}{\partial \tau} = \tfrac{1}{2}\,\Delta_g P\,; \quad P = P(0, \overline{x}, \tau), \tag{2A.7-1}$$

where P is the transition density (or Heat Kernel) of the probability of being at $\overline{x}$ at time τ, given that the source is at or near the origin at time zero. The *Allometric Laplacian* is given by

$$\Delta_g = g^{ij}\left(\frac{\partial^2}{\partial x^i\,\partial x^j} - \Gamma^k_{ij}\,\frac{\partial}{\partial x^k}\right), \tag{2A.7-2}$$

and the Riemannian metric is given by (2A.6-7) or by the *perturbation metrics* of essays 2B, 2C, 2D, and 2E in this chapter. We are concerned with the asymptotic evaluation of the probability that a given sample path of the equation (2A.7-1) remains inside a small tube of (Riemannian) radius $\rho = \varepsilon$, about a geodesic curve $\phi_t : [0, T] \to H^n$. Recently, Watanabe and Takahashi (1980) have obtained the important geometric result:

THEOREM. The probability that a sample path x_t of (2A.7-1), starting at $x_0 = \phi_0$ in H^n, remains inside a tube of radius $\rho = \varepsilon$ for all $t \in [0, T]$, i.e. that $\mathrm{Prob}_{\phi_0} \{ \rho(x_t, \phi_t) < \varepsilon, \ \forall_t \in [0, T] \}$ is asymptotically, $c \circ \exp(-(\lambda_1 T)/(\varepsilon^2)) \exp(- \int_0^T L(\phi_t, \dot{\phi}_t)\, dt)$ as $\varepsilon \to 0$. Here,

$$L(\underline{x}, \dot{\underline{x}}) = \tfrac{1}{2} g_{ij} \frac{dx^i}{dt} \frac{dx^j}{dt} + \frac{1}{12} R(\underline{x}), \qquad (2A.7\text{-}3)$$

and the dot denotes d/dt. **R** is the Riemann scalar curvature of the metric (watch the sign convention in the definition of **R** !) The constant c is given by

$$c = \psi_1(0) \int_{\mathbf{D}^n} \psi_1(\underline{x})\, d\underline{x}, \qquad (2A.7\text{-}4)$$

where $(\psi_m(\underline{x}), \lambda_m)$ $m = 1, 2, \ldots$ is the eigensystem for $-\tfrac{1}{2}\Delta_e$ in the unit n-ball $\mathbf{D}^n$, (Δ_e is the euclidean Laplacian), with Dirichlet boundary conditions.

From this theorem of Watanabe and Takahaski (1980) it is transparent that *the more negative* $\mathbf{R}(\underline{x})$ *is in the tubular region around* ϕ_t, *the greater the probability of a sample path of* (2A.7-1) *remaining inside it up to time* T. For passive Volterra-Hamilton systems, $n \geqslant 3$, $\mathbf{R}(\underline{x})$ is negative everywhere (see equation 3B.1-29). For, $n = 2$, $\mathbf{R}(\underline{x})$ is negative for perfectly symbiotical systems (see essay 2F). These have Riemannian metrics perturbed from the metric (2A.6-7), and are interpreted as explicit chemical influences in the ecology of the system. The above theorem has important statistical implications for allelochemical experiments. Moreover, if one interprets (2A.7-3) as a *cost functional* for growth it is not difficult to show that the dosage dependent strategy discussed in essay 2D in this chapter, is costly to the plant community and increased plant response causes decreased $|\mathbf{R}|$. It is tempting to interpret, $-\mathbf{R}$, as a measure of *vigour.* For the perturbed metrics discussed in essay 3B in the next chapter, the magnitude of $\mathbf{R}(x)$ decreases with increasing x. Thus canalization decreases with time for the transient Allometric Brownian motion in this nonequilibrium theory.

Acknowledgement

This research was partially supported by NSERC-A-7667.

References

Alexander, R.M. 1979. *The Invertebrates.* Cambridge University Press.

Huxley, J. 1932. *Problems of Relative Growth,* 2nd Edition. 1972. Dover Press.

Kittredge, J. 1944. Estimation of the amount of foliage of trees and stands. *J. Forestry,* 42: 905-912.

Jackson, J., and L. Buss. 1975. Allelopathy and spacial competition among coral reef invertebrates. *Proc. Nat. Acad. Sci. USA.* 72, no. 12: 5160-5163.

Lotka, A.J. 1956. *Elements of Mathematical Biology.* Dover Press.

Maragos, J.E. 1978. Coral growth: geometrical relationships. In D.R. Stoddard and R.E. Johannes, eds., *Coral Reefs: Research Methods.* Unesco monographs in oceanographic methodology, pp. 543-550.

Needham, J.A. 1934. Heterogony, a chemical ground-plan for development. *Biol. Rev.,* 9: 79-109.

Patefield, W.M., and R.B. Austin. 1971. A model for the simulation of the growth of *Beta Vulgaris L. Ann. Bot.,* 35: 1227-1250.

Porter, J. 1976. Autotrophy, heterotrophy and resource partitioning in Caribbean reef-building corals. *Amer. Nat.,* 110: 731-742.

Rhunau, E. 1977. Spaces of locally constant connection. Master's Thesis, University of Alberta, 94 pages.

Stearn, C.W., and J.P. Scoffin. 1977. Carbonate budget of a fringing reef, Barbados. In *Proc. Third International Coral Reef Symposium.* University of Miami.

Volterra, V. 1936. Principles de Biologie Mathématique. *Acta Biotheoretica III:* 1-36.

Watanabe, S., and Y. Takahashi. 1980. The probability functional (Onsager-Machlup functions) of diffusion processes. In D. Williams, ed., *Stochastic Integrals,* Proc. LMS, Durham Symposium, pp. 433-463.

Whittaker, R.H., and P.P. Feeny. 1971. Allelochemics: chemical interactions between species. *Science,* 171: 757-768.

2B

Starfish Predation of a Growing Coral Reef Community

P.L. Antonelli and N.D. Kazarinoff

2B.1 Introduction

In the 1960's and 1970's some Indo-Pacific coral reefs were devastated by large "crown-of-thorns" starfish (Endean, 1973; Pearson and Endean, 1969). This has occurred especially on portions of the Great Barrier Reef of Australia. The single species involved in the recent population "outbreaks" of coral-eating starfish is *Acanthaster planci.* An excellent and thorough review of the status of *A. planci* as a possible pest has been written by D.C. Potts (1981). In his review Potts summarized and analyzed the then available scientific data. He assessed these data and theories based upon them in relation to three basic questions: "Is crown-of-thorns a pest? Are its population outbreaks natural phenomena? Are they caused by human disturbance of tropical ecosystems?" We discuss these basic questions briefly in section 2B.2. Since we believe that Potts' review is both valuable and not easily accessible, we quote from it several times in section 2B.2.

Potts concludes that of the possible alternatives "the outbreaks appear to be responses to natural disturbances" and that recovery of coral polyps from *A. planci* caused damage is likely. In section 2B.3 we develop a model whose theoretical and numerical behavior supports Potts' conclusion. Indeed, we view predation of autotropic coral polyps by crown-of-thorns to be a natural cyclic phenomena that is like predation of plant communities by gregarious insects. The aggregation behavior of *A. planci* (Campbell, 1976) has motivated us to introduce a quadratic cooperative term (an autocatalytic term in the language of chemistry) into the predator equation of our model. Such a term is probably the simplest representation of the aggregation effect. As is expected in analogy with chemical and biochemical models such as the Oregonator and Brusselator, it is this quadratic cooperative term that makes possible the existence of stable limit cycles.

In sections 2B.4-6 we prove existence of periodic solutions of our model and study their stability using Hopf bifurcation theory and B.D. Hassard's code BIFOR2 (Hassard, Kazarinoff and Wan, 1981). Although Hopf bifurcating periodic solutions have small amplitudes, a property which may contrast with the apparent large amplitude behavior of *A. planci,* more realistic parameter values than those we have used may give rise to large amplitude oscillations. We have not proved that our model supports such solutions; however, the periods of the bifurcating solutions we have found can be of moderate size ($\approx 2\pi$). Data collected over a sufficiently long interval of time will confirm or deny the existence of a natural cycle.

A further feature of our model for two optimally producing coral and one starfish species (sections 2B.3-6) is that if the two corals react identically to the starfish (the starfish do not discriminate between the coral polyps), then the bifurcation parameters μ_2 and β_2 are identically zero. Here μ_2 gives the direction of bifurcation and β_2 determines the critical Floquet exponent that governs stability. The symmetry that causes μ_2 and β_2 to vanish seems to be due to the optimal growth condition to which our model conforms; see (2B.3-5) below. As the corals become asymmetric in their preference (or as the starfish increases its preference for one coral over the other but without change in benefit to starfish population growth), μ_2 and β_2 become nonzero (positive and negative, respectively) and increase in magnitude together with the amount of preference asymmetry. The increase in magnitude is especially rapid for β_2.

2B.2 Alternative Theories

One theory, advanced mainly by the biologist R. Endean (1977), to account for the outbreaks of *A. planci* is that a natural predator, namely, the giant triton *Charonia tritonis,* is removed by man for its beautiful shell. *C. tritonis* preys on adult *A. planci* by cutting through the starfish's body wall with its radula, inserting its proboscis into the arms, and removing gonads and digestive glands (Pearson and Endean, 1969). *C. tritonis* eats juvenile starfish whole. But *C. tritonis* is not a specialist, and *A. planci* has other predators. For example, territorial pairs of the painted shrimp *Hymenscera picta* feed exclusively on large starfishes (Wickler, 1973).

Other consumers of *A. planci* are the large puffer fish *Arothron bispidus* and the triggerfishes *Balistoides viridescens* and *Pseudobalistes flavimarginatus.* Finally, it appears from ongoing research of P. Antonelli and R. Endean that such a theory is unlikely to yield natural cycles in crown-of-thorns populations when quantified.

A second alternative explanation for outbreaks of *A. planci* is enhanced recruitment during periods of physical conditions which are particularly favorable for *A. planci* larvae (Pearson, 1975). However, "there is no direct evidence to support the juvenile recruitment hypothesis, and no outbreak composed mainly of juveniles has ever been detected" (Potts, 1981).

Finally, Potts writes, "The adult aggregation hypothesis proposes that primary outbreaks are formed by aggregation of adults from normal, low-density populations after abnormal storms have destroyed most of their food through increased sedimentation over polyp bodies. The hypothesis accounts for the observation that all known primary populations consist only of adults... . The adult aggregation hypothesis predicts the possibility of previous "outbreaks" and explains the general distribution, local sites, timing and age distributions of post-World War II outbreaks. It also explains why only *A. planci* was affected: this is the only starfish which feeds almost entirely on the structurally dominant organisms, and so is the only species likely to be left without alternative foods. Since it is composed of existing adults, a primary outbreak can form within a few weeks of the initial disturbance without having to survive the vicissitudes of planktonic and juvenile life for up to two years. Because larval stages are not involved, the adult aggregation model is the only one which not only explains the existence of primary outbreaks on isolated oceanic islands, but also explains why secondary outbreaks have not occurred on these islands" (Potts, 1981).

Whatever the cause, there are both major and minor outbreaks of *A. planci.* Indeed, Dana, Newman, and Fager (1972) conclude that frequency distribution of population densities of *A. planci* forms a single continuous distribution with no obvious separation between numerous low density, normal populations and a few large outbreak populations. This is in harmony with the possibility that a model such as ours may support periodic solutions whose amplitudes vary from small to large with continuous changes in a parameter. Potts concludes that outbreaks of *A. planci* in which large aggregations persist for several years and cause extensive coral mortality over large areas of reef surface are real and widespread but that "the Great Barrier Reef is a unique situation with outbreaks of *A. planci* and consequent coral mortality on a much greater scale than anywhere else." He further concludes that "outbreaks appear to be responses to natural disturbances" and that "recovery of coral faunas from *A. planci* damage seems to involve the same ecological processes as recovery from other causes of extensive mortality."

2B.3 Optimal Growth of the Coral Skeleton

Elsewhere, a nonlinear generalization of the classical theory of allometric growth, due to Julian Huxley, has been developed (Antonelli, 1980a; Antonelli, 1980b; and Antonelli and Voorhees, 1983). We apply this theory, which is especially appropriate for nonequilibrium growth of plant communities, to n species of reef building corals (Harper, 1977).

The sessile coral polyps produce calcium carbonate, in the form of aragonite, through a symbiotic relationship with a brown-yellow algae, known as, *zooxanthellae,* which inhabits the endodermic tissue layer of polyp guts (Alexander, 1979). This production of aragonite is evidently the main component of the reef skeleton growth process (Stearn and Scoffin, 1977). It is largely due to photosynthesis. Coral polyps are animals; and they do compete for food as well as sunlight and space. But, the competition for food is not intense for a large number of autotropic species, which live off sunlight at the courtesy of *zooxanthellae* (Porter, 1976). Furthermore, corals, like plants, proliferate their biomass by asexual budding. Some species even do this at a constant rate in clear sunlight (Alexander, 1979). We shall model the competition for space in sunlight of autotropic coral polyps subject to predation by starfish. The plant-like behavior of these coral polyps and the swarming behavior of crown-of-thorns play key roles in the model.

In order to keep track of the total aragonite production, a variable x is introduced to denote the natural logarithm of the *total accumulation* of aragonite at a site specified on the reef surface. In practice, x is estimated by core sampling procedures (Maragos, 1978). Therefore, letting $C_i(t)$ denote the number of autotropic coral polyps of species $i = 1, ..., n$, we write

$$\frac{dx^i}{dt} = k_{(i)} C_i \quad (i = 1, ..., n), \qquad \text{(2B.3-1)}$$

where $k_{(i)}$ is a constant and where the index i is not summed. Without loss of generality, we shall normalize the $k_{(i)}$ to unity for each $i = 1, ..., n$.

We further use a constant coefficient Volterra equation

$$\frac{dC^i}{dt} = - \sum_{j,k} \Gamma^i_{jk} C_j C_k + \sum_j \lambda^i_j C_j , \qquad (2B.3\text{-}2)$$

to describe the time change of polyp numbers. It will be (temporarily) assumed that

$$\lambda^i_j = \lambda \delta^i_j \quad (\lambda > 0), \text{ where } \delta^i_j \text{ is the Kronecker delta.} \qquad (2B.3\text{-}3)$$

Our next hypothesis is that the $x^i(t)$ solve the second order system

$$\frac{d^2x^i}{dt^2} = - \sum_{j,k} \Gamma^i_{jk} \frac{dx^j}{dt} \frac{dx^k}{dt} + \lambda \frac{dx^i}{dt} , \qquad (2B.3\text{-}4)$$

with all indices running from 1 to n . Our optimal growth condition is that the equations (2B.3-4) are the Euler-Lagrange equations for the extremal

$$\delta \int_{t_1}^{t_2} F(x^1, \ldots, x^n, \frac{dx^1}{dt}, \ldots, \frac{dx^n}{dt}, t)\, dt = 0 , \qquad (2B.3\text{-}5)$$

where F is C^∞ in all variables and is first order homogeneous in the dx^i/dt . A fairly difficult theorem in classical Finsler differential geometry is that this is possible *if and only if* there are n positive constants $\alpha_1, \ldots, \alpha_n$ such that the n^3 quantities Γ^i_{jk} are given by

$$\begin{cases} \Gamma^i_{ii} = + \alpha_i \\ \Gamma^i_{ij} = \Gamma^i_{ji} = + \alpha_j , \quad i \neq j \\ \Gamma^i_{jj} = - \alpha_i , \quad i \neq j \\ \Gamma^i_{jk} = 0 , \quad i \neq j \neq k . \end{cases} \qquad (2B.3\text{-}6)$$

The condition (2B.3-5) may be interpreted as minimizing the "cost of Aragonite production," $F(x, \dot{x})$, to the coral community.

The interested reader may consult (Antonelli, 1980a; Antonelli, 1980b; Antonelli and Voorhees, 1983) and essay 3B.2 for further details.

The *predator equation* we employ is

$$\frac{dF}{dt} = \left(\sum_{i=1}^{\mu} \beta_i C_i\right) F + \gamma F^2 - \varepsilon F, \tag{2B.3.7}$$

where γF^2 is the *cooperative term* of special interest. The influence of the predator upon dC_i/dt is given by including terms $-\delta_i F C_i$ in the prey equations (2B.3-2).

2B.4 Existence of Small Amplitude Periodic Solutions

The coupled system we analyze is for *two* coral species and *one* starfish species. Our model is

$$\Gamma_F: \begin{cases} \dfrac{dC_1}{dt} = -2\alpha_2 C_1 C_2 + \alpha_1(C_2^2 - C_1^2) - \delta_1 F C_1 + \lambda_1 C_1 \\ \dfrac{dC_2}{dt} = -2\alpha_1 C_1 C_2 - \alpha_2(C_2^2 - C_1^2) - \delta_2 F C_2 + \lambda_2 C_2 \\ \dfrac{dF}{dt} = \beta F(C_1 + C_2) + \gamma F^2 - \varepsilon F . \end{cases} \tag{2B.4-1}$$

We suppose all Greek letters in (2B.4-1) represent positive constants. We also suppose in this section and in section 2B.5 that $\delta_1 = \delta_2 = \delta$ and $\lambda_1 = \lambda_2 = \lambda$. Further, to simplify notation we assume that $\alpha_2 = k\alpha_1$ $(k > 0)$ and $\alpha_1 = \alpha > 0$.

The unique equilibrium of (2B.4-1) in the positive orthant is then (C_{01}, C_{02}, F_0), where

$$C_{01} = \frac{\lambda - \delta F_0}{\alpha(1+k^2)} \equiv C_0, \quad C_{02} = kC_0 \text{ and}$$

$$F_0 = \frac{\alpha\varepsilon(1+k^2) - \beta\lambda(1+k)}{\alpha\gamma(1+k^2) - \beta\delta(1+k)} . \tag{2B.4-2}$$

(There is no need to assume $\delta_1 = \delta_2$ and $\lambda_1 = \lambda_2$ for such an equilibrium to exist.) Because α and k are positive, we must have $\lambda_0 - \delta F_0 > 0$. Since $F_0 > 0$, either

$$\beta\lambda(1+k) > \alpha\varepsilon(1+k^2) \text{ and } \beta\delta(1+k) > \alpha\gamma(1+k^2), \tag{2B.4-3}$$

or

$$\beta\lambda(1+k) < \alpha\varepsilon(1+k^2) \text{ and } \beta\delta(1+k) < \alpha\gamma(1+k^2). \qquad (2B.4\text{-}4)$$

To analyze (2B.4-1) for Hopf bifurcation we follow the Recipe-Summary given in Chapter 2 of *Theory and Applications of Hopf Bifurcation* (Hassard, Kazarinoff, Wan, 1981). We choose γ to be the bifurcation parameter.

We next perform a translation so as to bring the equilibrium (C_0, kC_0, F_0) of the system Γ_F to the origin. We define $x_i = C_i - C_{0i}$ $(i = 1, 2)$, and we let $x_3 = F - F_0$. Using the equilibrium relations

$$(1+k)\beta C_0 + \gamma F_0 - \varepsilon = 0 \quad \text{and}$$

$$-(1+k^2)\alpha C_0 + \lambda - \delta F_0 = 0, \qquad (2B.4\text{-}5)$$

we find that in terms of the x_i the system Γ_F becomes

$$\Gamma_\gamma \begin{cases} \dfrac{dx_1}{dt} = -2k\alpha(kC_0x_1+x_1x_2) + \alpha(x_2^2-x_1^2) - 2\alpha C_0x_1 \\ \qquad - \delta(C_0x_3+F_0x_1+x_1x_3) + \lambda x_1 \\ \dfrac{dx_2}{dt} = -2\alpha(C_0x_2+x_1x_2) - k\alpha(x_2^2-x_1^2) - 2k^2\alpha C_0x_2 \\ \qquad - \delta(kC_0x_3+F_0x_2+x_2x_3) + \lambda x_2 \\ \dfrac{dx_3}{dt} = \beta F_0x_1 + \beta F_0x_2 + [2\gamma F_0 - \varepsilon + \beta(1+k)C_0]x_3 \\ \qquad + \beta(x_1+x_2)x_3 + \gamma x_3^2. \end{cases} \qquad (2B.4\text{-}6)$$

The linear part of this new system is

$$\dot{x} = A(\gamma)x, \qquad (2B.4\text{-}7)$$

where

$$A(\gamma) = \begin{pmatrix} -(1+k^2)\alpha C_0 & 0 & -\delta C_0 \\ 0 & -(1+k^2)\alpha C_0 & -k\delta C_0 \\ \beta F_0 & \beta F_0 & \gamma F_0 \end{pmatrix},$$

and where (2B.4-5) has been used.

The characteristic equation of $A(\gamma)$ is $[\nu + (1+k^2)\alpha C_0]\,[\{\nu + (1+k^2)\alpha C_0\}(\nu - \gamma F_0) + (1+k)\delta\beta F_0 C_0] = 0$. Therefore, $\nu_3 = -(1+k^2)\alpha C_0$ is an eigenvalue of $A(\gamma)$, and the remaining two eigenvalues, ν_1 and ν_2 are roots of

$$\nu^2 + [(1+k^2)\alpha C_0 - \gamma F_0]\,\nu + [(1+k)\delta\beta - (1+k^2)\alpha\gamma]\,F_0 C_0 = 0\,. \qquad (2B.4\text{-}8)$$

We have $\mathrm{Re}\,\nu_1 = \mathrm{Re}\,\nu_2 = 0$ and $\mathrm{Im}\,\nu_1 \neq 0$ if and only if

$$\gamma = \gamma_c = (1+k^2)\alpha C_0 / F_0 = \frac{\alpha\,\delta\,\varepsilon(1+k^2)}{\beta\lambda(1+k) + \alpha(\lambda-\varepsilon)(1+k^2)} \qquad (2B.4\text{-}9)$$

and

$$(1+k)\beta\delta - (1+k^2)\alpha\,\gamma_c > 0\,,$$

which is the second inequality in (2B.4-3). It follows from (2B.4-8) that

$$\frac{d}{d\gamma}\mathrm{Re}\,\nu_1 = \tfrac{1}{2}\frac{d}{d\gamma}(\mathrm{Re}\,\nu_1 + \mathrm{Re}\,\nu_2) = \frac{F_0}{2}\left\{1 - (\delta + \gamma)\left[\frac{\alpha(1+k^2)}{\alpha\gamma(1+k^2) - \beta\delta(1+k)}\right]\right\}.$$

Hence, by (2B.4-3),

$$\left.\frac{d\,\mathrm{Re}\,\nu_1}{d\gamma}\right|_{\gamma=\gamma_c} > 0\,, \qquad (2B.4\text{-}10)$$

and the desired transversality condition holds. By (2B.4-10) the equilibrium (0, 0, 0) of Γ_γ is stable if $\gamma < \gamma_c$ and (2B.4-3) holds.

But it (2B.4-4) holds, then

$$(1+k)\beta\delta - (1+k^2)\alpha\gamma_c < 0$$

and ν_1 and ν_2 are real and have opposite sign at γ_c and for γ in a neighborhood of γ_c.

THEOREM I. If one of the conditions (2B.4-3) and (2B.4-4) holds, the system Γ_F has a unique equilibrium in the positive orthant. If (2B.4-3) holds, this equilibrium is unstable for $\gamma > \gamma_c$, which is defined by (2B.4-9), and is stable for $\gamma < \gamma_c$. If (4.4) holds, this equilibrium is unstable in a neighborhood of γ_c including γ_c. Further, if (2B.4-3) holds, there is a Hopf bifurcation at γ_c.

In the next section we study the asymptotic orbital stability of the bifurcating periodic solutions whose existence is guaranteed by Theorem I. The calculation required is complicated because so many parameters occur in the nonlinear terms of the model.

2B.5 Stability of Periodic Solutions

We now investigate the asymptotic orbital stability of the bifurcating periodic solutions and the direction of the bifurcation. We also discuss the periods of these solutions. To do these things we follow closely the Recipe-Summary in Chapter 2 of *Theory and Applications of Hopf Bifurcation* (Hassard, Kazarinoff and Wan, 1981). At γ_c the matrix A in (2B.4-7) becomes

$$A(\gamma_c) = \begin{pmatrix} \delta F_0(\gamma_c) - \lambda & 0 & -\delta C_0(\gamma_c) \\ 0 & \delta F_0(\gamma_c) - \lambda & -k\delta C_0(\gamma_c) \\ \beta F_0(\gamma_c) & \beta F_0(\gamma_c) & \lambda - \delta F_0(\gamma_c) \end{pmatrix} \quad (2B.5\text{-}1)$$

where γ_c is given by (2B.4-9) and F_0 and C_0 are given by (2B.4-2). The eigenvalues of $A(\gamma_c)$ are $\nu_3 = \delta F_0(\gamma_c) - \lambda$ and $\nu_1 = \overline{\nu}_2 = i\omega_0$, where

$$\omega_0 = \{[(1+k)\delta\beta - (1+k^2)\alpha\gamma_c]F_0(\gamma_c)C_0(\gamma_c)\}^{1/2} > 0 \quad (2B.5\text{-}2)$$

by the second of the inequalities (2B.4-3). Eigenvectors v_i $(i = 1,2,3)$ corresponding to the ν_i are:

$$v_1 = \begin{pmatrix} 1 \\ k \\ \dfrac{\delta F_0(\gamma_c) - \lambda - i\omega_0}{\delta C_0(\gamma_c)} \end{pmatrix}, \quad v_2 = \bar{v}_1,$$

and $v_3 = (1, -1, 0)^{tr}$, where the superscript tr denotes transpose and where $\bar{v}_1$ is the complex conjugate of v_1. We transform the system (2B.4-6) at γ_c replacing x by Py, where $P = (\mathrm{Re}\, v_1, -\mathrm{Im}\, v_1,\ v_3)$. Then $\dot{y} = P^{-1}A(\gamma_c)Py + F(y)$. To compute F we apply P^{-1}, to the nonlinear part of (2B.4-6) with x replaced by Py.

Let $\psi = (\delta F_0(\gamma_c) - \lambda)/\omega_0$. Then $F = (F^1, F^2, F^3)^{tr}$, in which

$$F^1(y) = \frac{1}{k+1}\,[N_1(y) + N_2(y)],$$

$$F^2(y) = \psi\left[F_1(y) + \frac{\delta}{\alpha(1+k^2)}\, N_3(y)\right],$$

$$F^3(y) = \frac{1}{k+1}\,[kN_1(y) - N_2(y)],$$

and

$$N_1(y) = 2k\alpha y_3^2 - \alpha(1+k^2)[y_1y_3 + (1/\psi)y_2(y_1 + y_3)],$$

$$N_2(y) = \alpha(1+k^2)[y_1y_3 - (1/\psi)\,y_2(ky_1 - y_3)] + 2\alpha y_3^2,$$

$$N_3(y) = \tau[\gamma_c\tau - \beta(1+k)]\,y_1^2 + \gamma_c[\omega_0/\delta C_0(\gamma_c)]^2 y_2^2$$

$$+ \frac{\omega_0}{\delta C_0(\gamma_c)}\,[\beta(1+k) - 2\gamma_c\tau]\,y_1y_2,$$

and $\tau = \alpha(1+k^2)/\delta$.

The next task is to compute certain derivatives of the F^i. These are:

$$\frac{\partial^2 F^1}{\partial y_1^2} = \frac{\partial^2 F^1}{\partial y_2^2} = 0,$$

$$\frac{\partial^2 F^2}{\partial y_1^2} = \frac{2}{\omega_0}\,[\tau\gamma_c - \beta(1+k)]\,(\delta F_0(\gamma_c) - \lambda)\,,$$

$$\frac{\partial^2 F^2}{\partial y_2^2} = -\frac{2\gamma_c\omega_0}{\delta C_0(\gamma_c)}\,, \qquad \frac{\partial^2 F^1}{\partial y_1 \partial y_2} = +\frac{2\omega_0}{(1+k)C_0(\gamma_c)}\,,$$

$$\frac{\partial^2 F^2}{\partial y_1 \partial y_2} = -\beta(1+k) + 2\tau(\gamma_c - \delta/(1+k))\,,$$

$$\frac{\partial^2 F^3}{\partial y_1^2} = \frac{\partial^2 F^3}{\partial y_2^2} = 0\,, \qquad \frac{\partial^2 F^3}{\partial y_1 \partial y_2} = 0\,,$$

and

$$\frac{\partial^2 F^i}{\partial y_j \partial y_3} = 0 \quad (i = 1, 2;\ j = 1, 2)\,.$$

All third order partial derivatives of the components of F with respect to the components of y are zero.

A certain constant $c_1(\gamma_c)$ determines the stability, periods, and direction of bifurcation of the bifurcating periodic solutions. This constant is defined in terms of the partial derivatives computed above as follows:

$$c_1(\gamma_c) = \frac{i}{2\omega_0}\,[g_{20}g_{11} - 2|g_{11}|^2 - \tfrac{1}{3}|g_{02}|^2] + g_{21}/2\,,$$

where

$$g_{11} = \tfrac{1}{4}\,[\Delta F^1 + i\Delta F^2]\,,$$

$$g_{02} = \tfrac{1}{4}\,[\Box F^1 - 2\partial^2 F^2/\partial y_1 \partial y_2] + \tfrac{i}{4}\,[\Box F^2 + 2\partial^2 F^1/\partial y_1 \partial y_2]\,,$$

$$g_{20} = \tfrac{1}{4}\,[\Box F^2 + 2\partial^2 F^1/\partial y_1 \partial y_2] + \tfrac{i}{4}[\Box F^2 - 2\partial^2 F^2/\partial y_1 \partial y_2]\,,$$

$$g_{21} = 2G_{110}w_{11} + G_{101}w_{20}\,,$$

where

$$G_{110} = \tfrac{1}{2}\,[\partial^2 F^1/\partial y_1 \partial y_3 + \partial^2 F^2/\partial y_2 \partial y_3] + \tfrac{i}{2}\,[\partial^2 F^2/\partial y_1 \partial y_3 - \partial^2 F^1/\partial y_2 \partial y_3]\,,$$

$$G_{101} = \frac{1}{2}\,[\partial^2F^1/\partial y_1\partial y_3 - \partial^2F^2/\partial y_2\partial y_3] + \frac{i}{2}\,[\partial^2F^1/\partial y_2\partial y_3 + \partial^2F^2/\partial y_1\partial y_3]\,,$$

$$w_{11} = \frac{1}{4}\,\Delta F^3\,, \quad w_{20} = \frac{1}{4}\left(\frac{\Box F^3 - 2i\partial^2F^3/\partial y_1\partial y_2}{1 + 2i\omega_0}\right),$$

and where Δ denotes the Laplace operator in y_1 and y_2, and $\Box$ denotes the wave operator $\partial^2/\partial y_1^2 - \partial^2/\partial y_2^2$.

We may now compute $\mathrm{Re}\ c_1(\gamma_c) = (-1/2\omega_0)\,\mathrm{Im}(g_{20}g_{11}) + \tfrac{1}{2}\,\mathrm{Re}\ g_{21}$. We find, using (2B.4-2), (2B.4-5), (2B.5-2), that $g_{11} \equiv G_{110} \equiv G_{101} \equiv 0$. Thus $g_{21} \equiv 0$ and $\mathrm{Re}\ c_1(\gamma_c) \equiv 0$.

Now, either the real parts of all subsequent c_i $(i > 1)$ in the Poincaré normal form corresponding to the system (2B.4-6) are zero, in which case there is true neutral stability or one of them is not zero. In any case analytical computation of the higher order terms is out of reach.

The result $\mathrm{Re}\ c_1(\gamma_c) = 0$ has been established under the hypotheses that $\delta_1 = \delta_2$ and $\lambda_1 = \lambda_2$ in (2B.4-1). Fortunately, it is biologically highly unlikely that the preferences of *A. planci* for the two coral species are equal, although each coral species may be equally beneficial to population increases of starfish. However, if, for example, $\delta_1 \neq \delta_2$ but $\lambda_1 = \lambda_2$, analytical computation of $\mathrm{Re}\ c_1(\gamma_c)$ is still out of reach. In the next section we present results obtained in this case by applying B.D. Hassard's computer code BIFOR2 (Hassard, Kazarinoff, Wan, 1981) to the model (2B.4-1).

2B.6 Bifurcation Results if $\delta_1 \neq \delta_2$

Suppose $\delta_2 = \ell\delta_1$ $(0 < \ell < 1)$, but $\lambda_1 = \lambda_2 = \lambda$. Then the analysis of sections 2B.4 and 2B.5 cannot be carried out "by hand." We have carried it out numerically on a computer, a CDC Cyber 170 Dual 730, using the code BIFOR2 with an error criterion which yields estimated errors in the results that are less than 1.0E-6. For the model (2B.4-1) we computed the equilibrium point (C_{01}, C_{02}, F_0), the basic frequency ω_0, the value γ_c of γ at which a Hopf bifurcation occurs, the direction of bifurcation μ_2, the correction coefficient τ_2 to the basic period $2\pi/\omega_0$, and the stability exponent β_2 for the following ranges of parameters: $\ell = 0.98(0.02)0.90$, $\beta = 0.5(0.5)3.5$, $\alpha = 0.1(0.1)0.5$ and $\alpha = 0.05$ with $\delta_1 = 0.5$, $\varepsilon = 1.0$, $\lambda_1 = \lambda_2 = 2.0$, $\alpha_1 = \alpha_2 = \alpha$. We present the results in Tables 2B.6-I - 2B.6.V . In each "cell" eight numbers appear. Those in the first column are, in order, values of C_{01}, C_{02}, F_0, and $10^2 \gamma_c$. Those in the second column are, in order, values of ω_0, $10^3 \mu_2$, τ_2 , and $10^3 \beta_2$.

The results in the Tables 2B.6-I - 2B.6-V show that *as* ℓ *decreases, the stability of the periodic, bifurcating solutions rapidly strengthens. For fixed* ℓ*, their periods lengthen as either* α *decreases or* β *increases.* These results are not unexpected. They accord with the principle that diversity of species is more stable than a large homogenous population. However, we emphasize that the preferences expressed through our assumption $\delta_1 = \ell\delta_2$ $(0 < \ell < 1)$ do not affect the benefits to the starfish population received from each species of coral. We have assumed that these benefits are equal throughout this paper. This equality of benefit is expressed by the single coefficient β common to the FC_1 and FC_2 terms in the predator equation in (2B.4-1).

2B.7 Conclusions

J. Porter's often cited dictum that *A.planci* acts to maintain high diversity of corals (thereby stabilizing the system) by preferentially feeding on competitively dominant species is apparently incorrect (Potts, 1981, p. 61). In fact, it is frequently found that *A. planci* feeds on rarer species, preferentially (Potts, 1981). The model developed here shows that *preferential feeding of any sort enhances stability* of the system as a whole!

Our model (2B.4-1) may be tested and fitted to real data. First, it is necessary to determine λ_1, α_1 and λ_2, α_2 from tank studies. If each of two coral species, known to occur together in nature and believed to compete for space and sunlight on the reef surface, is grown separately in a tank, a logistic curve may be fitted to polyp number increases (Harper, 1977 and Alexander, 1979). (P. Antonelli witnessed this procedure at the Hearon Island Research Station on the Great Barrier Reef at the courtesy of R. Endean.) In fact, this procedure may be carried out directly on the reef site using glass enclosures, provided small portions of the corals are naturally isolated from each other (Alexander, 1979).

If we set $\delta_1 = \delta_2 = 0$ in (2B.2-1), the equilibrium values are $C_{0i} = \alpha_i((\lambda_i)/(\alpha_1^2 + \alpha_2^2))$ $(i = 1,2)$, while the single species equilibria are $C_{0i}^S = (\lambda_i)/(\alpha_i)$ $(i = 1,2)$. If $\lambda_2 > \lambda_1$ for the separate intrinsic

Table 2B.6-I: $\ell = 0.98$

α \ β	3.5		3.0		2.5		2.0		1.5		1.0		0.5	
0.05	0.053	1.405	0.067	1.403	0.089	1.400	0.124	1.394	0.186	1.386	0.320	1.368	0.733	1.318
	0.230	0.180	0.262	0.130	0.305	0.0882	0.367	0.0545	0.463	0.0288	0.637	0.0110	1.09	0.00163
	4.05	21.6	4.04	13.4	4.03	7.70	4.02	4.00	4.00	1.78	3.97	0.608	3.87	0.119
	0.219	-0.736	0.272	-0.523	0.349	-0.363	0.469	-0.225	0.673	-0.119	1.09	-0.0461	2.28	-0.00697
0.1	0.075	1.391	0.093	1.386	0.119	1.379	0.160	1.368	0.229	1.351	0.366	1.318	0.744	1.227
	0.204	0.163	0.231	0.115	0.267	0.0755	0.319	0.0441	0.399	0.0211	0.545	0.00650	0.925	-0.0000957
	4.01	10.9	4.00	7.11	3.99	4.34	3.97	2.43	3.93	1.20	3.87	0.476	3.71	0.122
	0.556	-0.672	0.673	-0.477	0.838	-0.314	1.09	-0.184	1.49	-0.0888	2.28	-0.0279	4.46	+0.00435
0.2	0.094	1.361	0.114	1.351	0.142	1.338	0.183	1.318	0.249	1.286	0.372	1.227	0.671	1.070
	0.177	0.126	0.199	0.0842	0.230	0.0509	0.273	0.0260	0.340	+0.00926	0.462	-0.000383	0.757	-0.00522
	3.95	6.96	3.93	4.79	3.91	3.13	3.87	1.90	3.81	1.05	3.71	0.489	3.47	0.168
	1.26	-0.529	1.49	-0.355	1.81	-0.216	2.28	-0.112	3.04	-0.0405	4.46	+0.00174	8.24	+0.0272
0.3	0.102	1.332	0.122	1.318	0.149	1.299	0.187	1.271	0.248	1.227	0.355	1.145	0.597	0.936
	0.161	0.0937	0.182	0.0585	0.209	0.0315	0.248	0.0120	0.308	-0.000861	0.414	-0.00875	0.652	-0.0155
	3.90	6.02	3.87	4.29	3.84	2.92	3.79	1.87	3.71	1.10	3.58	0.565	0.329	0.241
	1.95	-0.399	2.28	-0.251	2.74	-0.137	3.40	-0.0529	4.46	+0.00391	6.44	+0.0424	11.4	+0.0945
0.4	0.106	1.304	0.125	1.286	0.150	1.262	0.186	1.227	0.241	1.171	0.336	1.070	0.536	0.817
	0.151	0.0667	0.170	0.0371	0.195	+0.0146	0.231	-0.00153	0.286	-0.0126	0.379	-0.0209	0.575	-0.0332
	3.85	5.74	3.81	4.19	3.77	2.94	3.71	1.95	3.62	1.21	3.47	0.672	3.15	0.349
	2.61	-0.288	3.04	-0.162	3.62	-0.0648	4.46	+0.00696	5.80	+0.0599	8.24	+0.109	14.1	+0.244
0.5	0.107	1.277	0.125	1.256	0.149	1.227	0.182	1.105	0.232	1.119	0.317	1.001	0.484	0.707
	0.143	0.0426	0.161	0.0170	0.185	-0.00239	0.218	-0.0168	0.268	-0.0277	0.350	-0.0382	0.515	-0.0611
	3.80	5.71	3.76	4.25	3.71	3.05	3.64	2.09	3.54	1.35	3.37	0.805	3.03	0.510
	3.25	-0.187	3.76	-0.0759	4.46	+0.0109	5.47	+0.0787	7.05	+0.137	9.89	+0.215	16.5	+0.556

Table 2B.6-II: $\ell = 0.96$

α \ β	3.5		3.0		2.5		2.0		1.5		1.0		0.5	
0.05	0.033	1.408	0.043	1.406	0.059	1.403	0.084	1.399	0.132	1.392	0.239	1.375	0.604	1.325
	0.250	1.48	0.287	1.07	0.337	0.735	0.408	0.462	0.520	0.253	0.723	0.107	1.23	0.0219
	4.13	51.7	4.13	30.8	4.12	16.9	4.11	8.25	4.09	3.39	4.05	1.04	3.94	0.168
	0.157	- 6.17	0.199	- 4.48	0.262	- 3.07	0.363	- 1.94	0.543	- 1.07	0.928	- 0.454	2.11	- 0.0957
0.1	0.052	1.396	0.066	1.392	0.087	1.385	0.200	1.375	0.179	1.358	0.302	1.325	0.664	1.232
	0.229	1.40	0.260	1.01	0.302	0.688	0.361	0.427	0.453	0.227	0.615	0.0875	1.01	0.0113
	4.10	21.7	4.09	13.6	4.07	7.87	4.05	4.14	4.01	1.88	3.94	0.671	3.77	0.147
	0.439	- 5.87	0.543	- 4.26	0.694	- 2.91	0.928	- 1.82	1.32	- 0.974	2.11	- 0.383	4.33	- 0.0522
0.2	0.072	1.368	0.089	1.358	0.114	1.345	0.151	1.325	0.213	1.293	0.332	1.232	0.629	1.072
	0.201	1.28	0.226	0.907	0.260	0.598	0.307	0.350	0.379	0.165	0.505	0.0451	0.800	- 0.0153
	4.03	11.4	4.01	7.53	3.98	4.68	3.94	2.68	3.88	1.38	3.77	0.589	3.52	0.184
	1.10	- 5.45	1.32	- 3.89	1.64	-2.59	2.11	- 1.53	2.88	- 7.37	4.33	- 0.209	8.12	0.0810
0.3	0.083	1.339	0.101	1.325	0.126	1.305	0.162	1.277	0.221	1.232	0.327	1.148	0.596	0.937
	0.183	1.14	0.205	0.788	0.234	0.495	0.275	0.266	0.337	0.101	0.443	- 0.00180	0.679	- 0.0609
	3.97	8.83	3.94	6.04	3.91	3.93	3.85	2.39	3.77	1.33	3.63	0.641	3.33	0.255
	1.78	-4.96	2.11	- 3.45	2.58	- 2.19	3.25	- 1.20	4.33	- 0.470	6.32	+0.00887	1.13	+0.378
0.4	0.089	1.311	0.107	1.293	0.131	1.268	0.166	1.232	0.220	1.175	0.315	1.072	0.515	0.817
	0.169	0.999	0.190	0.662	0.216	0.389	0.252	0.180	0.307	0.0333	0.400	- 0.0612	0.593	- 0.136
	3.92	7.82	3.88	5.50	3.83	3.71	3.77	2.36	3.68	1.39	3.52	0.736	3.19	0.362
	2.44	- 4.41	2.88	- 2.95	3.47	- 1.76	4.33	- 0.835	5.68	- 0.161	8.12	+0.324	14.0	+1.02
0.5	0.092	1.284	0.109	1.261	0.133	1.232	0.166	1.189	0.216	1.122	0.300	1.002	0.468	0.707
	0.159	0.851	0.178	0.534	0.202	0.282	0.235	0.0908	0.285	- 0.0446	0.367	- 0.140	0.529	- 0.252
	3.86	7.38	3.82	5.32	3.77	3.68	3.70	2.43	3.59	1.50	3.42	0.862	3.07	0.522
	3.09	- 3.81	3.62	- 2.43	4.33	- 1.30	5.35	- 0.434	6.94	+0.225	9.77	+0.799	16.3	+2.33

Table 2B.6-III: $\ell = 0.94$

α \ β	3.5		3.0		2.5		2.0		1.5		1.0		0.5	
0.05	0.024	1.409	0.032	1.408	0.044	1.406	0.064	1.402	0.102	1.395	0.192	1.381	0.514	1.332
	0.260	4.94	0.299	3.59	0.353	2.46	0.430	1.55	0.552	0.850	0.774	0.364	1.33	0.0821
	4.22	92.5	4.22	53.9	4.21	28.8	4.20	13.6	4.18	5.33	4.14	1.52	4.03	0.221
	0.129	-21.0	0.164	-15.3	0.217	-10.5	0.304	- 6.60	0.463	- 3.65	0.815	- 1.58	1.96	- 0.366
0.1	0.040	1.399	0.051	1.395	0.068	1.390	0.962	1.381	0.147	1.365	0.257	1.332	0.596	1.238
	0.241	4.68	0.276	3.40	0.322	2.32	0.387	1.46	0.487	0.794	0.664	0.329	0.108	0.0579
	4.19	35.0	4.18	21.3	4.16	12.0	4.14	6.09	4.10	2.64	4.03	0.882	3.84	0.174
	0.370	-20.0	0.463	-14.6	0.600	-10.0	8.15	- 6.31	1.19	- 3.47	1.96	- 1.46	4.17	- 0.274
0.2	0.059	1.374	0.074	1.365	0.095	1.352	0.129	1.332	0.186	1.300	0.298	1.238	0.589	1.076
	0.216	4.40	0.244	3.18	0.281	2.15	0.332	1.31	0.409	0.675	0.542	0.232	0.841	- 0.0128
	4.12	16.4	4.10	10.6	4.07	6.37	4.03	3.53	3.96	1.73	3.84	0.697	3.57	0.201
	0.973	-19.1	1.19	-13.9	1.49	- 9.46	1.96	- 5.86	2.72	- 3.07	4.17	- 1.09	7.99	+0.0690
0.3	0.070	1.346	0.086	1.332	0.109	1.313	0.143	1.284	0.199	1.238	0.302	1.153	0.541	0.939
	0.197	4.16	0.221	2.96	0.253	1.95	0.296	1.14	0.361	0.521	0.470	+0.107	0.704	- 0.128
	4.06	11.9	4.03	7.94	3.99	5.02	3.93	2.95	3.84	1.57	3.69	0.720	3.38	0.269
	1.63	-18.4	1.96	-13.2	2.41	- 8.79	3.09	- 5.21	4.17	- 2.46	6.18	- 0.537	11.2	+0.809
0.4	0.076	1.318	0.093	1.300	0.116	1.274	0.149	1.238	0.202	1.180	0.294	1.076	0.493	0.818
	0.183	3.88	0.205	2.70	0.233	1.71	0.271	0.927	0.327	0.341	0.420	- 0.0511	0.611	- 0.311
	4.00	10.1	3.96	6.91	3.91	4.53	3.84	2.79	3.74	1.58	3.57	0.803	3.23	0.375
	2.28	-17.4	2.72	-12.3	3.31	- 7.91	4.17	- 4.38	5.53	- 1.68	7.99	+0.276	13.8	+2.36
0.5	0.081	1.291	0.097	1.268	0.119	1.238	0.151	1.194	0.200	1.126	0.283	1.005	0.450	0.707
	0.172	3.56	0.191	2.41	0.217	1.45	0.251	0.690	0.301	0.131	0.383	- 0.253	0.542	- 0.583
	3.94	9.18	3.90	6.46	3.84	4.36	3.76	2.79	3.65	1.67	3.47	0.923	3.12	0.534
	2.93	-16.3	3.46	-11.1	4.17	- 6.84	5.20	- 3.36	6.80	- 0.674	9.64	+1.47	0.162	+5.50

Table 2B.6-IV: $\ell = 0.92$

α \ β	3.5		3.0		2.5		2.0		1.5		1.0		0.5	
0.05	0.019	1.410	0.025	1.409	0.035	1.407	0.051	1.404	0.083	1.398	0.161	1.385	0.448	1.338
	0.265	11.4	0.306	8.26	0.362	5.66	0.443	3.56	0.571	1.95	0.808	0.841	1.40	0.195
	4.32	143.	4.31	82.3	4.30	43.2	4.29	19.9	4.27	7.57	4.23	2.07	4.11	0.276
	0.115	-49.3	0.146	-35.9	0.193	-24.6	0.270	-15.5	0.431	-8.56	-0.737	-3.71	1.83	-0.886
0.1	0.032	1.402	0.042	1.398	0.056	1.393	0.080	1.385	0.125	1.370	0.224	1.338	0.540	1.244
	0.249	10.8	0.286	7.83	0.335	5.35	0.404	3.36	0.511	1.84	0.701	0.781	1.14	0.157
	4.28	50.6	4.27	30.3	4.25	16.7	4.23	8.26	4.19	3.46	4.11	1.10	3.92	0.202
	0.329	-47.1	0.413	-34.2	0.538	-23.5	0.737	-14.8	1.09	-8.20	1.83	-3.54	4.02	-0.756
0.2	0.049	1.378	0.062	1.370	0.082	1.357	0.112	1.338	0.164	1.306	0.270	1.244	0.551	1.080
	0.226	10.2	0.256	7.38	0.295	5.03	0.350	3.12	0.433	1.66	0.572	0.627	0.878	0.0227
	4.21	22.0	4.19	13.9	4.16	8.18	4.11	4.41	4.04	2.09	3.92	0.808	3.64	0.218
	0.885	-45.1	1.09	-32.8	1.38	22.5	1.83	-14.2	2.57	-7.69	4.02	-3.02	7.85	-0.125
0.3	0.060	1.352	0.075	1.338	0.955	1.319	0.127	1.291	0.180	1.244	0.279	1.158	0.513	0.941
	0.208	9.77	0.234	7.03	0.267	4.73	0.314	2.85	0.382	1.41	0.494	0.398	0.729	-0.202
	4.15	15.2	4.11	9.93	4.07	6.15	4.01	3.53	3.92	1.82	3.76	0.802	3.44	0.284
	1.51	-44.0	1.83	-31.9	2.27	-21.7	2.94	-13.3	4.02	-6.81	6.03	-2.05	11.0	+1.31
0.4	0.067	1.325	0.082	1.306	0.103	1.281	0.135	1.244	0.185	1.186	0.275	1.080	0.470	0.819
	0.194	9.35	0.216	6.64	0.246	4.36	0.286	2.51	0.344	1.09	0.439	0.0910	0.628	-0.552
	4.08	12.4	4.04	8.37	3.99	5.37	3.92	3.23	3.81	1.78	3.64	0.872	3.29	0.389
	2.15	-42.8	2.57	-30.7	3.16	-20.5	4.02	-12.1	5.38	-5.46	7.85	-0.502	13.7	+4.30
0.5	0.072	1.297	0.087	1.275	0.108	1.244	0.138	1.200	0.186	1.132	0.267	1.008	0.431	0.707
	0.182	8.87	0.202	6.18	0.229	3.92	0.265	2.09	0.316	0.685	0.398	-0.306	0.554	-1.06
	4.02	11.0	3.98	7.63	3.92	5.05	3.84	3.16	3.72	1.84	3.53	0.986	3.17	0.549
	2.78	-41.3	3.30	-29.2	4.02	-18.9	5.05	-10.4	6.66	-3.61	9.51	+1.82	16.0	+10.3

Table 2B.6-V: $\ell = 0.90$

α \ β	3.5		3.0		2.5		2.0		1.5		1.0		0.5	
0.05	0.015	1.411	0.021	1.410	0.029	1.408	0.043	1.405	0.070	1.400	0.138	1.387	0.395	1.343
	0.269	21.2	0.311	15.4	0.368	10.6	0.452	6.67	0.585	3.66	0.833	1.57	1.46	0.368
	4.41	203.	4.41	116.	4.40	59.9	4.39	27.2	4.36	10.1	4.32	2.66	4.21	0.333
	0.109	-94.0	0.137	-68.5	0.180	-47.0	0.250	-29.6	0.383	-16.3	0.684	-7.07	1.72	-1.70
0.1	0.027	1.403	0.035	1.400	0.048	1.395	0.069	1.387	0.108	1.373	0.198	1.343	0.492	1.251
	0.255	20.2	0.293	14.7	0.344	10.0	0.416	6.29	0.529	3.45	0.730	1.47	1.20	0.314
	4.38	68.4	4.36	40.4	4.35	21.9	4.32	10.6	4.28	4.34	4.21	1.33	4.00	0.230
	0.305	-89.9	0.383	-65.4	0.498	-44.8	0.684	-28.3	1.02	-15.6	1.72	-6.80	3.88	-1.55
0.2	0.043	1.382	0.054	1.373	0.071	1.362	0.099	1.343	0.147	1.312	0.246	1.251	0.515	1.085
	0.233	19.0	0.265	13.8	0.307	9.42	0.365	5.88	0.452	3.17	0.598	1.26	0.913	0.105
	4.31	27.9	4.28	17.4	4.25	10.1	4.21	5.33	4.13	2.46	4.00	0.920	3.71	0.236
	0.824	-85.9	1.02	-62.5	1.30	-43.0	1.72	-2.72	2.45	-15.0	3.88	-6.18	7.71	-0.591
0.3	0.053	1.356	0.066	1.343	0.085	1.325	0.115	1.297	0.164	1.251	0.259	1.164	0.486	0.944
	0.216	18.3	0.243	13.2	0.279	8.97	0.328	5.51	0.399	2.83	0.515	0.912	0.752	-0.269
	4.24	18.6	4.21	12.0	4.16	7.31	4.10	4.11	4.00	2.07	3.84	0.885	3.50	0.300
	1.42	-84.0	1.72	-61.2	2.16	-42.0	2.81	-26.2	3.88	-1.39	5.89	-4.79	10.9	+1.79
0.4	0.60	1.330	0.074	1.312	0.093	1.287	0.123	1.251	0.171	1.192	0.258	1.085	0.448	0.820
	0.202	17.7	0.226	12.7	0.257	8.47	0.299	5.03	0.360	2.35	0.456	0.419	0.644	-0.852
	4.17	14.8	4.13	9.86	4.08	6.23	4.00	3.68	3.89	1.98	3.71	0.944	3.35	0.406
	2.04	-82.6	2.45	-59.9	3.03	-40.6	3.88	-24.7	5.24	-1.21	7.71	-2.36	13.5	+6.81
0.5	0.065	1.303	0.079	1.281	0.098	1.251	0.127	1.206	0.173	1.137	0.251	1.012	0.411	0.706
	0.190	17.0	0.212	12.1	0.239	7.85	0.276	4.41	0.329	1.72	0.411	-0.237	0.565	-1.69
	4.11	12.9	4.07	8.82	4.00	5.72	3.92	3.54	3.80	2.02	3.60	1.05	3.24	0.567
	2.66	-80.9	3.17	-58.1	3.88	-38.6	4.91	-22.4	6.51	-9.25	9.37	1.45	15.8	+16.9

growth rates, then $C^s_{01} > C^s_{02}$ implies $C_{02} > C_{01}$. This we term a *reversal.* Such reversals are known to occur in aquatic plant systems (Harper, 1977). If $\lambda_1 > \lambda_2$, no reversal occurs necessarily. However, the appearance of a reversal would strongly corroborate our basic model, as a reversal is entirely a consequence of our optimality condition for aragonite production. (See 2C.4-5.)

Once a reversal has been established we can proceed to analyze the deeper properties of the model, especially the limit cycle parameters. At present the only such parameter available is the period, which according to several authorities is of the order of twenty years (Campbell, 1976; Taylor and Taylor, 1981; Potts, 1981; and a personal communication from R. Endean, 1983).

Clearly, further numerical work will have to be done, as well as statistical analysis of experimental design before calibration of our full model can be achieved. The measurement of the preference coefficients δ_1 and δ_2 is perhaps the most pressing of these problems.

Acknowledgement

The authors acknowledge support from the following grants: NSERC-A-7667 (Antonelli) and N.S.F. Grant MCS-8106657 (Kazarinoff).

References

Alexander, R.M. 1979. *The Invertebrates.* Cambridge University Press.

Antonelli, P.L. 1980a. Optimal growth of an ideal coral reef. *Acta Cientifica Venezolana,* 31: 521-525.

Antonelli, P.L. 1980b. Huxley's allometric space for an ideal coral reef or forest. In G.E. Lasker, ed., *Proc. Int. Cong. on Systems Research,* IV: 1973-1978. Acapulco, Mexico.

Antonelli, P.L. and B.H. Voorhees. 1983. Nonlinear growth mechanics I. *Bull. Math. Biol.,* 45, no. 1: 103-116.

Campbell, A.C. 1976. *The Coral Seas.* Orbis Publishing, London.

Dana, T.F., W.A. Newman, and E.W. Fager. 1972. *Acanthaster* aggregations: interpreted as primarily responses to natural phenomena. *Pacific Sci.,* 26: 355-372.

Endean, R. 1973. Population explosions of *Acanthaster planci* and associated destruction of hermatypic corals in the Indo-West Pacific region. In O.A. Jones and R. Endean, eds., *Biology and Geology of Coral Reefs,* Vol. II, Biology 1, pp. 389-438. Academic Press, New York.

Endean, R. 1977. *Acanthaster planci* infestations of reefs of the great barrier reef. Proc. 3rd Internat. Coral Reef Symp., Vol. 1, pp. 185-191.

Endean, R. 1983. Personal communication.

Harper, J. 1977. *The Population Biology of Plants.* Academic Press, New York.

Hassard, B.D., N.D. Kazarinoff, and Y.-H. Wan. 1981. *Theory and Applications of Hopf Bifurcation.* London Math. Soc. Lect. Note Series, No. 41, Cambridge University Press, London - New York.

Maragos, J.E. 1978. Coral growth: geometrical relationships. In D.R. Stoddard and R.F. Johannes, eds., *Coral Reefs,* pp. 543-550. UNESCO Monographs in Oceanographic Methodology.

Pearson, R.G. 1975. Coral Reefs. *Unpredictable Climatic Factors and Acanthaster.* Crown-of-thorns Starfish Seminar Proc., pp. 131-134. A.G.P.S., Canberra.

Pearson, R.G. and R. Endean. 1969. *A Preliminary Study of the Coral Predator Acanthaster planci (L.) (Asteroidea) on the Great Barrier Reef.* Qld. Vol. 3, 27-55. Fish. Notes Dept. Harbours and Marine.

Porter, J. 1976. *Autotrophy, Heterotrophy and Resource in Caribbean Reef Building Corals, Amer. Nat.,* Vol. 110, pp. 731-742.

Potts, D.C. 1981. Crown-of-Thorns Starfish—man-induced pest or natural phenomena. Chapter 4 in R.L. Kitching and R.E. Jones, eds., *The Ecology of Pests (Some Australian Case Histories),* pp. 55-86. CSIRO, Melbourne, Australia.

Stearn, C.W. and J.P. Scoffin. 1977. Carbonate budget of a fringing reef, Barbados. In *Proc. 3rd Int. Coral Reef Symposium.*

Taylor, R. and V. Taylor. 1981. Paradise beneath the sea. *Natl. Geog. Mag.,* 159: 5.

Wickler, W. 1973. Biology of *Hymenocera picta* dana. *Micronesica,* 9: 225-230.

2C
Competition and Productivity in Aquatic Plant Communities: Experiments and Theories

P.L. Antonelli

2C.1 Introduction

The G.F. Gause type competition experiments for plants were executed by J.L. Harper and J.N. Clatworthy in 1962 (Harper and Clatworthy, 1962). They used four species of simple aquatic plants known to live closely in nature. These were grown in beakers under excellent conditions, singly and two at a time. The populations were only space limited and then only towards climax (i.e. a steady state in which biomass and net frond number are constant). A total of three combinations were statistically and biologically analyzable. For both singles and doubles, three phases of growth were observed. Phase I was an exponential increase of biomass, phase II an equilibrium of numbers of new fronds budded off the parents per day and phase III was a climax phase in which both biomass and frond numbers were constant. Actually, Harper and Clatworthy used biomass instead of frond numbers to detect phase II, assuming effectively that frond size was normally distributed in all populations. This could be a step away from the general idea of Harper that *a plant should be considered a population of parts, or more precisely, a set of populations of essentially identical modular units* like fronds, roots, or leaves. However, there is a real need to verify *Harper's population paradigm* experimentally by showing that the production rate of *frond numbers* is constant in phase II. To accomplish this Professor M.R.T. Dale and the author are beginning similar experiments at the University of Alberta. If such an equilibrium of frond numbers is not clearly forthcoming from our data, then Harper's idea can not be taken seriously. We do not expect this.

The Harper/Clatworthy growth data for singles and doubles were statistically fitted to Gompertz curves by P.H. Leslie of Oxford University. In the case of doubles, the loser population does not fit a Gompertz curve, but the *total* biomass in the combination does fit. Moreover, it was noticed in two out of three cases of the doubles, that the single species phase II averages did *not* predict the winners at phase III. I shall call these cases *phase II reversals.* Also, for doubles, the phase II outcomes were the same as phase III outcomes. Perhaps these results may have contributed to Harper's opinion that models should not be used except as crude guides in the study of *even the simplest* plant communities. It seems to me, however, that this may be somewhat presumptuous, for while it is true that the usual Volterra competition equations offers no help in understanding the two (2) reversals, this classical theory does *not* incorporate optimal growth criteria. Yet, this criteria is experimentally invoked! I intend to show how *optimality of production* may be brought into the mathematical modelling in such a way (essentially unique) that a new set of Volterra equations applies to the present situation and, in fact, *predicts* the very *phase II reversals* mentioned. This theory is rooted in the Gompertz growth of single species. I combine this with *Harper's population paradigm* applied to phase II equilibrium of frond numbers mentioned above. In this way the unique intrinsic optimality of single species growth is revealed. The step to the multiple species theory is also unique, but is difficult to prove mathematically. The uniqueness is proved in essay 3B, part II. The theorem is that there is only one optimality principle (based on a second order Homogeneous Lagrangian) which yields a Volterra equation with constant coefficients. It is also a theorem that the induced ecology is of competitive/cooperative type as opposed to predator/prey. The n-species theory does not exhibit a phase III steady state because of the constancy of coefficients. However, a simple perturbation method has recently been used by the author to remedy this, and appears in essay 2F of this chapter.

In section 2C.2, I discuss the single species system within the optimality framework. Section 2C.3 employs the *theory of sprays* and the theory of *Volterra-Hamilton systems with constant coefficients* from essay 3B in the next chapter. In section 2C.3 the 2-species Gompertzian growth equations are derived directly. The solution to the n-species system are given explicitly in section 2C.4, using the Voorhees solution of the constant coefficient Volterra-Hamilton spray. The phase II reversals are discussed with explicit formulas as is the remaining case. I am led to believe the notion that if the species are not on par with each other then no reversals can occur.

2C.2 The Single Species Gompertzian

One could interpret the system

$$\Gamma_y : \begin{cases} \dfrac{dy}{dt} = \ell \cdot L \\ \dfrac{dL}{dt} = \lambda L(1 - \alpha L), \end{cases} \tag{2C.2-1}$$

as describing the increase of $y(t)$, the natural logarithm of the biomass at time t, and the logistic behavior of $L(t)$, the net number of fronds generated per unit time, at time t.† However, $y(t)$ from (2C.2-1) is *not* Gompertzian. Indeed, $y(t)$ is essentially linear as numerical studies easily show. However, the system

$$\Gamma_{Gomp} : \begin{cases} \dfrac{dy}{dt} = \ell \cdot \left(\dfrac{(1/\alpha) - L}{L}\right) \\ \dfrac{dL}{dt} = \lambda L(1 - \alpha L), \end{cases} \tag{2C.2-2}$$

is solved by

$$y(t) = \ell n\, q(t) = \ell n\, a - ce^{-\lambda t} \tag{2C.2-3}$$

where,

$$q(t) = ae^{-ce^{-\lambda t}}, \tag{2C.2-4}$$

is the Gompertz function for the biomass $q(t)$. If we set

$$\frac{(1/\alpha) - L}{L} = L_* , \tag{2C.2-5}$$

Then (2C.2-2) becomes,

$$\left. \begin{aligned} \frac{dy}{dt} &= \ell \cdot L_* \\ \frac{dL_*}{dt} &= -\lambda L_* . \end{aligned} \right\} \tag{2C.2-6}$$

This system may, by an astute coordinate change, be converted into the Hamiltonian form of (2C.2-1). Namely, I get

†This definition is different than that used in essays 2D and 2E.

$$\Gamma_\phi : \begin{cases} \dfrac{d\phi(y)}{dt} = \ell \cdot L \\ \dfrac{dL}{dt} = -\lambda L(1 - \alpha L), \end{cases} \tag{2C.2-7}$$

by defining the nonsingular differentiable map

$$x(t) \equiv \phi(y(t)) = x_{max} - \frac{\ell}{\lambda\alpha} \ell n(1 + \frac{1}{b_0} e^{-\lambda t}). \tag{2C.2-8}$$

Here, x_{max} is arbitrary except it must exceed the log term in order that $x(t) > 0$ for all $t > 0$. However, $\phi(y_{max}) = x_{max}$ and

$$\alpha L(0) = (1 + b_0)^{-1}, \tag{2C.2-9}$$

defines b_0 in terms of the *initial frond number* $L(0)$ and the *coupling constant* α.

As we are interested in the approach to steady state, it will be convenient to define $\hat{y} = y_{max} - y(t) = \ell n\, a - y(t)$ and $\hat{x} = x_{max} - x(t)$. One sees clearly that

$$\frac{d\hat{x}}{d\hat{y}} > 0, \tag{2C.2-10}$$

from (2C.2-3) and (2C.2-8), so that $\hat{x}$ is a monotone increasing function. In terms of $\hat{x}$, the system (2C.2-7) becomes

$$\Gamma_{\hat{x}} : \begin{cases} \dfrac{d\hat{x}}{dt} = \ell \cdot (-L) \\ \dfrac{d(-L)}{dt} = \lambda L(1-\alpha L).^{\dagger} \end{cases} \tag{2C.2-11}$$

This system *defines* phase II as *both* an equilibrium of new frond numbers *and* as a constant addition of biomass, as reflected in the constant decrease of $\hat{x}$. Note that $\hat{x} = 0$, if and only if $\hat{y} = 0$, and that

$$\hat{x}(t) = \frac{\ell}{\lambda\alpha} \ell n(1 + \frac{1}{b_0 c} \hat{y}), \tag{2C.2-12}$$

†Forward running time is measured by negative reals from 0 to $-\infty$ in the second equation of $\Gamma_{\hat{x}}$. The number of fronds per unit time, at time t, is given by the solution of the second equation at $-t$.

defines the nonsingular differentiable *coordinate transformation.* Use of (2C.2-12) converts (2C.2-11) into

$$\Gamma_*: \begin{cases} \dfrac{d\hat{y}}{dt} = -\ell \cdot L_* \\ \dfrac{d(L_*)}{dt} = -\lambda L_* , \end{cases} \tag{2C.2-13}$$

which is obviously the same as (2C.2-6). Furthermore, it should be stressed that $x(t)$ and $L(t)$ are independent variables in our treatment, as in any classical Hamiltonian approach.

In what is to follow we suppress the constant ℓ in $\hat{\Gamma}$. This may be accomplished by linearly scaling the time t to $t' = \ell \cdot t$. (I drop the prime.) Having assumed this, we now write (2C.2-6) as

$$\frac{d^2y}{dt^2} = -\lambda \frac{dy}{dt} , \tag{2C.2-14}$$

or equivalently,

$$\frac{d^2\hat{y}}{dt^2} = -\lambda \frac{d\hat{y}}{dt} . \tag{2C.2-15}$$

Using (2C.2-12), (2C.2-15) transforms to

$$-\frac{d^2\hat{x}}{dt^2} = \lambda \frac{d\hat{x}}{dt} + \lambda\alpha \left(\frac{d\hat{x}}{dt}\right)^2 . \tag{2C.2-16}$$

This is the Euler-Lagrange equation of the second order homogeneous Lagrangian

$$L_{\hat{x}} = e^{2\lambda\alpha\hat{x}+\lambda t} \left(\frac{d\hat{x}}{dt}\right)^2 , \tag{2C.2-17}$$

whereas,

$$L_{\hat{y}} = e^{+\lambda t} \left(\frac{d\hat{y}}{dt}\right)^2 , \tag{2C.2-18}$$

is the Lagrangian of (2C.2-14). Since

$$L_{\hat{x}} = \frac{1}{(\lambda\alpha b_0 c)^2} L_{\hat{y}} , \tag{2C.2-19}$$

it is clear that the optimality principles for (2C.2-15) and (2C.2-16) are equivalent. However, I shall find it convenient to work with $L_{\hat{x}}$ for the n-species cases. These single species Lagrangians are also positive definite forms, so the extremals solving the variational problem

$$\delta \int_{t_1}^{t_2} L_{\hat{x}}^{1/2}\, dt = 0 \tag{2C.2-20}$$

are actually minimal time paths and likewise for $L_{\hat{y}}$.

In section 2C.3, I use the product Lagrangian

$$L_{\text{Prod}}^{1/2} = e^{\alpha_1 \lambda \hat{x}^1} \cdot e^{\alpha_2 \lambda \hat{x}^2} \cdot [(L^1)^2 + (L^2)^2]^{1/2}, \tag{2C.2-21}$$

in the form of a Riemannian metric tensor, g_{ij}, on biomass space. The main theorem of essay 3B in the next chapter shows this to be the only choice yielding Euler-Lagrange equations which reduce to Volterra equations with constant coefficients.

2C.3 The Gompertzian Community for Two

In order to obtain the Gompertzian dynamics we need to compute the new Riemannian metric from the old one using now the same formula as (2C.2-12) for each $(\hat{x}^i, \hat{y}^i)$. Via the classical formula (see Appendix of essay 2C)

$$g'_{ij}(\hat{y}) = g_{k\ell} \frac{\partial \hat{x}^k}{\partial \hat{y}^i} \frac{\partial \hat{x}^\ell}{\partial \hat{y}^j}, \tag{2C.3-1}$$

I obtain

$$g'_{ij} = e^{2(\lambda_1 \alpha_1 \hat{x}^1 + \lambda_2 \alpha_2 \hat{x}^2)} \delta_{k\ell} \left(\frac{\partial \hat{x}^k}{\partial \hat{y}^i} \circ \frac{\partial \hat{x}^\ell}{\partial \hat{y}^j} \right), \tag{2C.3-2}$$

or, after differentiation,

$$g'_{ij} \doteq (Q_1^2 + Q_2^2)\, \delta_{ij}, \tag*{(2C.3-2)$'$}$$

where

$$\left.\begin{aligned} Q_1 &= \frac{\alpha_1 \lambda_1 b_{01} c_1}{d_0}\left(1 + \frac{\hat{y}^1}{b_{01} c_1}\right) \\ Q_2 &= \frac{\alpha_2 \lambda_2 b_{02} c_2}{d_0}\left(1 + \frac{\hat{y}^2}{b_{02} c_2}\right) \\ d_0 &= (\alpha_1 \lambda_1 b_{01} c_1)(\alpha_2 \lambda_2 b_{02} c_2) \end{aligned}\right\} \tag{2C.3-3}$$

Now, I compute the new Γ'^{i}_{jk} from the *formula of Levi-Cività*

$$\Gamma'^{i}_{jk} = \frac{1}{2} g'^{i\lambda}\left(\frac{\partial g'_{j\lambda}}{\partial \hat{y}^k} + \frac{\partial g'_{k\lambda}}{\partial \hat{y}^j} - \frac{\partial g'_{kj}}{\partial \hat{y}^\lambda}\right). \tag{2C.3-4}$$

I obtain the formulas

$$\left.\begin{aligned} \Gamma'^{1}_{11} &= \frac{\lambda_1 \alpha_1}{d_0} \cdot \frac{Q_1}{Q_1^2 + Q_2^2} \\ \Gamma'^{2}_{22} &= \frac{\lambda_2 \alpha_2}{d_0} \cdot \frac{Q_2}{Q_1^2 + Q_2^2} \\ \Gamma'^{1}_{12} &= \frac{\lambda_2 \alpha_2}{d_0} \cdot \frac{Q_1}{Q_1^2 + Q_2^2} \\ \Gamma'^{2}_{21} &= \frac{\lambda_1 \alpha_1}{d_0} \circ \frac{Q_2}{Q_1^2 + Q_2^2} \\ \Gamma'^{1}_{22} &= -\Gamma'^{1}_{11}, \quad \Gamma'^{2}_{11} = -\Gamma'^{2}_{22} \end{aligned}\right\} \tag{2C.3-4}$$

Let me now write out the Gompertzian equations for two species. First, I do this for natural parameter s and then using $\lambda_1 = \lambda_2 = \lambda$, I convert these to *real time form.* The s-equations are

$$\frac{d^2\hat{y}^1}{ds^2} = \frac{-Q_1}{d_0(Q_1^2 + Q_2^2)} \cdot \left[-2\lambda_2 \alpha_2 \left(\frac{d\hat{y}^1}{ds}\right)\left(\frac{d\hat{y}^2}{ds}\right) + \lambda_1 \alpha_1 \left(\left(\frac{d\hat{y}^2}{ds}\right)^2 - \left(\frac{d\hat{y}^1}{ds}\right)^2\right)\right], \tag{2C.3-5}$$

and

$$\frac{d^2\hat{y}^2}{ds^2} = \frac{-Q_1}{d_0(Q_1^2+Q_2^2)}\left[-2\lambda_1\alpha_1(\frac{d\hat{y}^1}{ds})(\frac{d\hat{y}^2}{ds}) + \lambda_2\alpha_2\left((\frac{d\hat{y}^1}{ds})^2 - (\frac{d\hat{y}^2}{ds})^2\right)\right]. \qquad (2C.3\text{-}6)$$

The transformation of time parameter given by $s = e^{-\lambda t}$ converts (2C.3-5) and (2C.3-6) to

$$\frac{d^2\hat{y}^1}{dt^2} = \frac{-Q_1}{d_0(Q_1^2+Q_2^2)}\left\{-2\lambda_2\alpha_2(\frac{d\hat{y}^1}{dt})(\frac{d\hat{y}^2}{dt}) + \lambda_1\alpha_1\left((\frac{d\hat{y}^2}{dt})^2 - (\frac{d\hat{y}^1}{dt})^2\right) + \lambda_1\cdot\frac{Q_1^2+Q_2^2}{Q_1Q_2}\cdot\frac{d\hat{y}^1}{dt}\right\}, \qquad (2C.3\text{-}7)$$

and

$$\frac{d^2\hat{y}^2}{dt^2} = \frac{-Q_2}{d_0(Q_1^2+Q_2^2)}\left\{-2\lambda_1\alpha_1(\frac{d\hat{y}^1}{dt})(\frac{d\hat{y}^2}{dt}) + \lambda_2\alpha_2\left((\frac{d\hat{y}^1}{dt})^2 - (\frac{d\hat{y}^2}{dt})^2\right) + \lambda_2\cdot\frac{Q_1^2+Q_2^2}{Q_1Q_2}\cdot\frac{d\hat{y}^2}{dt}\right\}. \qquad (2C.3\text{-}8)$$

These two equations (2C.3-7) and (2C.3-8) are the *Growth Equations for a 2-species Gompertzian community* with $\lambda_1 = \lambda_2$. The n-species equations are exactly analogous. What is extremely important is that upon transformation from $\hat{y}_i$ to $\hat{x}^i$, the n-species Gompertzian becomes

$$\Gamma_{\hat{x}}: \begin{cases} \dfrac{d\hat{x}^i}{dr} = -L^i \\ \dfrac{d(-L^i)}{dr} = +\Gamma^i_{jk}L^jL^k, \end{cases} \qquad (2C.3\text{-}9)$$

where

$$\Gamma : \begin{cases} \Gamma^i_{ii} = \alpha_i \lambda , & \\ \Gamma^i_{ji} = \Gamma^i_{ij} = \alpha_i \lambda , & i \neq j \\ \Gamma^i_{jj} = -\alpha_j \lambda , & i \neq j \\ \Gamma^i_{jk} = 0 , & i \neq j \neq k , \end{cases} \qquad (2C.3\text{-}10)$$

and

$$r = b_0 (e^{-\lambda t} - b_0) = b_0 s - b_0 = as + b . \qquad (2C.3\text{-}11)$$

The Lagrangian for (2C.3-9) takes the product form

$$L^{1/2}_{\mathrm{Prod}} = e^{\alpha_1 \lambda \hat{x}^1} \cdots e^{\alpha_n \lambda \hat{x}^n} \left(\sum_{i,j} \delta_{ij} L^i L^j \right)^{1/2} . \qquad (2C.3\text{-}12)$$

I shall use (2C.3-9) in the next section to give the solution of the n-species Gompertzian.

2C.4 Solution of the n-Species Gompertzian Growth Equations and Conclusion

In essay 3A in the next chapter, B.H. Voorhees has given the solution of the system (2C.3-9) (see Antonelli and Voorhees, 1983). Unfortunately, this Voorhees solution uses a *peculiar parametrization* and extreme care must be exercised in matching these parameters to those used in this paper, which are biologically motivated. Following Voorhees' notation I use $\underline{\alpha}$ to denote $(\alpha_1, \ldots, \alpha_n)$, $|\underline{\alpha}|$ for the Euclidean norm and $\underline{K} = (L^1(0), \ldots, L^n(0))$ for the initial *frond vector.*

The solution of the Gompertzian growth equations in the x variables and time parameter r is

$$\hat{x}(r) = \underline{x}(0) + \frac{1}{|\underline{\alpha}|^2} \Big\{ \tfrac{1}{2} \ln(1 + \tau^2)\, \underline{\alpha} - \left[\underline{\alpha}(\cot \theta_0) - \frac{|\underline{\alpha}|}{|\underline{K}| \sin \theta_0} \underline{K} \right] \tan^{-1} \tau \Big\}, \qquad (2C.4\text{-}1)$$

if $\theta_0 \neq 0, \pi$. If $\theta_0 = 0$ or π,

$$\hat{\underline{x}}(t) = \underline{x}(0) + \frac{(\underline{\alpha})}{|\underline{\alpha}|^2} \ln(1 + |\underline{\alpha}||\underline{K}| r), \tag{2C.4-2}$$

where for either case,

$$\tau = \cot\theta_0 + c_0 |\underline{\alpha}| r,$$

$$r = b_0 s - b_0$$

$$s = e^{-\lambda t},$$

$$c_0 = \frac{|\underline{K}|}{\sin\theta_0}, \tag{2C.4-3}$$

and θ_0 is an angle between $\underline{\alpha}$ and $\underline{K} = L(0)$.

The equilibrium of (2C.3-7) and (2C.3-8), or its n-species analogue is of phase II type. It is given by

$$-\left.\frac{dy^i}{dt}\right]_{II} = \alpha_i \circ \frac{1}{\sum_{j=1}^{n} (\alpha_j)^2} \cdot \frac{Q}{\prod_{i=1}^{n} Q_i} \tag{2C.4-4}$$

where $Q = \sum_{i=1}^{n} Q_i^2$ and $\prod_{i=1}^{n} Q_i = Q_1 \ldots Q_n$. Note that we have assumed in our derivation that $\lambda_1 = \ldots = \lambda_n = \lambda$. Now it is easy to see how phase II reversals come about from (2C.4-4). Recall that the phase II equilibria in single species (2C.2-11) are equal to $1/\alpha_i$. If we generalize (2C-2-11) at (2C.4-4) to include $\lambda_i \neq \lambda_j$, we obtain hypothetical equilibria

$$-\left.\frac{dy^i}{dt}\right]_{II} = \alpha_i \cdot \frac{\lambda_i^2}{\sum_{j=1}^{n} (\alpha_j \lambda_j)^2} \cdot \frac{Q}{\prod_{i=1}^{n} Q_n}. \tag{2C.4-5}$$

From (2C.4-5) we see that a single species phase II loser will be a winner in the 2-species system provided $\lambda_i > \lambda_j$. But this system is not an optimal one! It is necessary to have $\lambda_i = \lambda_j$ for this. However, rough analysis of the original Harper/Clatworthy data indicates that the variation in the λ values between the species may, in fact, be negligible. For using the relation $b_0 = \lambda c$, one finds that λ has values .07, .08, .09 while the phase I rate values (b_0) are .21, .25, .28 . Analysis of the data gathered in our experiments with M.R.T. Dale shall hopefully confirm this result.

Another point to recognize about our phase II equilibria is that they depend on the log biomass $\hat{y}^i$ explicitly. However, the $\hat{y}^i$ should not be treated as constants close to zero, near or at phase II.

This brings up another point. For the system (2C.3-7), (2C.3-8) or its n-species analogue, there is no phase III steady state. However, I have developed a linear perturbation technique which allows phase III to occur in a finite time after phase II. This method has been employed in the study of starfish and coral and appears in *Math. Model.* (Antonelli, 1983a; 1983b).

For the case where no reversal occurs the winner seems to have a strong biological advantage. For example, the aquatic fern, *Salvinia natans,* beat out *Lemna polyrrhiza* and *Lemna minor* and its single species phase II equilibrium predicts just this. The strong advantage *Salvinia* has over the others is described by Harper in Chapter I of *The Population Biology of Plants* (1977). *Salvinia* produces its new fronds in the air and then lowers them onto the water surface. The others form theirs at the surface itself. Thus, *Salvinia* is bound to overtop and shade the others. This advantage allows *Salvinia* to escape coupling with either of the *Lemna* species and consequently its growth dynamics is of single species type (2C.2-2).

Acknowledgement

This research was partially supported by NSERC-A-7667 at the University of Alberta.

References

Antonelli, P.L. 1983a. Nonlinear allometric growth I. Perfectly cooperative systems. *Math. Model.,* 4: 367-372.

Antonelli, P.L. 1983b. Nonlinear allometric growth II. Allelopathic plant communities. *Math. Model.,* 4: 373-380.

Antonelli, P.L., and B.H. Voorhees. 1983. Nonlinear growth mechanics I. Volterra-Hamilton systems. *Bull. Math. Biology* 45, no. 1: 103-116.

Douglas, J. 1927. The geometry of paths. *Ann. Math.,* 29: 143-168.

Eisenhart, L.P. 1927. *Non-Riemannian Geometry.* Princeton University Press.

Harper, J.L., and J.N. Clatworthy. 1962. The comparative biology of closely related species living in the same area V. Inter- and intra-specific interference within cultures of *Lemna spp.* and *Salvinia natans. J. Exp. Bot.,* 13: 307-324.

Harper, J.L. 1977. *The Population Biology of Plants.* Academic Press.

2D

A Mathematical Model of Chemical Defense of Apparent Plants Against Insects

P.L. Antonelli

2D.1 Introduction

P.P. Feeny and D.F. Rhoades hypothesize that the proportion of metabolic resources allocated by plants to chemical defense against herbivores tends to be greater in "apparent" species, and that these species use dosage dependent chemical defenses. For example, tannins are known to act as quantitative barriers (i.e. dosage dependent) for oaks *(Quercus robur)* against many species of *Lepidoptera.* Their apparency is reflected in their dominance in climax forests. Using ideas of J. Harper, oak leaves may be thought of as modular units whose population ecology is influenced by production of tannins. In addition, it is becoming clear that trees can increase the content of tannins in direct response to attack by insects (Rhoades, 1979). A mathematical model, in which the amount of produced tannin is allometrically related to leaf biomass for a one oak species/one insect species system will be presented. Because some caterpillars aggregate while foraging, the existence of a periodic cycle whose (small) period increases with the percent of tannin in the leaves, can be established. Stability holds for a range of parameter values, but the first Hopf bifurcation indices τ_2 and β_2 for asymptotic orbital stability are zero, so this is neutral stability only (see essay 2B). Also, *production stability* holds and decreases with increased tannin content. A delightfully readable account of gregarious cooperative foraging is given in *The Life of the Caterpillar,* by the eminent French entomologist J.H. Fabre (1925). (See also, Frost's *Insect Life and Insect Natural History* (1959).)

2D.2 Derivation of the Production Model

In a classic 1944 paper in *The Journal of Forestry,* J. Kitteridge discovered that Huxley's allometric law holds, via linear regression analysis, for many species of conifer trees and oaks. We shall use this fact now. Let Q denote the average total leaf biomass (dry weight) for a stand of oaks or conifers and let B.H.D. denote the average breast height diameter. Then,

$$\ell n\, Q = \rho\, \ell n(\mathrm{B.H.D.}) + C\,, \tag{2D.2-1}$$

holds with specific values of ρ and C for specific species. Further, let T denote the average total tannin content of the leaves. We shall employ a simple formula describing the response of the plants to attack by herbivores due to myself and D.F. Rhoades. Thus, *Rhoades' Plant Response Mechanism* is expressed as

$$d \cdot Q^g = T\,, \qquad 0 \leqslant g \leqslant 1\,, \tag{2D.2-2}$$

for some portionality constant d . Note that if $g = 1$, then $T/Q = d$ (no response). While, if $g = 0$, $T/Q = d/Q$, so that for small Q (defoliation implies small Q), T/Q is large and we obtain maximal response.

Suppose that B.H.D. and Q follow Gompertz curves

$$\mathrm{B.H.D.} = a e^{-c e^{-\lambda t}}\,, \tag{2D.2-3a}$$

for given positive constants a, c, λ, and likewise, suppose

$$Q = a_1 e^{-c_1 e^{-\lambda_1 t}} \tag{2D.2-3b}$$

holds for positive constants a_1, c_1, λ_1. (See for example Harper and Clatworthy (1963). P.H. Leslie of Oxford University fitted their data to Gompertz curves (2D.2-3).) Clearly, the Kitteridge/Huxley allometric relation

$$\ell n\, Q = \rho\, \ell n(\mathrm{B.H.D.}) + B\,, \tag{2D.2-4}$$

holds, if and only if $\lambda_1 = \lambda$ and $\rho = c/c_1$.

But, from (2D.2-2)

$$g \cdot \ell n\, Q = \ell n\, T - \ell n\, d\,, \tag{2D.2-5}$$

so that (2D.2-4) yields

$$\ell n\, T = g\rho\, \ell n(\text{B.H.D.}) + (B + \ell n\, d)\,. \tag{2D.2-6}$$

Now,

$$\frac{d^2(\ell n(\text{B.H.D.}))}{dt^2} = -\lambda\, \frac{d(\ell n(\text{B.H.D.}))}{dt}\,, \tag{2D.2-7}$$

follows from (2D.2-3). Then, if we introduce N = average number of leaves per tree, we are able to rewrite (2D.2-7) as

$$\begin{cases} \dfrac{d(\ell n(\text{B.H.D.}))}{dt} = \dfrac{(1/\alpha) - N}{N}, \quad \alpha > 0\,, \\[2ex] \dfrac{dN}{dt} = \lambda N(1 - \alpha N) \end{cases} \tag{2D.2-8}$$

where

$$N(t) = \frac{1/\alpha}{1 + b_0 e^{-\lambda t}}\,, \tag{2D.2-9}$$

solves the *logistic* equation of (2D.2-8), provided the *initial time relation*

$$\alpha N(0) = \frac{1}{\lambda c + 1} = \frac{1}{b_0 + 1}\,, \tag{2D.2-10}$$

is satisfied. Given the asymptote determining constant, a , in (2D.2-3), knowledge of $\ell n(\text{B.H.D.})$ at $t = 0$ and $\left.\frac{d\,\ell n(\text{B.H.D.})}{dt}\right]_{t=0}$ determines α uniquely in terms of N(0), and conversely. Finally, using (2D.2-4) and (2D.2-5) we obtain the *tannin production equations*

$$\begin{cases} \dfrac{d(\ell n\, T)}{dt} = \rho g \cdot \left(\dfrac{(1/\alpha) - N}{N}\right) \\[2ex] \dfrac{dN}{dt} = \lambda N(1 - \alpha N)\,, \end{cases} \tag{2D.2-11}$$

where ρ is the allometric constant for oak in relation to B.H.D. Obviously, decreasing g will increase the yearly time average relative amount of tannin in the typical modular unit. The coefficient α may be interpreted as the *self-shading effect,* here, and in what is to follow.

2D.3 A Perturbation Yields Stable Production

The system (2D.2-7) is unstable. We remedy this defect by perturbation of the underlying optimality principle. Clearly, if tannin production is to provide a reliable defense against attack by herbivores, stability of the production process itself will increase reliability. However, the perturbed system must make biological sense on its own.

We must perturb (2D.2-7) in a way which does not destroy the ecological meaning of (2D.2-8). This is just an application of what I call *Harper's criterion,* that a plant should be considered as a population (indeed several populations!) of leafy modular units (Harper, 1977). We suppose that simple systems can be treated in classical Volterra fashion using the "encounters method." Hence, the logistic has much meaning in our model.

Let us begin by writing (2D.2-11) in the form

$$\begin{cases} \dfrac{d(\ell n\, T)}{dt} = g\rho \cdot N_* \\ \dfrac{dN_*}{dt} = -\lambda N_* , \end{cases} \tag{2D.3-1}$$

where N_* is the so-called *relative proportion variable* of classical literature (see 2A.3). Further, rewrite (2D.3-1) in terms of

$$\hat{y} = (\ell n\, T)_{max} - \ell n\, T , \tag{2D.3-2}$$

so that we obtain

$$\begin{cases} \dfrac{d\hat{y}}{dt} = -g\rho N_* \\ \dfrac{dN_*}{dt} = -\lambda N_* . \end{cases} \tag{2D.3-3}$$

Now define $\hat{x} = (g\rho/\lambda\alpha)\, \ell n(1 + \frac{1}{b_0 c}\hat{y})$, and note that (2D.3-3) may be written

$$\begin{cases} \dfrac{d\hat{x}}{dt} = g\rho\,(-N) \\ \dfrac{d(-N)}{dt} = \lambda N(1 - \alpha N) , \end{cases} \tag{2D.3-4}$$

or in the second order form, after rescaling the rate λ by setting $\lambda = g\rho \cdot \lambda'$ and then dropping the prime,

$$\frac{d^2\hat{x}}{dt^2} + \lambda\alpha\left(\frac{d\hat{x}}{dt}\right)^2 + g\rho\lambda\left(\frac{d\hat{x}}{dt}\right) = 0\,.^{\dagger} \tag{2D.3-5}$$

The *cost functional* or Lagrangian for (2D.3-5) is

$$L_{\hat{x}} = \frac{1}{g^2\rho^2}\, e^{2\lambda\alpha\hat{x}+g\rho\lambda t}\left(\frac{d\hat{x}}{dt}\right)^2 . \tag{2D.3-6}$$

The perturbed Lagrangian we seek is obtained by addition of higher powers of $\hat{x}$. Thus,

$$L_{\text{pert}} = \frac{1}{g^2\rho^2}\, e^{2\lambda\alpha\hat{x}+\nu(\hat{x})^2+g\rho\lambda t}\left(\frac{d\hat{x}}{dt}\right)^2 , \tag{2D.3-7}$$

is the cost functional for the Euler-Lagrange equation

$$\frac{d^2\hat{x}}{dt^2} + \lambda\left(\alpha + \frac{\nu}{\lambda}\,\hat{x}\right)\left(\frac{d\hat{x}}{dt}\right)^2 + g\rho\lambda\left(\frac{d\hat{x}}{dt}\right) = 0\,. \tag{2D.3-8}$$

After scaling time using $t' = g\rho\, t$ and $\hat{x} = (g\rho/\lambda\alpha)\,\ell n(1 + \frac{1}{b_0 c}\,\hat{y})$, we get

$$\begin{cases} \dfrac{d\hat{x}}{dt} = -N \\[2ex] \dfrac{d(-N)}{dt} + \lambda\left(\alpha + \dfrac{\nu\rho g}{\lambda\alpha}\,\ell n\left(1 + \dfrac{1}{b_0 c}\,\hat{y}\right)\right)N^2 - \lambda N = 0\,, \end{cases} \tag{2D.3-9}$$

after dropping the prime on t .

Expanding the log term in a Taylor series for $\hat{y}$, we obtain for the time reversed second equation of (2D.3-9)

$$\frac{dN}{dt} = \lambda N\left[1 - \left(\alpha + \frac{\nu\rho g}{\lambda\alpha}\cdot\frac{1}{b_0 c}\circ\hat{y}\right)N\right], \tag{2D.3-10}$$

as first order approximation assuming ν is small. The corresponding second order equation (2D.3-9) or its first order approximation (2D.3-10) are *production stable.* My proof of this fact is a consequence of the method contained in essay 2F, and the fact that $\hat{x} \to 0$ as $t \to \infty$, for small ν . Note

† Scaling λ in this way is biologically valid as inspection of (2D.2-2) and (2D.2-1) shows; λ is independent of g and λ' may be measured in units of g .

that increased plant response (i.e. decreasing g) results in diminished self-shading. This is consistent with the common field observation that leaves are smaller in the year following the initial attack by insects (Rhoades, 1979).

2D.4 The Optimally Foraging Cooperative Herbivore

Let $\hat{q} = \ell n(Q)_{max} - \ell n\, Q$ and write

$$\begin{cases} \dfrac{d\hat{q}}{dt} = \ell_f \cdot F \\ \dfrac{dF}{dt} = -\varepsilon F + \varepsilon\gamma F^2 . \end{cases} \tag{2D.4-1}$$

This system describes cooperative foraging by the herbivore, F. The $\hat{q}$ represents the *amount consumed,* measured against the maximum of $\ell n(Q)$ achieved when F is not present. The term F^2 represents cooperation while foraging. We can normalize to unity the positive constant ℓ_f. The resulting

$$\frac{d^2\hat{q}}{dt^2} = -\varepsilon\frac{d\hat{q}}{dt} + \varepsilon\gamma\left(\frac{d\hat{q}}{dt}\right)^2, \tag{2D.4-2}$$

represents *optimal* foraging because of the fact that (2D.4-2) is the Euler-Lagrange equation for the variational principle

$$\delta\int_{t_0}^{t_1} e^{-\varepsilon\gamma\hat{q}+\varepsilon t}\, dt = 0 . \tag{2D.4-3}$$

Note that $\ell n(Q_{max}/Q) = \hat{q}$, so that (2D.4-3) can be written as

$$\delta\int_{t_0}^{t_1}\left(\frac{Q_{max}}{Q}\right)^{-\varepsilon\gamma} e^{\varepsilon t}\, dt = 0 . \tag{2D.4-4}$$

We now perturb the Lagrangian to obtain

$$\hat{L}_{pert} = e^{-2\varepsilon\gamma(\hat{q} - \frac{1}{2}\frac{\mu}{\varepsilon}(\hat{q})^2) + \varepsilon t}\left(\frac{d\hat{q}}{dt}\right)^2 . \tag{2D.4-5}$$

The resulting *Herbivore cooperative foraging equation* is

$$\frac{dF}{dt} = -\ \varepsilon F + [\varepsilon\gamma - \mu\hat{q}]F^2 . \tag{2D.4-6}$$

This system is (consumption) *unstable* (see essay 2F). Furthermore, it is clear from (2D.4-6) that as more and more leaves are consumed (i.e. $\hat{q}$ increasing), the cooperative term is less and less beneficial to the herbivore. But, we should also expect the coefficient ε in (2D.4-6) to be altered. We shall mention this again in the next section.

2D.5 The Periodic Cycle

The complete plant/herbivore system is obtained from (2D.3-9) and (2D.4-6). It is

$$\frac{dN}{dt} = \lambda N - (\lambda\alpha + g(\frac{\nu\rho}{\lambda\alpha})\,\ell n(1 + \frac{1}{b_0 c}\hat{y}))N^2 - \hat{\delta}NF$$

$$\frac{dF}{dt} = -\ \hat{\varepsilon}F + (\hat{\varepsilon}\gamma - \frac{\mu}{g}\hat{y})F^2 + \hat{\beta}NF . \tag{2D.5-1}$$

There is good evidence that $\hat{\varepsilon}$ increases and $\hat{\beta}$, $\hat{\delta}$ decreases as g decreases (i.e. as plant response increases) (Feeny, 1968; Rhoades, 1979; and Rhoades and Cates, 1976). Note that from (2D.2-5)

$$g \cdot \hat{q} = \hat{y} , \quad \text{(tannin in leaves)}. \tag{2D.5-2}$$

Thus, we see that plant response via Rhoades' mechanism results in decreasing the cooperative coefficient in (2D.5-1) and increases the instability of consumption. This still holds when the log term above is replaced by its first order approximation.

When considering cyclic behavior in ecological time, we shall always treat $\hat{y}$ as a constant and call it the *slow variable.* The N and F variables are referred to as *fast variables.* One does orbital stability analysis holding the slow variable fixed. Therefore, it will suffice to investigate orbital stability of the system with constant coefficients,

$$\begin{cases} \dfrac{dN}{dt} = \lambda N - \hat{\alpha} N^2 - \hat{\delta} FN \\ \dfrac{dF}{dt} = \hat{\beta} FN + \hat{\gamma} F^2 - \hat{\varepsilon} F , \end{cases} \tag{2D.5-3}$$

where $\hat{\delta} = \delta - m(\mu, g, \hat{y})$ and $\hat{\beta} = \beta - n(\nu, g, \hat{y})$ are to be identified from systems analysis and biological data. We suppose m, n and $\hat{\varepsilon}$ increase if g decreases while $\hat{\gamma}$ decreases if g decreases. This is biologically reasonable.

The existence of a neutrally stable periodic cycle for (2D.5-3) has been established. However, if the F^2 term is replaced by $F^{1.999}$, then stability holds for a certain range of parameters using Hopf bifurcation theory (see essay 2B). The author and N. Kazarinoff have also proved results for a related three species system and these may be applied to (2D.5-3) by a method of dimension reduction. Analysis of asymptotic orbital stability requires use of Hopf bifurcation methods, and a computer. The reader should see related paper "Nonlinear allometric growth I" (Antonelli, 1983) and essays 2A, 2B and 2F of this chapter. The resulting formula for the asymptotic period is

$$\mathrm{Per}(\nu, \rho) \approx \frac{2\pi}{\sqrt{\lambda\alpha + (\nu g\rho / b_0 \alpha c)\hat{y}}} \circ \frac{\sqrt{\hat{\delta} - \lambda\alpha - (\nu g\rho / b_0 \alpha c) y}}{(\delta - \lambda)} \tag{2D.5-4}$$

($\hat{\beta} = \hat{\gamma}/2 = \hat{\varepsilon}/2$ in the calculation.) From this we see that the period lengthens as $g\rho$ decreases (ν and μ are held fixed). This models the well-known "dosage dependent" effect as an optimal defense strategy (Rhoades and Cates, 1976; and Varley et al., 1973). Finally, another dimension reduction argument can be used to prove that *production stability* decreases with small $g\rho$ values.

2D.6 Conclusions

This model gives a mathematical description of the well-known periodic defoliation of oaks by the caterpillars of the moth *Operophtera brumata* and especially that of the larch *Tortrix* devastation of the Engadine Valley of Switzerland (Feeny, 1968; Baltensweiler et al., 1977). Evidently, the *O. brumata* cycle has not been explained completely in terms of parasites with caterpillar hosts, or environmental stress, like drought (Varley et al., 1973).

The model presented uses *Rhoades' mechanism* and *cooperative foraging.* It predicts that plant/herbivore interactions which are *not* of dosage dependent type (g = 1) and are cooperative, should have shorter periods. Voles and lemmings are known to forage cooperatively on herbs (unapparent plants) and cycle with periods of 3 or 4 years while larch budmoths, forest tent caterpillars and scotch grouse (feeding on heather -- an apparent plant) cycle with periods of 9 or 10 years. These facts, therefore, corroborate our model.

Acknowledgements

The author is grateful to Professor N. Kazarinoff and Professor D.F. Rhoades for their interest and assistance in the preparation of this paper. The research was supported in part by NSERC-A-7667.

References

Antonelli, P.L. 1983. Nonlinear allometric growth I. Perfectly cooperative systems. *Math. Model.*, 4: 367-372.

Baltensweiler, W., G. Benz, P. Bovey, and V. DeLucchi. 1977. Dynamics of larch budmoth populations. In *Annual Rev. Ent.*, 22.

Fabre, J.H. 1925. *The Life of the Caterpillar.* Dodd, Mead and Company, New York. (Translation by A.T. DeMattos.)

Feeny, P.P. 1968. Effect of oak leaf tannins on larval growth of winter moth *Operophtera brumata. J. Insect Physiol.*, 14: 805-817.

Frost, S.W. 1959. *Insect Life and Insect Natural History* (2nd Ed.). Dover.

Harper, J.L. 1977. *The Population Biology of Plants.* Academic Press.

Harper, J.L., and J.N. Clatworthy. 1963. The comparitive biology of closely related species living in the same area V. Inter- and intra-specific interference within cultures of *Lemna spp.* and *Salvinia ratans,* 13: 307- .

Kittredge, J. 1944. Estimation of the amount of foliage of trees and stands. *J. Forestry,* 42: 905-912.

Rhoades, D.F. 1979. Evolution of plant chemical defense against herbivores. In Rosenthal and Janzen, eds., *Herbivores, Their Interaction with Secondary Plant Metabolites.* Academic Press.

Rhoades, D.F., and R.G. Cates. 1976. A general theory of plant anti-herbivore chemistry, in Biochemical interactions between plants and insects. In Wallace and Mansell, eds., *Recent Adv. Phytochem.*, Vol. 10. Plenum, New York.

Varley, G.C., G.R. Gradwell, and M.P. Hassell. 1973. *Insect Population Ecology, An Analytical Approach.* Blackwell Scientific Publ., Oxford, England.

2E
A Mathematical Model of Multiple Tannin Defense of Apparent Plants Against Insects

P. L. Antonelli

2E.1 Introduction

In the present paper, I wish to extend the mathematical results for the (one plant/one herbivore)–system previously treated. This model, although ecologically motivated, takes into account tannin chemistry as discussed by Haslam (1974) and Zucker (1983). There are two large classes of tannins, the so-called *condensed* sort and the *hydrolyzable* sort. The latter class splits into two distinct subclasses known as the *gallotannins* and *ellagitannins*. The latter variety of molecules are relatively small spherically shaped with protruding carboxyl or carbonyl groups, which are the *binding sites* for complexing with proteins, celluloses, pectins and alkyloids. The gallotannins are larger, usually flat and disc-like with phenolic groups for binding, distributed on the periphery of the disc. Yet, the condensed tannins are much larger than either of the hydrolyzable classes and often take the form of very regular *double Helices.* Furthermore, both hydrolyzable types typically have a glucose core to which a polygallyol chain of variable length is esterified onto all or some of the available bonding sites. It is the sugar that allows the hydrolyzable tannins to be safely sequestered in cell vacuoles. Naturally, these kinds of tannin molecules are small enough to pass through cell membranes so that, presumably, they can be released into the guts of insect herbivores (Varley et al., 1973; Baltensweiler et al., 1977; Fabre, 1925; Frost, 1959; Schultz, 1983; and Zucker, 1983).

The condensed tannins are not as mobile as the other types and make good candidates for *dosage dependent defense.* The hydrolyzable sort may exhibit another mode of defense. This mode is to be described herein.

According to Haslam, "protein complex formation ..., although broadly related to molecular size is primarily dependent on the number of separate sites in the molecule able to associate with the protein." For gallotannins, Zucker states, "the ability to precipitate a protein increases nonlinearly with increasing numbers of separate galloyl groups, up to a maximum of five per residue," and also, "... the stability of complex formation may be a simple function of the actual number of hydrogen bonds that can be formed."

Based on a plethora of chemical facts and theories, the hypothesis that plants have evolved a condensed tannin defense against microbes and pathogens, while they have evolved the extreme stereo-specific diversity of hydrolyzable tannins to fend off insects has been recently advanced (Zucker, 1983). I intend to show here that it is possible to provide a unified mathematical model, which removes some of the recent objections with the *Optimal Defense Theory* of Feeny and Rhoades/Cates, arising from the chemistry of tannins (Rhoades, 1979; Rhoades and Cates, 1976; and Zucker, 1983).

From my perspective the *multiple tannin defense system* is simply a result of the plant's having one more tannin to mutate. However it should be stressed that *this model is based on ecological theory of* Feeny, Rhoades and Cates *and extends to multiple tannins by a simple "reductionist" argument.* I thereby obtain an ecologically motivated "tannin kinetics" which, in the presence of a herbivore who prefers one type of tannin (or leaf) over another, and who forages cooperatively, results in a *stable periodic cycle whose period increases with divergence of tannin (or leaf) preference* (see essay 2B). It is *therefore not necessary for the plant to use a dosage dependent mechanism for its defense. Rather, it can choose to evolve different tannins (or leaves) of divergent palatability to the herbivore!* (Schultz, 1983).

2E.2 The Reductionist Argument

Let N = *average* number of leafy modular units over a set of plants and let x(t) = ℓn (average total amount of complexed tannin of a particular type). Just as in the dosage dependent model, we let

$$\frac{dx}{dt} = \frac{N_{max} - N}{N}, \qquad (2E.2\text{-}1)$$

and *assume* allometry between the tannin total and the average breast height diameter, B.H.D. (Kitteridge, 1944). Let M = *average,* over a set of trees, of total number of *binding sites* on *all* molecules of the tannin. We *assume* binding sites are distributed in the leafy biomass with a *constant uniform density,* σ, in cubic millimeters, say. J.L. Harper (1977) tells us that plants have modular units which vary little in size. Accordingly, *we suppose a normal distribution* of modular shape parameters and weights, of small variance.

Let V denote the volume of an average leafy modular unit. Then M will be given by the simple relation

$$(\sigma \cdot V) \cdot N = k \circ M , \qquad (2E.2\text{-}2)$$

where k is a (dimensionless) proportionality constant. Substitution of (2E.2-2) into (2E.2-1) yields

$$\frac{dx}{dt} = \frac{M_{max} - M}{M} . \qquad (2E.2\text{-}3)$$

Now, the logistic associated with Gompertz growth, derived from growth experiments of Harper and Clatworthy (1963), is

$$\frac{dN}{dt} = \lambda N(1 - \alpha N), \text{ with } \alpha N(0) = \frac{1}{\lambda c + 1} , \qquad (2E.2\text{-}4)$$

which becomes, using (2E.2-2),

$$\frac{dM}{dt} = \lambda M \left(1 - \frac{\alpha \cdot k}{\sigma \cdot V} M\right), \text{ with } \frac{\alpha \circ k \cdot N(0)}{\sigma \cdot V} = \frac{1}{\lambda c + 1} \qquad (2E.2\text{-}5)$$

The above argument allows us to equivalently formulate the original dosage dependent, ecologically-based model for one plant and one herbivore in terms of numbers of bonding sites for precipitation of protein by (condensed) tannin. However, as we quoted above, precipitation may be produced by varying numbers of sites in combination, on the molecular surface. Thus, there is a variation in the *chemical production unit* just as there is with the leafy modular units (which may have changing productivity depending on their ages). Thus, the equation (2E.2-3) should be viewed as characterizing the accumulation of precipitate by the totality of functioning *site combinations.* The latter are the *production modules.*

2E.3 Volterra's "Encounter Method"

We postulate the "encounter kinetics,"

$$\frac{dM^i}{dt} = \sum_{j,k=1}^{2} C^i_{jk} M^j M^k + \lambda M^i , \quad i = 1,2, \qquad (2E.3\text{-}1)$$

where, apriori, C^i_{jk} can depend on x^1, x^2, the *accumulated* tannin precipitates. We further require that this results in an *optimal production* of x^1 and x^2 which is *production stable* and which generalizes the production equation for the (one plant/one herbivore)-system. We use the same type of auxillary variables,

$$\hat{x}^i = \frac{\rho_i g_i}{\lambda \alpha_i} \ln\left(1 + \frac{1}{b_{0i} c_i} \hat{y}^i\right), \quad i = 1, 2. \tag{2E.3-2}$$

Here,

$$\hat{y}^i = (\ln T_i)_{max} - \ln T_i, \ \rho_1 = \rho_2 = \rho, \ g_1 = g_2 = g, \ i = 1, 2 \tag{2E.3-3}$$

and T_i denotes "biomass of average total of complexed tannin sites." Thus, the production equations would be of the general time reversed form

$$\begin{cases} \dfrac{d\hat{x}^i}{dt} = -M^i \\ \dfrac{d(-M^i)}{dt} = \displaystyle\sum_{j,k=1}^{2} C^i_{jk} M^j M^k + \lambda M^i. \end{cases} \tag{2E.3-4}$$

Note that the coefficient λ is the *same* for $i = 1$ and $i = 2$! This is because of (2E.2-5) and the plant growth logistic (2E.2-4). There may be *two* tannins but there is only *one* plant. That is,

$$\begin{cases} \dfrac{dM^i}{dt} = \lambda M^i(1 - \alpha_{(i)} M^i), \quad i = 1, 2 \\ \alpha_{(i)} = \dfrac{\alpha k_{(i)}}{\sigma_{(i)} \cdot V} . \end{cases} \tag{2E.3-5}$$

It would seem that σ_1 and σ_2 might be *very* different, as for condensed versus hydrolyzed tannins. I do not know if there would be a measurable difference in these densities for two hydrolyzable tannins. However, gallotannins *can* be long chains.

Now, we assume a herbivore cooperative feeding equation essentially as in the (one plant/one herbivore) case. Thus,

$$\frac{dF}{dt} = \beta F(M^1 + M^2) + \gamma F^2 - \varepsilon F. \tag{2E.3-6}$$

It is a *hard* theorem of my own that if C^i_{jk} are assumed constant (to begin with) then

$$\begin{cases} C^i_{ii} = \beta_i \\ C^i_{ij} = C^i_{ji} = \beta_i, \quad i \neq j \\ C^i_{jj} = -\beta_j, \qquad i \neq j \end{cases} \tag{2E.3-7}$$

must hold for some constants β_1 and β_2, provided optimal production is defined in terms of a *cost functional*

$$L(\hat{x}^i, \dot{\hat{x}}^i, t)$$

which is second order homogeneous in $\dot{\hat{x}}^i = dx^i/dt$ (a *very* general assumption). Therefore, it is *not possible for the system* (2E.3-5) *to respond with an optimal defense (which is a consequence of optimal production)* against the herbivore (2E.3-6). For in (2E.3-5)

$$\begin{cases} C^1_{11} = -\lambda\alpha_1 \\ C^1_{12} = C^1_{21} = 0 \\ C^2_{22} = -\lambda\alpha_2 \\ C^2_{12} = C^2_{21} = 0 \end{cases} \tag{2E.3-8}$$

contradicts (2E.3-7), unless $\alpha_1 = \alpha_2 = 0$!

Therefore, the two tannin types must "evolve" to a *coupled state of interdependence* as (2E.3-7) describes. So, the constant coefficients allowed for Volterra's "encounter method" result in the time reversed production equations

$$\begin{cases} \dfrac{d^2\hat{x}^1}{dt^2} = -2\lambda\alpha_2 \dfrac{d\hat{x}^1}{dt}\dfrac{d\hat{x}^2}{dt} + \lambda\alpha_1\left(\left(\dfrac{d\hat{x}^2}{dt}\right)^2 - \left(\dfrac{d\hat{x}^1}{dt}\right)^2\right) + \lambda\dfrac{d\hat{x}^1}{dt} \\ \dfrac{d^2x^2}{dt^2} = -2\lambda\alpha_1 \dfrac{d\hat{x}^1}{dt}\dfrac{d\hat{x}^2}{dt} + \lambda\alpha_2\left(\left(\dfrac{d\hat{x}^1}{dt}\right)^2 - \left(\dfrac{d\hat{x}^2}{dt}\right)^2\right) + \lambda\dfrac{d\hat{x}^2}{dt} \end{cases} \tag{2E.3-9}$$

The cost functional is

$$L_M = e^{2\lambda(\alpha_1 \hat{x}^1 + \alpha_2 \hat{x}^2 + g\rho\lambda t)} \left((\frac{d\hat{x}^1}{dt})^2 + (\frac{d\hat{x}^2}{dt})^2 \right), \tag{2E.3-10}$$

and its perturbation, after the method of the dosage dependent model is

$$L_M^{pert} = e^{2\lambda(\hat{\alpha}_1 \hat{x}^1 + \hat{\alpha}_2 \hat{x}^2 + g\rho\lambda t)} \left((\frac{d\hat{x}^1}{dt})^2 + (\frac{d\hat{x}^2}{dt})^2 \right), \tag{2E.3-11}$$

where

$$\begin{cases} \hat{\alpha}_1 = \alpha_1 + \frac{1}{2\lambda} \nu_1 \hat{x}^1 \\ \hat{\alpha}_2 = \alpha_2 + \frac{1}{2\lambda} \nu_2 \hat{x}^2 , \end{cases} \tag{2E.3-12}$$

for perturbation parameters ν_1 , ν_2 .

The perturbed cost functional for the foraging herbivore is essentially the same as for the (one plant/one herbivore)-system. The resulting *fast variable equations are:*

$$\begin{cases} \frac{dM^1}{dt} = -2\lambda\hat{\alpha}_2 M^1 M^2 + \lambda\hat{\alpha}_1 \ (M^2)^2 - (M^1)^2 \ - \hat{\delta}_1 M^1 F + \lambda M^1 \\ \frac{dM^2}{dt} = -2\lambda\hat{\alpha}_1 M^1 M^2 + \lambda\hat{\alpha}_2 \ (M^1)^2 - (M^2)^2 \ - \hat{\delta}_2 M^2 F + \lambda M^2 \\ \frac{dF}{dt} = -\hat{\varepsilon}F + \hat{\varepsilon}\hat{\gamma}F^2 + \beta(M_1 + M_2)F , \end{cases} \tag{2E.3-13}$$

where

$$\begin{cases} \hat{\gamma} = \gamma - \frac{\mu}{\hat{\varepsilon}} \left(\frac{\hat{T}_1}{g_1} + \frac{\hat{T}_2}{g_2} \right), \quad \hat{T}_i = T_{i_{max}} - T_i , \quad i = 1, 2 \\ \hat{\delta}_1 = \delta_1 - m_1(\mu, g_1, g_2) \\ \hat{\delta}_2 = \delta_2 - m_2(\mu, g_1, g_2) \\ \hat{\beta} = \beta - n_1(\nu, g_1, g_2) \\ \hat{\varepsilon} = \varepsilon + n_2(\nu, g_1, g_2) . \end{cases} \tag{2E.3-14}$$

Here, n_1, n_2, m_1, m_2 are naturally *assumed* to be monotone increasing functions of $1/g_1$ and $1/g_2$ and monotone decreasing functions of μ, ν, the perturbation parameters (assumed fixed), and δ_1 , δ_2 , β, ε are interpreted as in 2B.4-1.

For *fast* variable orbital stability analysis, we fix all *slow* parameters and study the constant coefficient system. As it happens, this type of system has been studied by myself and N. Kazarinoff. For a range of parameter values we have shown stability of the periodic cycle using Hopf bifurcation analysis (see essay 2B).

Finally, the two tannin system extends to any number of tannins (Antonelli, 1983; see essay 2A). But, stability of any possible limit cycles has not as yet been investigated. Nevertheless, production stability does hold. The proof is similar to that in essay 2F in this chapter.

Acknowledgements

I would like to thank N. Kazarinoff and D.F. Rhoades for their kind assistance in the preparation of this paper. This research was supported in part by NSERC-A-7667.

References

Antonelli, P.L. 1983. Nonlinear allometric growth I. Perfectly cooperative systems. *Math. Model.*, 4: 367-372.

Baltensweiler, W., G. Benz, P. Bovey, and V. DeLucchi. 1977. Dynamics of larch budmoth populations. In *Annual Rev. Ent.*, 22.

Fabre, J.H. 1925. *The Life of the Caterpillar.* Translation by A.T. DeMattos. Dodd, Mead and Company, New York.

Feeny, P.P. 1968. Effect of oak leaf tannins on larval growth of winter moth *Operophtera brumata. J. Insect Physiol.*, 14: 805-817.

Frost, S.W. 1959. *Insect Life and Insect Natural History,* (2nd Ed.). Dover.

Harper, J.L. 1977. *The Population Biology of Plants.* Academic Press.

Harper, J.L., and J.N. Clatworthy. 1963. The comparative biology of closely related species living in the same area V. Inter- and intra-specific interference within cultures of *Lemna spp.* and *Salvinia ratans. J. Exp. Bot.*, 13: 307-324.

Haslam, E. 1974. Polyphenol-protein interaction. *Biochem. J.*, 139: 285-287.

Kittredge, J. 1944. Estimation of the amount of foliage of trees and stands. *J. Forestry*, 42: 905-912.

Rhoades, D.R. 1979. Evolution of plant chemical defense against herbivores. In Rosenthal and Janzen, eds., *Herbivores, Their Interaction with Secondary Plant Metabolites.* Academic Press.

Rhoades, D.F., and R.G. Cates. 1976. A general theory of plant anti-herbivore chemistry, in biochemical interactions between plants and insects. In Wallace and Mansell, eds., *Recent Adv. Phytochem.*, Vol. 10. Plenum, New York.

Schultz, J.C. 1983. Impact of variable plant defensive chemistry on susceptibility of insects to natural enemies. In Paul A. Hedin, ed., *Plant Resistance to Insects.* Amer. Chem. Soc. Symp. 208.

Varley, G.C., G.R. Gradwell, and M.P. Hassell. 1973. *Insect Population Ecology, An Analytical Approach.* Blackwell Scientific Publ., Oxford, England.

Zucker, W.V. 1983. Tannins: does structure determine function? An ecological perspective. *Amer. Naturalist.,* 121, no. 3: 335-365.

2F

Autotoxin Production in Some Symbiotical Plant Communities Is Unstable

P.L. Antonelli

2F.1 Introduction

In a mutually symbiotic system, it is not surprising that autotoxin production, a widespread phenomenon in plant communities, results in the *instability of productivity* when each autotoxin is toxic to all other species. However, in the present paper, we show that, *for a large class of symbiotic plant communities, instability holds even when the autotoxins are beneficial to all other species,* i.e. even when they are *kairomones* in the terminology of R.H. Whittaker. This suggests, for example, that two plant species living off each other's leachates may have *unstable productivity,* in the sense that systems with close initial conditions can be vastly disparate in a very short time. In such systems, the leachates accumulate rapidly and yet, for large times, the number of production modules (i.e. cell, leaves with axillary bud, etc. ...) may reach stable equilibrium. This result, for systems the author has considered, hinges on the fact that, for large times, the actual amounts of accumulated leachates do not change much. It follows that the notion of "large time" is system specific and other systems, even with initial conditions close to those of the given system, may require a much longer period to settle down into equilibrium for the numbers of modular production units.

All the results reported on in this paper hold for mutually symbiotic communities of any number of species defined by the appropriate differential equations and variational principle. We shall, for ease of mathematical exposition, restrict our mathematical theorizing to two species systems only in what is to follow.

2F.2 Ecological Equations Parameterized by Allomones and Kairomones

Each of the two species of plant considered here possesses C_1 and C_2 numbers of *production modules.* These may be cells or leaves with axillary bud or roots, for example. These are considered the producers of the secondary substances which enter into competition, succession and growth. The basic notion of *modular unit* is due to Professor John Harper, for whom they form the basis of a *proper* ecological theory of plant communities (Harper, 1977). Each module is assumed to produce a secondary substance of one of the *five* chemical types (Whittaker and Feeny, 1977). We suppose the production rate to be constant, per module, and hence implicitly suppose *ideal environmental conditions.* Mathematically, we write

$$\frac{dx^i}{dt} = k_{(i)}\, C_i\,, \qquad i = 1, \ldots, n. \tag{2F.2-1}$$

Here, x^i denotes the *total amount of* ith *substance produced by all* ith *type production units,* of which there are C_i. In what follows, we will suppose the positive proportionality constant $k_{(i)}$ are each set to unity, without loss of real generality in the mathematical work. We further suppose a given *ideal geographical transect* is selected or that a given volume of water from an *ideal aquatic community* is specified. It is from these portions of the community that we make *ideal measurements* of x^i and C_i, as functions of time. From now on, we let $i = 1, 2$. Furthermore, we shall assume for convenience that the *intrinsic growth rates* λ_1 and λ_2 are *commensurate, or equal,* because community interactions are most easily studied using this *normal form.* Indeed, in the laboratory, it may be convenient to add phosphate to aquatic cultures to *normalize* the initial growth rates (Harper, 1977, Chapter 1).

The ecological equations we study are

$$\left.\begin{aligned} \frac{dC_1}{dt} &= -2(\alpha_2 + \nu_2 x^2 + \nu_3 x^1)C_1 C_2 + (\alpha_1 + \nu_1 x^1 + \nu_3 x^2)(C_2^2 - C_1^2) + \lambda C_1 \\ \frac{dC_2}{dt} &= -2(\alpha_1 + \nu_1 x^1 + \nu_3 x^2)C_1 C_2 + (\alpha_2 + \nu_2 x^2 + \nu_3 x^1)(C_1^2 - C_2^2) + \lambda C_2 \end{aligned}\right\},$$

where $\alpha_1 > 0,\ \alpha_2 > 0.$ (2F.2-2)

Here, λ denotes the (normalized) intrinsic growth rates. If we *treat* x^1 and x^2 *as parameters* in (2F.2-2), we find a unique nontrivial (stable) equilibrium (Antonelli, 1983a; 1983b).

$$\left.\begin{aligned} C_{01} &= \frac{\lambda(\alpha_1+\nu_1 x^1+\nu_3 x^2)}{(\alpha_1+\nu_1 x^1+\nu_3 x^2)^2+(\alpha_2+\nu_2 x^2+\nu_3 x^1)^2} \\ C_{02} &= \frac{\lambda(\alpha_2+\nu_2 x^2+\nu_3 x^1)}{(\alpha_1+\nu_1 x^1+\nu_3 x^2)^2+(\alpha_2+\nu_2 x^2+\nu_3 x^1)^2} \end{aligned}\right\} . \tag{2F.2-3}$$

We now use the convenient parameter change defined by $s = e^{+\lambda t}$, so that the linear terms of (2F.2-2) disappear and t is replaced by s in (2F.2-1) and (2F.2-2). Now consider the variational problem

$$\delta \int L(\dot{x}, x)ds = 0\,, \tag{2F.2-4}$$

where $x = dx/ds$ and the Lagrangian L is given by

$$\left.\begin{aligned} L(\dot{x},x) &= \frac{1}{2}\Big(\sum_{i,j}\delta_{ij}\,C_i\,C_j\Big)e^{2\phi}\,, \\ \phi &= \alpha_1 x^1+\alpha_2 x^2+\frac{1}{2}\nu_1(x^1)^2+\frac{1}{2}\nu_2(x^2)^2+\nu_3 x^1 x^2 \end{aligned}\right\} . \tag{2F.2-5}$$

Here the summation runs over from 1 to 2 and $\boldsymbol{\delta}_{ij}$ denotes the identity matrix. Note that ϕ is the *most general quadratic polynomial in two variables* satisfying $\phi(0) = 0$. The *Euler-Lagrange equations* for (2F.2-4) and (2F.2-5) are now written as

$$\left.\begin{aligned} \frac{d^2x^1}{ds^2} &= -2(\alpha_2+\nu_2 x^2+\nu_3 x^1)\frac{dx^1}{ds}\frac{dx^2}{ds} \\ &\quad +(\alpha_1+\nu_1 x^1+\nu_3 x^2)\left(\left(\frac{dx^2}{ds}\right)^2-\left(\frac{dx^1}{ds}\right)^2\right) \\ \frac{d^2x^2}{ds^2} &= -2(\alpha_1+\nu_1 x^1+\nu_3 x^2)\frac{dx^1}{ds}\frac{dx^2}{ds} \\ &\quad +(\alpha_2+\nu_2 x^2+\nu_3 x^1)\left(\left(\frac{dx^1}{ds}\right)^2-\left(\frac{dx^2}{ds}\right)^2\right) \end{aligned}\right\} . \tag{2F.2-6}$$

Note that using (2F.2-1) and (2F.2-6) and resubstituting t to s we obtain (2F.2-2). Thus, (2F.2-2) is the (reparameterized) *reduced form* of (2F.2-6).

If we choose $\nu_1 = \nu_2 = \nu_3 < 0$ in (2F.2-2), we obtain a *perfectly cooperative system* as studied by the author (1983a). It is easily seen, directly from (2F.2-2) that the *production of* x^1 *and* x^2 *is beneficial to* C_1 *and* C_2. Hence, we have a (symbiotical) system with x^1 and x^2 *allomones.*

On the other hand, if we choose $\nu_1 = \nu_2 > 0$ and $\nu_3 = -\nu_1 = -\nu_2$, we then obtain a *perfectly symbiotical* system, as studied by the author (1983b). In this case, the *production* of *secondary substances* x^1 and x^2 *benefit* the *receivers* C_2 and C_1, respectively, *so that* x^1 *and* x^2 *are kairomones.*

2F.3 Stability of the Production Process

The variables of (2F.2-2) are (x^i, C_i), $i = 1, 2$. The solution curves require *initial conditions* $(x^i(t_0), C_i(t_0))$, $i = 1, 2$, where t_0 is some *initial time*, to be specified. We say the system is *stable* if, given a solution $(x^i(t), C_i(t))$ and any $\varepsilon > 0$, there is $\delta(\varepsilon) > 0$ such that, for any other solution $(\bar{x}^i(t), \bar{C}_i(t))$ with absolute values $|\bar{x}^i(t_0) - x^i(t_0)| < \delta(\varepsilon)$ and $|\bar{C}_i(t_0) - C_i(t_0)| < \delta(\varepsilon)$, it follows that $|\bar{x}^i(t) - x^i(t)| < \varepsilon$ for all t. If a system is *not stable,* it is said to be *unstable.* We now state the *stability of production theorem. If* (2F.2-2) *is perfectly cooperative, it is stable. If* (2F.2-2) *is perfectly symbiotical, it is unstable.* In order to prove this result, we use the method of Riemannian geometry. The basic reference we use is Laugwitz's *Differential and Riemannian Geometry* (1965).

First, note that a Riemannian metric on the first quadrant of Cartesian (x^1, x^2)-space is defined by

$$g_{ij} = e^{2\phi}\delta_{ij}, \quad i, j = 1, 2. \tag{2F.3-1}$$

Moreover, the *production equations* (2F.2-6) are *geodesic equations* (written in the *natural* parameter s) for the metric (2F.3-1). The so-called *metric connection* associated with these geodesics are the *eight* quantities

$$\left\{ {i \atop jk} \right\} = \frac{1}{2} \sum_{\lambda} g^{\lambda i} \left(\frac{\partial g_{j\lambda}}{\partial x^k} + \frac{\partial g_{k\lambda}}{\partial x^j} - \frac{\partial g_{jk}}{\partial x^\lambda} \right), \tag{2F.3-2}$$

where g^{ij} is the inverse matrix of g_{ij}. By substituting (2F.3-1) into (2F.3-2), we obtain

$$\left\{ {i \atop ii} \right\} = \frac{\partial \phi}{\partial x^i},$$

$$\left\{ {i \atop ji} \right\} = \left\{ {i \atop ij} \right\} = \frac{\partial \phi}{\partial x^j}, \quad i \neq j,$$

$$\left\{ {i \atop jj} \right\} = -\frac{\partial \phi}{\partial x^i}, \quad i \neq j, \tag{2F.3-3}$$

where i, j is 1 or 2. The so-called *Riemannian Curvature Tensor* $R^i_{jk\ell}$ for the metric g_{ij} is defined, using the summation convention on repeated indices, by

$$R^{\ell}_{ijk} = \frac{\partial}{\partial x^k}\left\{{\ell \atop i\,j}\right\} - \frac{\partial}{\partial x^j}\left\{{\ell \atop i\,k}\right\} + \left\{{m \atop i\,j}\right\}\left\{{\ell \atop m\,k}\right\} - \left\{{m \atop i\,k}\right\}\left\{{\ell \atop m\,j}\right\}. \tag{2F.3-4}$$

Using (2F.3-3) and two-dimensionality, we obtain for $i \neq j$

$$R^i_{jij} = \frac{\partial^2 \phi}{(\partial x^i)^2} + \frac{\partial^2 \phi}{(\partial x^j)^2}, \tag{2F.3-5}$$

with all other R-quantities vanishing. Letting g denote the determinant of 2×2 matrix g_{ij}, the *Gaussian Curvature* of g_{ij} is the real number

$$K = -\frac{R_{1212}}{g}, \tag{2F.3-6}$$

where,

$$R_{1212} = \sum_{i=1}^{2} g_{1i}\, R^i_{212} = e^{2\phi}\left(\frac{\partial^2 \phi}{(\partial x^1)^2} + \frac{\partial^2 \phi}{(\partial x^2)^2}\right), \tag{2F.3-7}$$

(see Laugwitz, 1965). Thus, we arrive at the important formula

$$K = -e^{-2\phi}(\nu_1 + \nu_2). \tag{2F.3-8}$$

It is a classical *theorem of Jacobi* that *geodesic systems are unstable if* $K \leqslant 0$ *and stable if* $K > 0$, *for two-dimensional Riemannian geometries* (Antonelli, 1983a). The theorem now follows by inspection of (2F.3-8).

If we consider the case ν_1, ν_2, ν_3 are *all positive,* we might term it *perfectly antagonistic.* It is no surprise that such a system is unstable in its production of toxins. It is now clear that *autotoxin production, for the class of symbiotical systems we consider here, is always unstable.*

Acknowledgement

This research was partially supported by NSERC-A-7667 in Limoges, France in 1982.

References

Antonelli, P.L. 1983a. Nonlinear allometric growth I. Perfectly cooperative systems. *Math. Model.*, 4: 367-372.

Antonelli, P.L. 1983b. Nonlinear allometric growth II. Perfectly symbiotical systems. *Math. Model.*, 4: 373-380.

Harper, J. 1977. *The Population Biology of Plants.* Academic Press.

Laugwitz, D. 1965. *Differential and Riemannian Geometry.* Academic Press.

Whittaker, R.H. and P.P. Feeny. 1977. Allelochemics: chemical interaction between species. *Science,* 171: 757-768.

3

ON THE MATHEMATICAL FOUNDATIONS OF GROWTH MECHANICS

3A

Solving Geodesic Equations in Space of Locally Constant Connections

B.H. Voorhees

3A.1 Conformally Flat Spaces of Locally Constant Connection

A differential manifold M with connection Γ denoted (M,Γ) will be called a space of locally constant connection if for every point p of M there exists a coordinate chart containing p such that with respect to this chart the connection coefficient of (M,Γ) are constant. The first example of such a space was given by Antonelli and Voorhees (1974).

Such spaces have proved useful in modeling certain biological systems such as coral reefs and plant communities. They can be employed for analysis of cooperative or competitive systems obeying Lotka-Volterra type equations

$$\frac{dc_i}{dt} = \varepsilon_i c_i - \sum_{j,k=1}^{n} \Gamma^i_{jk} c_j c_k . \qquad (3A.1\text{-}1)$$

The coefficients ε_i and Γ^i_{jk} in this equation are constants. Note that c_i is the population of species i, ε_i is the growth rate of this species, and Γ^i_{jk} tells how the interaction between species j and k contributes to the growth of species i. The term $\varepsilon_i c_i$ can be removed without loss of generality by appropriate reparameterization of time, *assuming all* ε_i *equal.*

Thus we study (3A.1-1) in the form

$$\frac{dx^i}{dt} = c_i , \qquad \frac{dc_i}{dt} = - \sum_{j,k=1}^{n} \Gamma^i_{jk} c_j c_k \qquad (3A.1\text{-}2)$$

where the Γ^i_{jk} are the connection coefficients for a space of locally constant connection. (3A.1-2) are the geodesic equations for this space.

E. Ruhnau (1977) had shown that sufficient conditions for a locally constant Levi-Cività connection to be a metric connection are that in coordinates in which the Γ^i_{jk} are constant there exists a vector field with constant components γ_k such that

$$\sum_{s=1}^{n} (\delta_{is}\Gamma^s_{kj} + \delta_{js}\Gamma^s_{ki}) = \gamma_k \delta_{ij} \,. \qquad (3A.1\text{-}3)$$

She conjectures that this condition is also necessary. A proof is given in Antonelli's essay C in Chapter 2. So it is true that the only possible metric for a space of locally constant connection is the conformally flat metric derived by Antonelli and Voorhees (1974):

$$g_{ij} = e^{2\phi}\delta_{ij} \,,$$

$$\phi = \sum_{k=1}^{n} \alpha_k x^k \,, \qquad (3A.1\text{-}4)$$

where $\gamma_k = (2/n)\alpha_k$ and the x^k are local (actually global) coordinates. The connection coefficients, Riemann and Ricci tensors, sectional curvatures and scalar curvature for the metric of (3A.1-4) are given in the appendix 3A. The geodesic equations (3A.1-2) now take the form

$$\frac{dx^i}{dt} = c_i$$

$$\frac{dc_i}{dt} = -2c_i \sum_{j\neq i} \alpha_j c_j - \alpha_i c_i^2 + \alpha_i \sum_{j\neq i} c_j^2 \,. \qquad (3A.1\text{-}5)$$

The main result of this paper is the general solution of these equations. (Fred Bookstein has pointed out that the solutions of (3A.1-5) form a parabolic pencil of semi-circles, in an elegant argument from classical synthetic geometry.)

3A.2 Dynamics

Let $\{\hat{e}_1, \ldots, \hat{e}_n\}$ be an orthonormal basis for tangent spaces of M. Write

$\underline{c} = c_1\hat{e}_1 + \ldots + c_n\hat{e}_n$, $\underline{\alpha} = \alpha_1\hat{e}_1 + \ldots + \alpha_n\hat{e}_n$. Addition and subtraction of $2\alpha_i e_i^2$ to (3A.1-5) together with identification of c_i as the rate of change of x^i yields

$$\dot{\underline{x}} = \underline{c}$$
$$\dot{\underline{c}} = |\underline{c}|^2\underline{\alpha} - 2(\underline{\alpha}\cdot\underline{c})\underline{c} \tag{3A.2-1}$$

where $\dot{u}$ denotes the time derivative of u, $|\underline{c}|^2 = \underline{c}\cdot\underline{c}$ and scalar products are computed with respect to the Euclidean metric δ_{ij}.

THEOREM I (Dynamics). The general solution of (3A.2-1) with initial conditions $\underline{c}(0) = \underline{k}, \underline{x}(0)$ is

$$\underline{c}(t) = \frac{c_0}{|\underline{\alpha}|}\left(\frac{1}{1+\tau^2}\right)\left[(\tau - \cot\theta_0)\underline{\alpha} + \frac{|\underline{\alpha}|}{|\underline{k}|\sin\theta_0}\underline{k}\right] \tag{3A.2-2}$$

$$\underline{x}(t) = \underline{x}(0) + \frac{1}{|\underline{\alpha}|^2}\left\{\frac{1}{2}\ln(1+\tau^2)\underline{\alpha} - \left[\cot\theta_0\,\underline{\alpha} - \frac{|\underline{\alpha}|}{|\underline{k}|\sin\theta_0}\underline{k}\right]\tan^{-1}\tau\right\} \tag{3A.2-3}$$

so long as $\theta_0 \neq 0, \pi$. If $\theta_0 = 0$ or π

$$\underline{c}(t) = \frac{|\underline{k}|}{|\underline{\alpha}|}\left[1 + (\underline{\alpha}\cdot\underline{k})t\right]^{-1}\underline{\alpha} \tag{3A.2-4}$$

$$\underline{x}(t) = \underline{x}(0) + \frac{1}{|\underline{\alpha}|^2}\ln(1 + |\underline{\alpha}||\underline{k}|t)\underline{\alpha} \tag{3A.2-5}$$

where $\tau = \cot\theta_0 + c_0|\underline{\alpha}|t$ and θ_0, c_0 are constants:

$$c_0 = \frac{|\underline{k}|}{\sin\theta_0}$$
$$\theta_0 = \cos^{-1}\left[\frac{\underline{\alpha}\cdot\underline{k}}{|\underline{\alpha}||\underline{k}|}\right]. \tag{3A.2-6}$$

PROOF follows by direct substitution into (3A.1-5).

In using these equations to model coral reef growth $c_i(t)$ represents the population of coral polyps of species i at time t and $x^i(t)$ the total amount of calcium carbonate laid down in the reef by coral species i up to time t. Thus the following theorem is of some biological interest:

THEOREM II (Generalized Allometric Law). Let $y_i(t)$ be defined by $x_i = \ell n\, y_i$. Then

$$y_i = e^{P_{ij}(\tau)} (y_j)^{\alpha_i/\alpha_j} \tag{3A.2-7}$$

where

$$P_{ij}(\tau) = -\frac{1}{\alpha_j}[\alpha_i x_j(0) - \alpha_j x_j(0)] + \frac{|\underline{\alpha}|}{\alpha_j |\underline{k}| \sin\theta_0}(\alpha_i k_j - \alpha_j k_i)\tan^{-1}\tau \tag{3A.2-8}$$

COROLLARY. In the limit $\tau \to \infty$ (3A.2-7) becomes the classical allometric law

$$y_i = P_{ij}(y_j)^{\alpha_i/\alpha_j} \quad (P_{ij} \text{ constant}). \tag{3A.2-9}$$

Thus, this model leads to the prediction that in older reefs, but not in younger ones, the amounts of calcium carbonate deposited by the different coral species will follow the classical allometric law.

3A.3 Kinematics

We can obtain a kinematic form for the $\underline{c}(t)$ of Theorem I. Such a form is useful under some circumstances for computing the optical properties of congruences of system orbits. For example if $\underline{c}(\underline{x})$ is known then expansion of the orbits is just $\underline{\nabla} \cdot \underline{c}$ while their rotation is $\frac{1}{2}((\partial/\partial x^i)c_j - (\partial/\partial x^j)c_i)$ and their shear $\frac{1}{2}((\partial/\partial x^i)c_j + (\partial/\partial x^j)c_i) - (1/n)\underline{\nabla} \cdot \underline{c}\,\delta_{ij}$.

Let A be the matrix

$$A_{ij} = \begin{cases} \alpha_i c_j - \alpha_j c_i & i \neq j \\ \phi & i = j. \end{cases} \tag{3A.3-1}$$

THEOREM III (Kinematics). The general solution $\underline{c}(\underline{x})$ of (3A.3.1) for $\underline{c}(0) = \underline{k}$ is

$$\underline{c}(\underline{x}(t)) = \left[e^{\int_0^t A(\underline{c}(t'))\, dt'} \right] \circ \underline{k} \qquad (3A.3\text{-}2)$$

PROOF: According to a theorem of Magnus (1954) any system of the form $\dot{\underline{c}} = A(t) \cdot \underline{c}$ has solution $\underline{c}(t) = e^{\Omega(t)} \cdot \underline{c}(0)$ where $\Omega(t)$ is a certain matrix. If $A(t)$ commutes with its integral $\Omega(t) = \int_0^t A(t')\, dt'$. By inspection (3A.2-1) can be written $\dot{\underline{c}} = A(\underline{c}(t)) \circ \underline{c}$ where $A(\underline{c})$ is given by (3A.3-1). Further, it is simple to show that this matrix commutes with its integral by use of (3A.2-2) or (3A.2-4). Thus the result follows.

The kinematic nature of the solution follows since

$$\int_0^t A_{ij}(\underline{c})\, dt = \begin{cases} \alpha_i x^j - \alpha_j x^i & i \neq j \\ -\phi & i = j\,. \end{cases} \qquad (3A.3\text{-}3)$$

If we define $K_{ij} = \frac{1}{2}(A_{ij} - A_{ji})$ then (3A.3-2) becomes

$$\underline{c}(\underline{x}) = e^{-\phi} e^{K} \cdot \underline{k}\,. \qquad (3A.3\text{-}4)$$

Use of the standard expansion of an exponential matrix together with the fact that any skew symmetric 3×3 matrix satisfies the equation $K^3 = -|K|^2 K$ yields

THEOREM IV. For a 3-space of locally constant connection

$$\underline{c}(\underline{x}) = e^{-\phi} [I + F(\underline{x})K + G(\underline{x})K^2] \cdot \underline{k} \qquad (3A.3\text{-}5)$$

where I is the identity, and $F(\underline{x})$ and $G(\underline{x})$ are defined by

$$F(\underline{x}) = \sum_{j=1}^{\infty} \frac{1}{(2j-1)!} (\phi^2 - |\underline{\alpha}|^2 |\underline{x}|^2)^{j-1}$$

$$G(\underline{x}) = \sum_{j=1}^{\infty} \frac{1}{(2j)!} (\phi^2 - |\underline{\alpha}|^2 |\underline{x}|^2)^{j-1}\,. \qquad (3A.3\text{-}6)$$

The explicit form of (3A.3-5) may be written

$$e^{\phi}\underline{c}(\underline{x}) = \underline{k} + \{F(\underline{x})(\underline{k}\cdot\underline{x}) + G(\underline{x})[(\underline{\alpha}\cdot\underline{x})(\underline{k}\cdot\underline{x}) - (\underline{\alpha}\cdot\underline{k})|\underline{x}|^2]\}\underline{\alpha}$$
$$- \{(\underline{\alpha}\cdot\underline{k})F(\underline{x}) - G(\underline{x})[(\underline{\alpha}\cdot\underline{k})(\underline{\alpha}\cdot\underline{x}) - |\underline{\alpha}|^2(\underline{k}\cdot\underline{x})]\}\underline{x} \tag{3A.3-7}$$

Finally, (3A.3-7) allows us to obtain a constant of motion:

THEOREM V. Let $\underline{k}$ be any constant vector field on a 3-space of locally constant connection. The function $E(\underline{x}, \underline{c})$ defined by

$$E(\underline{x}, \underline{c}) = |\underline{c}|^2 - [F^2 - 2G - (\phi^2 - |\alpha|^2|\underline{x}|^2)G^2]\,\psi^2(\underline{x}) \tag{3A.3-8}$$

where

$$\psi^2(\underline{x}) = (\underline{\alpha}\cdot\underline{k})^2|\underline{x}|^2 - 2(\underline{\alpha}\cdot\underline{k})(\underline{\alpha}\cdot\underline{x})(\underline{k}\cdot\underline{x}) + |\underline{\alpha}|^2(\underline{k}\cdot\underline{x})^2 \tag{3A.3-9}$$

has the constant value $|\underline{k}|^2$ along orbits of (3A.2-1) for which $\underline{c}(0) = \underline{k}$.

Note that the first term of (3A.3-8) is, formally, velocity squared. Thus the function $E(\underline{x}, \underline{c})$ has the form of kinetic plus potential energy if the second term is interpreted as a velocity independent potential. This suggests an interpretation of $E(\underline{x}, \underline{c})$ as an energy integral.

Acknowledgements

This paper owes much to conversations with graduate students James Brooke and Thomas Zannias. The paper of Magnus was brought to my attention by Professor Herbert Freedman. Theorem II, the generalized allometric law, was obtained in discussions with P.L. Antonelli. This research was supported by a NSERC-A-7667 grant to P.L. Antonelli.

References

Antonelli, P.L., and B.H. Voorhees. 1975. Sherbrooke Notes. Sherbrooke, Quebec.

Antonelli, P.L., and B.H. Voorhees. 1983. Non-linear growth dynamics: I. Volterra-Hamilton systems. *Bull. Math. Biol.*, 45: 103-116.
Ruhnau, E. 1977. Spaces of Locally Constant Connection. Master's Thesis, University of Alberta.
Magnus, W. 1954. On the exponential solution of differential equations for a linear operator. *Communications on Pure and Applied Mathematics,* Vol. VII, pp. 619-673.

3A. Appendix*

Connection Coefficients::

$$\Gamma^{i}_{ii} = \alpha_i$$

$$\Gamma^{i}_{ji} = \Gamma^{i}_{ij} = \alpha_j \qquad i \neq j$$

$$\Gamma^{i}_{jj} = -\alpha_i \qquad i \neq j$$

$$\Gamma^{i}_{jk} = 0 \qquad i \neq j \neq k .$$

Riemann Tensor:

$$R^{i}_{jk\ell} = -\delta^{i}_{k} L_{j\ell} + \delta^{i}_{\ell} L_{kj} - \delta_{j\ell} L_{ik} + \delta_{jk} L^{i}_{\ell}$$

$$L_{ij} = \alpha_i \alpha_j - \frac{1}{2} |\underline{\alpha}|^2 \delta_{ij} .$$

Ricci Tensor:

$$R_{ij} = (n-2)(\alpha_i \alpha_j - |\underline{\alpha}|^2 \delta_{ij}) .$$

Sectional Curvatures:

$$K(i,j) = -R_{ijij} = -e^{-2\phi} \sum_{k \neq i,j} (\alpha_k)^2, \quad i \neq j .$$

Scalar Curvature:

$$R = -e^{-2\phi}(n-1)(n-2)|\underline{\alpha}|^2 .$$

*See Sherbrooke Notes of Antonelli and Voorhees (1974) or Antonelli and Voorhees (1983) for detailed description of calculations.

3B

On the Mathematical Theory of Volterra-Hamilton Systems

P.L. Antonelli

PART I Some Geometrical Background

3B.1 The Tangent Bundle

The script letter H^n will always be used to denote a *smooth* (that is, c^∞ or c^ω) *manifold* of dimension n . It will be assumed Hausdorff, separable and connected. The so-called *smooth structure* is characterized by a maximal family F of coordinate charts (U_α, h_α), $\alpha \in F$, which cover H^n and which have smooth overlap maps

$$h_\beta \circ h_\alpha^{-1} : h_\alpha(U_\alpha \cap U_\beta) \to \mathbf{R}^n , \tag{3B.1-1}$$

where each $h_\alpha : U_\alpha \to \mathbf{R}^n$, $\alpha \in F$, is a homeomorphism (1-1, bicontinuous) onto an open subset of euclidean n-space, $\mathbf{R}^n$. If all overlap maps are real analytic (i.e. c^ω), then H is called a *real analytic manifold.*

A *smooth map* (i.e. c^∞ or c^ω) $f : H_1^n \to H_2^m$ between smooth manifolds is a continuous map for which the collection of functions

$$k_\beta \circ f \circ h_\alpha^{-1} : h_\alpha(U_\alpha) \to k_\beta(V_\beta) \tag{3B.1-2}$$

are smooth and $\alpha \in F_1$, $\beta \in F_2$.

Note. This article was written with a view to include the classic work of J. Douglas (1927) on sprays and modern nonlinear mechanics and stochastic Riemannian geometry. The latter topic, of such importance to the stochastic growth mechanics, is well expounded in the graduate text by Ikeda and Watanabe (1981).

A *tangent vector* ξ to H at point $p \in H$ is a map which assigns to each $\alpha \in F$ where $p \in U_\alpha$, an n-tuple, ξ_α, of real numbers, such that, if (U_β, h_β) is another chart containing p then,

$$\xi_\beta = D(h_\beta \circ h_\alpha^{-1})_{h_\alpha(p)} \cdot \xi_\alpha, \tag{3B.1-3}$$

where D denotes the Jacobian matrix of the overlap map and $\xi_\alpha = (\xi_\alpha^1, \ldots, \xi_\alpha^n)$. The numbers ξ_α^i $i = 1, 2, \ldots, n$ are called the *contravariant components* of ξ relative to the chart (U_β, h_β). The collection of tangent vectors at p is denoted T_pH. It carries the structure of an n-dimensional vector space over the real field, with addition and scaler multication induced from the componentwise operations. None of the concepts mentioned above depend on particular choices of coordinate charts.

Let us rewrite equation (3B.1-3) in index notation. Thus let $u_\alpha^i(p) = x^i(h_\alpha(p))$, $i = 1, \ldots, n$ define chart coordinates in U_α, where x^i are the usual cartesian coordinate functions. Then (3B.1-3) becomes

$$\xi_\beta^j = \sum_i \left(\frac{\partial u_\beta^j}{\partial u_\alpha^i}\right)_{h_\alpha(p)} \cdot \xi^i, \quad j = 1, \ldots, n. \tag{3B.1-4}$$

The partial derivative of f with respect to u^i at $p \in H$ is given by

$$\left(\frac{\partial f}{\partial u^i}\right)_p = \left(\frac{\partial(f \circ h_\alpha^{-1})}{\partial x^i}\right)_{h_\alpha(p)}. \tag{3B.1-5}$$

Again, the value does not depend on the particular chart. Indeed, the chain rule gives us

$$\sum_i \left(\frac{\partial(f \circ h_\alpha^{-1})}{\partial x^i}\right)_{h_\alpha(p)} \xi_\alpha^i = \sum_j \left(\frac{\partial(f \circ h_\beta^{-1})}{\partial x^j}\right)_{h_\beta(p)} \xi_\beta^j \tag{3B.1-6}$$

The term on the left in (3B.1-6) may be written more succinctly as $\xi_\alpha^i (\partial f/\partial u^i)_p$, using Einstein's summation convention. (Namely, summation on identical indices occurring up and down is understood to run from 1 to n .) The quantity in question is called the *directional derivative* of f, relative to (U_α, h_α), in direction ξ. The case in which all components of ξ relative to (U_α, h_α) vanish except the kth, which is unity, is of special interest. We denote this contravariant vector, $(\partial)/(\partial u_\alpha^k)$, and note that its components in (U_β, h_β) are given by $(\partial u^j_\beta)/(\partial u^k_\alpha)$, $j = 1, 2, \ldots, n$, so that,

$$\frac{\partial}{\partial u^k_\alpha} = \left(\frac{\partial u^j_\beta}{\partial u^k_\alpha}\right)_{h_\alpha(p)} \cdot \frac{\partial}{\partial u^j_\beta} \; . \tag{3B.1-7}$$

From this it follows that $\partial/(\partial u^k_\beta)$ are a basis for the T_pH relative to (U_β, h_β). Thus, for any tangent vector ξ at p and any choice of chart around p, we can write uniquely

$$\xi_p = \xi^i \left(\frac{\partial}{\partial u^i}\right)_p \; . \tag{3B.1-8}$$

The *dual vector space* to T_pH is denoted T^*_pH and has the dual basis du^i, $i = 1, \ldots, n$ given by

$$\left\langle \frac{\partial}{\partial u^i}, du^j \right\rangle_p = \delta^i_j \tag{3B.1-9}$$

where $\langle , \rangle$ denotes the evaluation of the linear functional du^j on the vector $\partial/(\partial u^i)$ and δ^i_j is the Kronecker delta or identity matrix. Elements of T^*_pH may be written uniquely as

$$\xi^\alpha = \xi^\alpha_j \, du^j \, . \tag{3B.1-10}$$

ξ_j are called the *covariant components* of the tangent vector ξ relative to $\alpha \in F$.

The transformation from one chart to another is given as

$$\xi^\beta_j = \sum_j \left(\frac{\partial u^i_\alpha}{\partial u^j_\beta}\right)_{h_\beta(p)} \xi^\alpha_i \tag{3B.1-11}$$

for covariant components of ξ.

The collection of all tangent vectors to H^n is denoted $TH = \bigcup_{p \in H} T_pH$. One provides a smooth structure on this set by first defining the projection map $\pi : TH \to H$ as

$$\pi(\xi) = p \text{ iff } \xi \in T_pH \tag{3B.1-12}$$

and then requiring this to be continuous, while simultaneously requiring the map

$$\tilde{h}_\alpha : \pi^{-1}(U_\alpha) \to \mathbf{R}^n \times \mathbf{R}^n , \tag{3B.1-13}$$

given by

$$\tilde{h}_\alpha(\xi) = (h_\alpha(\pi(\xi)), \xi_\alpha), \tag{3B.1-14}$$

to be a (c^∞, c^ω) diffeomorphism. The overlap maps are given by the formula

$$\tilde{h}_\beta \circ \tilde{h}_\alpha^{-1}(p, \xi) = (h_\beta \circ h_\alpha^{-1}(p'), D(h_\beta \circ h_\alpha^{-1})_{p'} \cdot \xi) \tag{3B.1-15}$$

where $p' = h_\alpha(p)$. Since $h_\beta \circ h_\alpha^{-1}$ are $(c^\infty$ or $c^\omega)$ maps it follows that $\tilde{h}_\beta \circ \tilde{h}_\alpha^{-1}$ are also. Therefore, the charts $(\pi^{-1}(U_\alpha), \tilde{h}_\alpha)$, $\alpha \in F$, generate a smooth structure on TH. One checks that TH is connected, Hausdorff, separable and of dimension $2n$, and that the (fiber bundle) projection map is smooth.

We shall need to recall the basic ideas of *fiber bundle theory.* Preliminary to this is the idea of a *topological transformation group* G acting on a space F. Firstly, a *topological group* G is a topological space for which the group operations $(g_1, g_2) \mapsto g_1 \cdot g_2$ and $g_1 \mapsto g_1^{-1}$ taking $G \times G \to G$ and $G \to G$, respectively, are continuous. G is said to *act on* F, if $(g_1 \cdot (g_2 \cdot f)) = (g_1 \cdot g_2) \cdot f$, for all $g_1, g_2 \in G$ and all $f \in F$. The symbol $g \cdot f$ denotes the *continuous* image of the *action map* $G \times F \to F$. We say G *acts effectively* provided $I \cdot f = f$ and $g \cdot f = f$ for all $f \in F$, implies $g = I$, where I denotes the identity element of G. The *structural group* of a fiber bundle acts effectively on the fiber F. It often occurs that G is a *Lie group.* This means that G is a real analytic manifold on which the group operations are real analytic. A standard example is $GL(n)$, the group of non-singular $n \times n$ real matrices. The group operation is matrix multiplication and the topology and real analytic structure is induced from $\mathbf{R}^{n^2}$ which contains $GL(n)$ as an open set with two connected components.

In reviewing the notions of fiber bundle we follow Steenrod (1951). A *fiber bundle* is a 5-tuple (E, π, H, F, G) where E, H, F, G are topological spaces and $\pi : E \to H$ is a continuous map onto H with the following additional requirements. There is a *trivializing cover* of H by open charts (V_α, h_α). This means that there is a homeomorphism

$$\phi_\alpha : V_\alpha \times F \to \pi^{-1}(V_\alpha) \tag{3B.1-16}$$

for each V_α and

$$\pi\,\phi_\alpha(p, f) = p \text{ for all } p \in V_\alpha \tag{3B.1-17}$$

and $f \in F$. The space F is called the *fiber.*

Furthermore, there are continuous maps

$$h_{\alpha\beta}: V_\alpha \cap V_\beta \to G \tag{3B.1-18}$$

defined by

$$h_{\alpha\beta}(p) = \phi^{-1}_{\alpha,p} \circ \phi_{\beta,p} \tag{3B.1-19}$$

with

$$\phi_{\alpha,p}(f) = \phi_\alpha(p, f)\,. \tag{3B.1-20}$$

One shows that

$$\begin{cases} h_{\gamma\alpha}h_{\alpha\beta} = h_{\gamma\beta} \\ \quad h_{\gamma\gamma} = I\,. \end{cases} \tag{3B.1-21}$$

Properties (3B.1-16), (3B.1-17), and (3B.1-21) characterize fiber bundles up to *bundle equivalence.* Two bundles with the same E, π, H, F, G are equivalent if their h-functions are conjugate in G. That is, if there is a continuous map $\lambda : V_\alpha \cap V_\beta \to G$ such that

$$\tilde{h}_{\alpha\beta}(p) = \lambda^{-1}(p) \cdot h_{\alpha\beta}(p) \cdot \lambda(p) \tag{3B.1-22}$$

for all $p \in V_\alpha \cap V_\beta$.

The 5-tuple $(TH, \pi, H, \mathbf{R}^n, GL(n))$ is a fiber bundle. Likewise, so is $(GL(H), \pi, H, GL(n), GL(n))$ which is an important bundle *associated with the tangent bundle* TH. The bundle $GL(H)$ has the same *base space,* structural group $GL(n)$, and h-functions. Instead of the group acting on the tangent space fiber, as in TH, it acts on itself by group multiplication, producing a so-called *principle bundle.* This is just a bundle in which the group and fiber are identical. We shall have need of a more explicit construction of the bundle $GL(H)$.

By a frame e at a point $p \in H$, we mean a set of n linearly independent tangent vectors $\{X_1, \ldots, X_n\}$ at p. An element $g \in G = GL(n)$ acts (on the left) on this frame according to

$$g \cdot e = \{g_1^i X_i, g_2^i X_i, ..., g_n^i X_i\}, \tag{3B.1-23}$$

where $g_1^i, ..., g_n^i$ are the n rows of the matrix g. If (U, h) is a chart in H, then a corresponding chart in $GL(H) = E$ is coordinatized by (u^j, a_i^k) where

$$X_i = a_i^k \left(\frac{\partial}{\partial u^k}\right)_p, \quad i, j, k = 1, ..., n. \tag{3B.1-24}$$

Using these coordinates, it is easy to verify that $GL(H) = E$ is a smooth manifold of dimension $n^2 + n$ and that the bundle projection map π is smooth. We call $GL(H)$ the *bundle of frames* over H^n.

One may also construct the so-called *orthonormal frame bundle* over H, directly. However, since the orthogonal group $O(n)$ is a compact Lie subgroup of $GL(n)$ which is a deformation retract of $GL(n)$, it is possible to apply the *bundle reduction theorem* to show that the orthonormal frame bundle is equivalent (in $GL(n)$) to the frame bundle $GL(H)$.

We now briefly review tensor bundles. These are bundles whose fibers are linear spaces of tensors. The tangent bundle itself is a tensor bundle. The transformation law (3B.1-4) shows that the fiber is a space of tensors of type $(1, 0)$. That is, a space of contravariant vectors. The dual bundle, or so-called *cotangent bundle,* has as fiber a linear space of *covariant* vectors or tensors of type $(0, 1)$ as (3B.1-11) shows. The h-functions for T^*H are related to those of TH via the adjoint map $ad: GL(n) \to GL(n)$, which sends each matrix into the transpose of its inverse, by

$$h^*_{\alpha\beta}(P) = ad(h_{\alpha\beta}(P)). \tag{3B.1-25}$$

A *tensor bundle of type* (t, s) may be defined as the ordered product

$$\underbrace{T^*H \otimes ... \otimes T^*H}_{\text{s copies}} \otimes \underbrace{TH \otimes ... \otimes TH}_{\text{t copies}} = T_s^t H, \tag{3B.1-26}$$

where $\otimes$ indicates tensor product of bundles. This is simply to say that the fibers are amalgamated by tensor product and that the action of G on the $(t + s)$-dimensional fiber is given by

$$(ad(a) \cdot Q)^{i_1 ... i_t}_{j_1 ... j_s} = a^{i_1}_{k_1} ... a^{i_t}_{k_t} Q^{k_1 ... k_t}_{\ell_1 ... \ell_s} \bar{a}^{\ell_1}_{j_1} ... \bar{a}^{\ell_s}_{j_s}, \tag{3B.1-27}$$

where

$$\bar{a} = \mathrm{ad}(a) = \mathrm{ad}(h_{\alpha\beta}(p)) \quad \text{and} \quad Q_{\ell_1 \dots \ell_s}^{k_1 \dots k_t},$$

is a tensor of type (t, s). A (t, s)-*tensor field* over H is a *smooth cross section* of $T_s^t H$. That is, it is a smooth map (i.e. c^∞ or $c^{(\omega)}$)

$$\theta(t, x) : H \to T_s^t H, \tag{3B.1-28}$$

for which

$$\pi\, \theta(t, s)(p) = p, \quad \text{for all} \quad p \in H. \tag{3B.1-29}$$

The transformation of coordinates in $V_\alpha \cap V_\beta$ is given by

$$
\begin{aligned}
{}_\beta Q_{j_1 \dots j_s}^{i_1 \dots i_t} = {} & \left(\frac{\partial u_\beta^{i_1}}{\partial u_\alpha^{k_1}} \right)_{h_\alpha(p)} \cdots \left(\frac{\partial u_\beta^{i_t}}{\partial u_\alpha^{k_t}} \right)_{h_\alpha(p)} {}_\alpha Q_{\ell_1 \dots \ell_s}^{k_1 \dots k_t} \\
& \times \left(\frac{\partial u_\alpha^{\ell_1}}{\partial u_\beta^{k_1}} \right)_{h_\beta(p)} \cdots \left(\frac{\partial u_\alpha^{\ell_s}}{\partial u_\beta^{k_t}} \right)_{h_\beta(p)}
\end{aligned}
\tag{3B.1-30}
$$

which generalizes (3B.1-4) and (3B.1-11). We see from (3B.1-25) that classical transformation laws for tensors, as (3B.1-30) is, one compatible with the action of G.

A *Douglas tensor* on H is a tensor field on TH with the property that its image, as a cross section of the natural bundle projection $T_s^t(TH) \to TH$, lies in the image of the so-called zero section of the induced projection $T_s^t(TH) \xrightarrow{\pi_s^t} T_s^t(H)$ arising from the (t, s)-tensor product of the natural map $\pi : TH \to H$. As a consequence, the transformation laws, $X \to \bar{X}$, which result for Douglas tensors are identical to the usual ones. The difference is that they depend explicitly on ξ as well as p. The general theory of sprays discussed in the next section makes use of Douglas Tensors.

3B.2 The Global Theory of Sprays

A vector field on H^n is a tensor field of type $(1, 0)$. It is clear that a vector field is a smooth cross section of TH. Thus, $\theta(1,0)(p) = \xi = (\xi^1, ..., \xi^n) \in \mathbf{R}^n$ is a tangent vector at p. Vector fields possess *local integral curves* $f(u, t)$. That is, the ordinary differential equations

$$\begin{cases} \dfrac{df^i}{dt} = \xi^i(f^1(t), ..., f^n(t)), \quad i = 1, ..., n, \\ f^i(u, 0) = u^i, \quad u = (u^1, ..., u^n), \\ u^i(p) = 0, \end{cases} \tag{3B.2-1}$$

have a unique smooth solution $\phi_t(u) = (f^1(u, t), ..., f^n(u, t))$ in a small enough neighborhood of the initial point p and for small enough values of the parameter t. It is also true that $\phi_t(\phi_s(u)) = \phi_{t+s}(u)$ for $t + s$ small enough. This is the *local* 1*-parameter group* property of the diffeomorphism $\phi_t(\cdot)$. Also $\phi_0(u) = u$ in a neighborhood of p.

A vector field S on TH is a *second order differential equation* on H, if the Jacobian map

$$D\pi : TTH \rightarrow TH, \tag{3B.2-2}$$

has the property

$$D\pi \circ S(\xi) = \xi, \text{ for all } \xi \in TH. \tag{3B.2-3}$$

A *spray* is a second order differential equation with a *special property.* Let λ be any nonzero real number and define the bundle automorphism

$$\lambda_* : TH \rightarrow TH, \tag{3B.2-4}$$

by scalar multiplication on each fiber. Thus,

$$\lambda_*(\xi) = \lambda\xi. \tag{3B.2-5}$$

If a given second order differential equation S, satisfies

$$S(\lambda\xi) = \lambda_* \lambda S(\xi), \tag{3B.2-6}$$

for any $\lambda \neq 0$, it is called a *spray.*

If ξ is a vector in TH, let β_ξ be the integral curve of S with $\beta_\xi(0) = \xi$. Let V be the set of ξ in TH for which β_ξ is defined at least for $0 \leqslant t \leqslant 1$. One shows V is open in TH.

The so-called *exponential map,* Exp is now defined by

$$\begin{cases} \mathrm{Exp} : V \to H \\ \mathrm{Exp}(\xi) = \pi \circ \beta_\xi(1) . \end{cases} \tag{3B.2-7}$$

This map can be shown to be a diffeomorphism whenever it is defined. It need not be defined on all of TH or even on all of T_pH. The gist of this preliminary discussion is to point out that a global spray gives a family of smooth curves through each point of H^n, with one in each direction. It is also true, that for any two sufficiently close points p, q in H, there is a unique spray curve joining them. Indeed, Douglas (1927) showed these two properties characterized sprays in the real analytic case. See Section 3B.7.

3B.3 Local Sprays

We now proceed to discuss the local theory. One shall suppose throughout that $u^1, \ldots, u^n$ are smooth (c^∞) coordinates in a trivializing neighborhood of TH and of T^*H. Fundamental to the discussion will be Euler's Theorem on smooth homogeneous functions. Suppose a smooth map $\phi : TH \to \mathbf{R}^1$ has the nth order homogeneous property. That is, for any nonzero real λ

$$\phi(u, \lambda\xi) = \lambda^n \phi(u, \xi) . \tag{3B.3-1}$$

Here, $u = (u^1, \ldots, u^n)$ and $\xi = (\xi^1, \ldots, \xi^n)$ is a contravariant vector. Euler's Theorem is that

$$\frac{\partial \phi}{\partial \xi^i} \xi^i = n\phi . \tag{3B.3-2}$$

Obviously, if $\phi : T^*H \to \mathbf{R}^1$ were used instead we would write (3B.2-9) as

$$\frac{\partial \phi}{\partial \xi_i} \xi_i = n\phi . \tag{3B.3-3}$$

Let us suppose we are given a spray in (U, h). Then

$$\frac{d^2u^i}{dt^2} = \underset{2}{H^i}(u, \frac{du}{dt}), \quad i = 1, \ldots, n, \tag{3B.3-4}$$

and the number two under the H^i indicated that H^i is 2nd order homogeneous in du/dt. Define

$$G^i_j(u, \xi) = \frac{1}{2} \frac{\partial \underset{2}{H^i}}{\partial \xi^j}, \tag{3B.3-5}$$

for each pair of i, j running from 1 to n. Taking one more derivative with respect to ξ^k, we define n^3 smooth functions,

$$\Gamma^i_{jk}(u, \xi) = \frac{\partial G^i_j}{\partial \xi^k}. \tag{3B.3-6}$$

It follows from Euler's Theorem that G^i_j is homogeneous of degree one in ξ and that the n^3 functions (3B.3-6) are homogeneous of degree zero in ξ.

By choosing the particular trivializing diffeomorphism corresponding to (U, h) in the fiber bundle TH, we are able to rewrite the 2nd order equation (3B.3-4) for the spray as

$$\begin{cases} \dfrac{du^i}{dt} = \xi^i \\ \dfrac{d\xi^i}{dt} = G^i_j \xi^j . \end{cases} \tag{3B.3-7}$$

We may also rewrite the second system of (3B.3-7) as

$$\frac{d\xi^i}{dt} = \Gamma^i_{jk}(u, \xi)\xi^j \frac{du^k}{dt}. \tag{3B.3-8}$$

If we change coordinate charts from u to $\bar{u}$, say, the n^3 Γ^i_{jk} transform according to

$$\frac{\partial \bar{u}^r}{\partial u^j} \frac{\partial \bar{u}^s}{\partial u^k} \bar{\Gamma}^i_{rs} = \frac{\partial \bar{u}^i}{\partial u^r} \Gamma^r_{jk} - \frac{\partial^2 \bar{u}^i}{\partial u^j \partial u^k}. \tag{3B.3-9}$$

This equation system is identical to the usual law of transformation of a *linear (affine) connection* on H^n. However, the Γ's of our discussion are allowed to depend on ξ through the ratios of ξ^i (or through the ratios of du^i) and only through those, because Γ's are of degree zero. The *classical linear connection does not allow dependence on* ξ. For this reason we will call $\Gamma(u, \xi)$ a *Douglas Connection.* Let us discuss the idea of a connection in more detail. In the linear case, a manifold H is affinely connected if, given any two nearby points p and q , and a smooth parametrized curve from p to q , there is determined a linear map from T_pH into T_qH . One says that a vector ξ_p is parallel translated to a vector $\overline{\xi}_q$, by this map. The differential equations (3B. 3-8) provide the definition of this map. It is a linear map only when the Γ's are independent of ξ. In the general case, the map is only *homogeneous of degree one.* We give an explicit example of a Douglas Connection that is not linear later on in this section. Right now, let's discuss Douglas's Theorem. In the real analytic case, he showed that the most general first order differential equation system

$$\frac{d\xi^i}{dt} = F^i\left(u, \frac{du}{dt}, \xi\right), \quad i = 1, ..., n, \tag{3B.3-10}$$

which is both linear in ξ and homogeneous of degree one in du/dt, is derivable from a unique spray S , according to (3B.3-4) through (3B.3-7), provided the *connection is symmetric,* that is, provided,

$$\Gamma^i_{jk} = \frac{\partial G^i_j}{\partial \xi^k} = \frac{\partial G^i_k}{\partial \xi^j} = \Gamma^i_{kj}. \tag{3B.3-11}$$

Therefore, a local spray S gives rise to a unique Douglas Connection and, conversely, any system (3B.3-10) with linearity in ξ and degree one homogeneity in du/dt, defines a unique local spray S, provided the connection is symmetric. The symmetry (3B.3-11) is nothing other than the integrability conditions for the existence $\underset{2}{H^i}(u, \xi)$, while the first order homogeneity insures the map induced on the tangent spaces does not depend on any particular parametrization. Following Douglas, we introduce the tensor

$$D^i_{jk\ell} = \frac{1}{2} \frac{\partial^3 \underset{2}{H}{}^i}{\partial \xi^j \partial \xi^k \partial \xi^\ell}, \tag{3B.3-12}$$

and note that $D = 0$, if and only if the connection is linear. Henceforth, we call D *the Douglas tensor for the spray* S.

The nonlinear Douglas Connection we now present is due to W. Barthel. We begin with a variational principal,

$$\delta \int_{t_0}^{t_1} F(u, \xi)\, dt = 0 , \tag{3B.3-13}$$

where F is a smooth real valued function on TH which satisfies

$$\begin{aligned} &F(u; \xi) > 0 \quad \text{for} \quad \xi \neq 0 , \\ &F(u; \lambda \xi) = |\lambda| F(u; \xi) \quad \text{for all real} \quad \lambda , \end{aligned} \tag{3B.3-14}$$

with nonvanishing absolute value $|\lambda|$. A quadratic form of degree zero in ξ is defined by

$$g_{ij}(u, \xi) = \frac{1}{2} \frac{\partial^2 (F \cdot F)}{\partial \xi^i \partial \xi^j} . \quad * \tag{3B.3-15}$$

If this (0, 2)-type tensor field is positive definite, g_{ij} is a *Finsler metric* on H^n. The spray equations are now

$$\frac{d^2 u^i}{dt^2} = - \Gamma^i_{jk}(u, \xi) \frac{du^j}{dt} \frac{du^k}{dt} , \tag{3B.3-16}$$

where

$$\Gamma^i_{jk}(u, \xi) = \frac{1}{2} g^{is}(u, \xi) \left(\frac{\partial g_{js}(u, \xi)}{\partial u^k} + \frac{\partial g_{ks}(u, \xi)}{\partial u^j} - \frac{\partial g_{jk}(u, \xi)}{\partial u^s} \right) , \tag{3B.3-17}$$

is the Douglas Connection for the Finsler metric and is none other than the *Levi-Cività Connection* for $g_{ij}(u, \xi)$. The formulas (3B.3-17) are also known as the *Christoffel symbols of* 2nd *kind* for the metric g. Usually, these terms are reserved for the case in which the Γ are independent of ξ. Then, the metric is said to be *Riemannian.*

Parallel translation in the Barthel case is given by the formulas,

*Dot indicates ordinary multiplication.

$$\begin{cases} \dfrac{d\xi^i}{dt} = G^i_s \dfrac{du^s}{dt} \quad \text{and,} \\ G^i_s = \Gamma^i_{sj}\xi^j - \dfrac{1}{2} g^{im} \Gamma^\ell_{jk} \dfrac{\partial g_{m\ell}}{\partial \xi^s} \xi^j \xi^k . \end{cases} \tag{3B.3-18}$$

The Γ's are those in (3B.3-17). It is clear that the G^i_s are only homogeneous of degree zero in ξ and that they are linear in ξ if and only if the Douglas Tensor for the spray S vanishes. The special feature of Barthel's example is that the length of a tangent vector ξ_p defined by,

$$L(\xi, p) = (g_{ij}(p, \xi)\xi^i\xi^j)^{1/2}, \tag{3B.3-19}$$

is preserved under Γ-parallel translation as defined by (3B.3-17) and (3B.3-18).* For the Riemannian case, it was proved by Levi-Cività that there is only one symmetric Linear Connection on a given *Riemannian manifold* (H^n, g), which preserves $L(\xi)$ under parallel translation defined by g_{ij} and (3B.3-17). It is usually called the *Levi-Cività Connection* on (H^n, g).

We have seen that the vanishing of the Douglas Tensor D, answers the easy question of when a Finsler metric is Riemannian. A more difficult question is to decide when a linear affine connection is Riemannian. That is, when the connection is Levi-Cività. One way of answering this question is with the notion of the *Holonomy Group,* Φ_Γ, *of the connection* Γ. Let $p \in H$ be chosen and consider the set of all smooth curves (i.e. loops) which begin and end at p, while remaining in a local coordinate neighborhood U. A *linear connection* provides a *linear* map of T_pH into itself via (3B.3-8) for *each* such smooth curve, $u^i = u^i(t)$, at p. Each map composes associatively with every other, as do loops. The parallel translation in the reverse direction provides the group inverse and the constant loop provides the group identity element. One shows that for U small enough any other point $p' \in U$, yields a group $\Phi_\Gamma(p')$ isomorphic to $\Phi_\Gamma(p)$. We dispense with p or p'. Therefore, an important theorem is that a given linear connection, Γ, arises from a Riemannian metric if and only if $\Phi_\Gamma \subseteq SO(n)$. Moreover, this connection is unique, if parallel translation on H^n is length preserving. It is none other than the Levi-Cività connection for the metric g.

3B.4 Global Linear Connections and Curvature

An alternate description of a linear connection is as follows. An *affine*

*(3B.3-19) defines a local Banach space norm on TH.

(linear) connection ∇ is a rule which assigns to each vector field on H^n a *linear mapping,* $\nabla_X = Vect(H) \to Vect(H)$, where *Vect*, indicates the linear space of vector fields, with the following additional properties:

1) $\nabla_X Y$ is linear in X *and* Y

2) $\nabla_{fX+gY} = f\nabla_X + g\nabla_Y$, where $fX_{(p)} = f(p) \cdot X_{(p)}$, and

3) $\nabla_X(fY) = f\nabla_X Y + (Xf) \circ Y$.

If $X = \xi^i(\partial/\partial u^i)$ and $Y = \eta^i(\partial/\partial u^i)$, in (U, h), then $Xf = \xi^i(\partial f/\partial u^i)$ is a real number independent of the particular (U, h) used to describe X. Moreover, in (U, h) we have the equation

$$\nabla_X Y = \xi^i \left(\eta^j \Gamma^k_{ij} + \frac{\partial \eta^k}{\partial u^i}\right) \frac{\partial}{\partial u^k}. \tag{3B.4-1}$$

For the basis field $X = \partial/\partial u_i$, we write ∇_i for ∇_X and note that,

$$\nabla_i\left(\frac{\partial}{\partial u^j}\right) = \Gamma^k_{ij} \frac{\partial}{\partial u^k}. \tag{3B.4-2}$$

We define a new vector field from X, Y denoted $[X, Y]$, and call it the *Lie bracket,* by the equation

$$[X, Y]f = X(Yf) - Y(Xf). \tag{3B.4-3}$$

Note this operation is *anticommutative.* Locally, the Lie bracket is given by

$$\xi^i \frac{\partial \eta^k}{\partial u^i} \frac{\partial f}{\partial u^k} - \eta^k \frac{\partial \xi^i}{\partial u^k} \frac{\partial f}{\partial u^i} = [X, Y]f. \tag{3B.4-4}$$

One defines the *Curvature Tensor* $\mathbf{R}(X, Y)$ by,

$$\nabla_X \nabla_Y Z - \nabla_Y \nabla_X Z - \nabla_{[X,Y]} Z = \mathbf{R}(X, Y)Z, \tag{3B.4-5a}$$

where locally we have,

$$\nabla_k \nabla_j Z^h - \nabla_j \nabla_k Z^h = \mathbf{R}^h_{kji} Z^i, \tag{3B.4-5b}$$

for $X = \partial_i$, $Y = \partial_j$, $Z = \partial_k$. One obtains curvature components,

$$\mathbf{R}^h_{kji} = \partial_i \Gamma^h_{kj} - \partial_j \Gamma^h_{ki} + \Gamma^s_{kj}\Gamma^h_{si} - \Gamma^s_{ki}\Gamma^h_{sj} . \qquad (3B.4\text{-}6)$$

It is easy to see from (3B.4-6) that,

$$\begin{cases} \mathbf{R}(X,Y) = \mathbf{R}(Y,X) \\ \mathbf{R}(X,Y)Z + \mathbf{R}(Y,Z)X + \mathbf{R}(Z,X)Y = 0 , \end{cases} \qquad (3B.4\text{-}7)$$

where $\mathbf{R}(X,Y)Z$ is linear in X, Y, Z and of type $(1,2)$, whereas $\mathbf{R}(X,Y)$ is a tensor of type $(1,3)$. One also has the so-called *Bianchi identities,*

$$\nabla_X \mathbf{R}(Y,Z) + \nabla_Y \mathbf{R}(Z,X) + \nabla_Z \mathbf{R}(X,Y) = 0 . \qquad (3B.4\text{-}8)$$

A theorem is that *if* $\mathbf{R}(X,Y) = 0$, *then there are suitable coordinates* x^i *for which the* n^3 *quantifies* Γ^k_{jk} *vanish identically in* (U, h). In these coordinates and the original parameter t , the spray equations (3B.3-4) have the *straight-line form*

$$\frac{d^2x^i}{dt^2} = 0 , \quad i = 1, ..., n . \qquad (3B.4\text{-}9)$$

3B.5 Local Nonlinear Connections and Curvature

In this section we review the various curvature tensor fields useful in dealing with nonlinear Douglas connections. First, we have *Berwald's* curvature tensor for a spray S ,

$$\mathbf{B}^i_{jk\ell} = \mathbf{R}^i_{jk\ell} + \xi^b(\Gamma^a_{b\ell} D^i_{jka} - \Gamma^a_{bk} D^i_{j\ell a}) . \qquad (3B.5\text{-}1)$$

The *basic theorem* is that *the spray equation for* Γ *can be transformed into straight line form via smooth arbitrary coordinate changes and affine parameter changes, if and only if*

$$\mathbf{B}^i_{jk\ell} = 0 \text{ and } D^i_{jk\ell} = 0 . \qquad (3B.5\text{-}2)$$

We now ask what happens to the spray equations (3B.3-4) under an arbitrary smooth parameter change $t = \phi(s, a)$. The system (3B.3-4) becomes

$$\frac{d^2u^i}{ds^2} = \underset{2}{H^i}\left(u, \frac{du}{ds}\right) + \psi\left(u, \frac{du}{ds}\right)\frac{du^i}{ds}, \qquad (3B.5\text{-}3)$$

where,

$$s = \int e^{-\int \psi(u,\, du)}\, dt\,. \qquad (3B.5\text{-}4)$$

Differentiation of the right-hand side of (3B.5-3) by the operators $\partial^2/\partial\xi^i\partial\xi^j$ leads to

$$\bar{\Gamma}^i_{jk} = \Gamma^i_{jk} + \delta^i_j\,\psi_k + \delta^i_k\,\psi_j + \xi^i\,\frac{\partial\psi_j}{\partial\xi^k}, \qquad (3B.5\text{-}5)$$

where,

$$\psi_j = \frac{1}{2}\,\frac{\partial\psi}{\partial\xi^j}, \qquad (3B.5\text{-}6)$$

defines a covariant vector (i.e. Douglas $(0,1)$-tensor) from which one derives the effect on D as

$$\bar{D}^a_{jka} = D^a_{jka} + (n+1)\,\frac{\partial\psi_j}{\partial\xi^k}\,. \qquad (3B.5\text{-}7)$$

The equations (3B.5-5) are called a *projective change of Douglas connection.*

One defines the set of n^3 quantities

$$\pi^i_{jk} = \Gamma^i_{jk} - \frac{1}{n+1}\,\delta^i_j\Gamma^a_{ak} - \frac{1}{n+1}\,\delta^i_k\,\Gamma^a_{aj} - \frac{1}{n+1}\,\xi^i\,D^a_{jka} \qquad (3B.5\text{-}8)$$

and determines they do *not* transform as a Douglas Connection (linear or nonlinear). However, π is a tensor of type $(1,2)$ relative to the *projective group* of coordinate transformations:

$$\bar{x}^i = \frac{a^i_j\,x^j + b^i}{c_k\,x^k + h}, \qquad (3B.5\text{-}9)$$

where,

$$\Delta = \det\left(\frac{\partial \bar{x}^i}{\partial \bar{x}^j}\right) \neq 0 . \qquad (3B.5\text{-}10)$$

Because of this we refer to π^i_{jk} as a *projective connection.* The transformation low for π is given as

$$\frac{\partial^2 x'^k}{\partial x^i \partial x^j} = \pi^h_{ij}\frac{\partial x'^k}{\partial x^h} - \pi'^k_{rs}\frac{\partial x'^r}{\partial x^i}\frac{\partial x'^s}{\partial x^j} + \frac{\partial x'^k}{\partial x^i}\frac{\partial \theta}{\partial x^j} + \frac{\partial x'^k}{\partial x^j}\frac{\partial \theta}{\partial x^i} , \qquad (3B.5\text{-}11)$$

where, $\theta = (1/n+1)\log \Delta$. Moreover, these quantities π have the same values for any connection projectivity related to Γ . For a fixed Douglas Connection Γ , the quantities (3B.5-8) determine the so-called *normal connection* in the class of all connections related projectively to Γ . If t is the spray parameter for Γ , then

$$s = \int e^{(2/n+1)\int_\gamma \Gamma^a_{ak}\,\xi^k du}\, dt , \qquad (3B.5\text{-}12)$$

is the so-called *natural parameter* for π , and is the spray parameter for the normal connection of π . That is, for the normal connection, also denoted π , the spray equations are of the form

$$\frac{d^2 u^i}{ds^2} + \pi^i_{jk}\frac{du^j}{ds}\frac{du^k}{ds} = 0 , \quad i = 1, \ldots, n . \qquad (3B.5\text{-}13)$$

The *natural parameter* s is *not invariant* under any coordinate transformation but those of constant Jacobian.

We now define the *Projective Douglas Tensor* K *for the spray* S , as

$$K^i_{jk\ell} = D^i_{jk\ell} - P\left(\frac{1}{n+1}\delta^i_j D^a_{ak\ell}\right) - \frac{1}{n+1}\xi^i \partial_a D^a_{jk\ell} , \qquad (3B.5\text{-}14)$$

where P indicates a sum of three terms obtained by cyclic permutation of j, k, ℓ, and ∂_a indicates the ordinary partial derivative with respect to u_a . One can show that $K = 0$, *if and only if there is a parameter change for which the new* Γ*'s are independent of* ξ *(i.e. are linear).*

Following Douglas, we define the *Projective Weyl Tensor,* W *,* for the spray S as

$$W^i_{jk\ell} = B^i_{jk\ell} - \frac{1}{n-1}\,\delta^i_k\, B^a_{ja\ell} - \frac{1}{n-1}\,\delta^i_\ell\, B^a_{jka}\,, \tag{3B.5-15}$$

where B is the Berwald curvature tensor formed with π in place of Γ and K in place of D. One can prove for dim > 2 that the local spray equation (3B.3-4) transform into straight line form, for some smooth reparametrization of S, if and only if, $W = 0$. In the linear case, $D = 0$, and so, $K = 0$. In this instance, W is the *usual* projective Weyl tensor of the linear affine connection Γ.

3B.6 Parallel Translation of Douglas Tensors and Linear Horizontal Vector Fields

Using the Douglas connection Γ of a spray S one is able to define, after Berwald, a process called *covariant differentiation* which in local coordinate notation is given by

$$a^{r_1 \cdots r_m}_{s_1 \cdots s_p;i} = \frac{\partial a^{r_1 \cdots r_m}_{s_1 \cdots s_p}}{\partial x^i} + \sum_{\alpha}^{1 \dots m} a^{r_1 \cdots r_{\alpha-1} j\, r_{\alpha+1} \cdots r_m}_{s_1 \cdots s_p}\, \Gamma^{r_\alpha}_{ij}$$

$$- \sum_{\beta}^{1 \dots p} a^{r_1 \cdots r_m}_{s_1 \cdots s_{\beta-1} \ell s_{\beta+1} \cdots s_p}\, \Gamma^{\ell}_{s_\beta i}$$

$$+ \xi^m \Gamma^{\ell}_{mi}\, \frac{\partial a^{r_1 \cdots r_m}_{s_1 \cdots s_p}}{\partial \xi^\ell}\,, \tag{3B.6-1}$$

where the semicolon denotes the covariant derivatives of a given Douglas Tensor $a = (a^{r_1 \cdots r_m}_{s_1 \cdots s_p})$ of type (m, p). The differentiated tensor is of type $(m, p+1)$.

Now let $x^i(t)$ be a continuous piecewise smooth curve in H. One says $a = a(t)$ *is parallel along* $x^i(t)$, if the following differential equation is satisfied

$$\frac{d}{dt}\, a^{r_1 \cdots r_m}_{s_1 \cdots s_p}(t) + \sum_{\alpha}^{1\ldots m} a^{r_1 \cdots r_{\alpha-1} j\, r_{\alpha+1} \cdots r_m}_{s_1 \cdots s_p}\, \Gamma^{r_\alpha}_{jk}(x(t))\, \frac{dx^k}{dt}$$

$$- \sum_{\beta}^{1\ldots p} a^{r_1 \cdots r_m}_{s_1 \ldots s_{\beta-1}\, \ell s_{\beta+1} \cdots s_p}\, \Gamma^{\ell}_{ks_\beta}(x(t))\, \frac{dx^k}{dt}$$

$$+ \xi^m \Gamma^{\ell}_{mk}(x(t))\, \frac{\partial a^{r_1 \cdots r_m}_{s_1 \cdots s_p}}{\partial \xi^\ell}\, \frac{dx^k}{dt} = 0\,. \qquad (3B.6\text{-}2)$$

For the case of a linear connection, (3B.6-2) is none other than

$$\nabla_{\frac{dx}{dt}} a(t) = 0\,, \qquad \frac{dx}{dt} = \left(\frac{dx^1}{dt}, \ldots, \frac{dx^n}{dt}\right). \qquad (3B.6\text{-}3)$$

We now describe a bundle theoretic approach to Linear Affine Connections ∇ which can be useful in the stochastic theory on manifolds. Consider a fixed point $p \in H$ and a trivializing neighborhood for TH. A *horizontal vector* at p relative to Γ is any element in the linear subspace of $T_{(p,e)}GL(H)$ defined by

$$H_{(p,e)} = \left\{ a^k \frac{\partial}{\partial u^k} - \sum_{j}^{1\ldots n} \Gamma^i_{k\ell}(p) e^\ell_j\, a^k \frac{\partial}{\partial e^i_j} \;\middle|\; a = (a^1, \ldots, a^k) \in \mathbf{R}^n \right\}$$

$$(3B.6\text{-}4)$$

where $e = \{e_1, \ldots, e_n\}$ is a frame at p. $H_{(p,e)}$ is a vector space of dimension $n^2 + n$. Its elements $\tilde{a}$ are called *horizontal vectors* at (p, e). They have the property

$$(D\pi)_{(p,e)}\tilde{a} = a \in \mathbf{R}^n\,. \qquad (3B.6\text{-}5)$$

Given a smooth curve $c(t)$ in U, the *horizontal lift* $\tilde{c}(t)$ of $c(t)$ is a unique smooth curve in $GL(H)$ whose velocity field $\tilde{x}$ is given by

$$\tilde{x}(t) = x^k(t)\,\frac{\partial}{\partial u^k} - \sum_{j}^{1\ldots n} \Gamma^i_{k\ell}(t) e^\ell_j(t) x^k(t)\, \frac{\partial}{\partial e^i_j(t)} \qquad (3B.6\text{-}6)$$

where $x(t)$ is the velocity vector for $c(t)$ at time t. $\{e^i_j(t)\}$ are obtained by parallel translation via Γ of the given initial frame $\{e^i_j(0)\}$ at p. Of particular interest, are the n *canonical horizontal vector fields*

$$\tilde{L}_q = e^i_j \frac{\partial}{\partial u^j} - \sum_j^{1\ldots n} \Gamma^i_{k\ell}\, e^\ell_j\, e^k_q \frac{\partial}{\partial e^i_j} \qquad \text{(3B.6-7)}$$

One can make use of these basic fields in the stochastic theories in Chapter V of *Stochastic Differential Equations and Diffusion Processes*, by Ikeda and Watanabe (1981).

3B.7 Normal Coordinates for Sprays

The method we present for introduction of normal coordinates at (p, ξ) is due to T.Y. Thomas and generalizes the usual method in the case of a linear affine connection.

Consider the system of differential equations

$$\begin{cases} \dfrac{d^2x^i}{dt^2} = -\Gamma^i_{jk}(p,\xi)\dfrac{dx^j}{dt}\dfrac{dx^k}{dt}, \\[2ex] \dfrac{d\xi^i}{dt} = -\Gamma^i_{jk}(p,\xi)\dfrac{dx^j}{dt}\xi^k, \end{cases} \qquad i,j,k = 1, \ldots, n. \qquad \text{(3B.7-1)}$$

These are equations for the time evolution of x^i and ξ^i. Suppose these can be solved for the *initial conditions,* $t = t_0$, $x^i = x^i_0$, $\xi^i = \xi^i_0$, at p, where $dx/dt = \xi$. The Taylor expansion of x can be computed to have the form

$$\begin{cases} x^i = x^i_0 + \xi^i(t-t_0) + \dfrac{1}{2!}\Gamma^i_{jk}(x_0,\xi_0)\,\xi^j\xi^k(t-t_0)^2 \\[2ex] \qquad + \dfrac{1}{3!}\Gamma^i_{jk\ell}(x_0,\xi_0)\,\xi^j\xi^k\xi^\ell(t-t_0)^3 + \ldots \end{cases} \qquad \text{(3B.7-2)}$$

where,

$$\begin{cases} \Gamma^i_{\alpha_1 \ldots \alpha_m} = \dfrac{1}{m!}\dfrac{\partial^m H^i_m(x,\xi)}{\partial\xi^{\alpha_1}\ldots\partial\xi^{\alpha_m}}, \\[2ex] \dfrac{d^m x^i}{dt^m} = H^i_m(x,\xi). \end{cases} \qquad \text{(3B.7-3)}$$

Setting, $y^i = \xi^i(t - t_0)$, we obtain

$$x^i = x_0^i + y^i + \frac{1}{2!}\Gamma^i_{jk}(x_0, \xi_0)y^j y^k + \frac{1}{3!}\Gamma^i_{jk\ell}(x_0, \xi_0)y^j y^k y^\ell + \ldots \tag{3B.7-4}$$

This series converges for small y values and defines an analytic transformation of x's to y's which induces a linear transformation on the y's for analytic transformations on the x's. The y^i are the *normal coordinates* for (Γ, x_0, ξ_0) at p. One proves that Γ vanish at (x_0, ξ_0) in normal coordinates and so generalizes the classical theorem. Obviously, spray curves through (p, ξ_0) have linear equations in normal coordinates.

3B.8 Finsler and Riemannian Geometries

Suppose that H^n provided with a positive-definite Douglas Tensor field g of type (0, 2). Then, for $X, Y \in Vect(H)$ we must have

$$\begin{cases} g(X, Y) = g(Y, X), \\ g(X, X) \geqslant 0, \end{cases} \tag{3B.8-1}$$

with $g(X, X) = 0$ if and only if $X = 0$. In a coordinate chart (U, h) we write

$$g(X, Y) = g(\xi^i \partial_i, \eta^j \partial_j) = g_{ij}\, \xi^i \eta^j \tag{3B.8-2}$$

where $\partial_i \equiv \partial/\partial u_i$ and $g_{ij} = g(\partial_i, \partial_j)$ is the *classical metric tensor* (or fundamental tensor or first fundamental form) of Riemannian geometry if and only if $D = 0$.

Arc length along any smooth curve $x^i = x^i(t)$ is given by the well-known formula

$$s = \int_{t_1}^{t_2} \sqrt{g_{ij}\left(x(t), \frac{dx}{dt}\right) \frac{dx^i}{dt} \frac{dx^j}{dt}}\; dt, \tag{3B.8-3}$$

where dx/dt is the *velocity vector* of the curve. The integrand is none other than the square root of $L(\xi, p)$. That is, it is identical with F in (3B.3-15).

If $D = 0$, then an *angle* between tangent vectors X_p, Y_p in T_pH is defined by

$$\cos\theta = \frac{g(X_p, Y_p)}{L(X_p, p)\, L(Y_p, p)} \quad . \tag{3B.8-4}$$

Now, for example, let W be a Douglas Tensor field of type (1, 2). The local components of Q (X, Y) in (U, h) are given by $Q^i_{jk}\, X^j X^k$. Similar expressions result for tensors of type (r, s).

For example, with this field Q of type (1, 2), we can associate a tensor field of type (0, 3) defined by

$$g\,(Q\,(X, Y), Z)\ , \tag{3B.8-5}$$

whose value is given by

$$g_{rh}\, Q^r_{ji}\, X^j Y^i Z^h\ . \tag{3B.8-6}$$

So, in local coordinates, the components of Q are given by

$$Q_{ijh} = Q^r_{ij}\, g_{rh}\ , \tag{3B.8-7}$$

and we say the components of Q_{ijk} are obtained from Q^r_{ij}, by *lowering indices.* Since g_{ij} has inverse g^{ij}, we are able to *raise indices,* as well. Thus,

$$Q^r_{ij} = g^{rh}\, Q_{ijh}\ . \tag{3B.8-8}$$

Of course, these ideas generalize to all types of tensor fields on *H*.

Using the metric tensor to lower indices of the Riemannian Curvature Tensor $\mathbf{R}^i_{jk\ell}$, we obtain the conjugate tensor

$$\mathbf{R}_{mjk\ell} = g_{mi}\, \mathbf{R}^i_{jk\ell}\ , \tag{3B.8-9a}$$

or in index free notation

$$\mathbf{R}(X, Y; U, V) = g\,(\mathbf{R}(X, Y)\, U, V)\ . \tag{3B.8-9b}$$

It is easy to verify the following identities for the *Purely Covariant Riemann Tensor*:

$$\mathbf{R}(X, Y; U, V) + \mathbf{R}(Y, X; U, V) = 0$$
$$\mathbf{R}(X, Y; U, V) + \mathbf{R}(X, Y; V, U) = 0$$
$$\mathbf{R}(X, Y; U, V) = \mathbf{R}(U, V; X, Y)$$
$$\mathbf{R}(X, Y; U, V) + \mathbf{R}(Y, U; X, V) + \mathbf{R}(U, X; Y, V) = 0 \qquad (3B.8\text{-}10)$$

Consider the global tensor field

$$\mathbf{R}(Y, Z) = \sum_{i}^{1\ldots n} \mathbf{R}(e_i, Y; Z, e_i), \qquad (3B.8\text{-}11)$$

where $e_1, \ldots, e_n$ are n orthonormal tangent fields on H. $\mathbf{R}$ is a field of type $(0, 2)$ and is called the *Ricci Curvature Tensor.* Locally, we have

$$\mathbf{R}_{ij} = \mathbf{R}_{ji} = g^{rs}\,\mathbf{R}_{rjis} = \mathbf{R}^{s}_{jis}, \qquad (3B.8\text{-}12)$$

and is said to result from *contraction* of $\mathbf{R}^{k}_{ji\ell}$ on k and ℓ. We can also define a global *scalar field* (a tensor of type (0,0)),

$$\mathbf{R} = \sum_{i}^{1\ldots n} \mathbf{R}(e_i, e_i), \qquad (3B.8\text{-}13)$$

where it is also true that

$$\mathbf{R} = g^{ij}\,\mathbf{R}_{ij}. \qquad (3B.8\text{-}14)$$

This field is known as the *Riemann Scalar Curvature* on H. $\mathbf{R}$ does not depend on the specific choice of e_i.

A Riemannian 2-manifold H^2 has *Gaussian Curvature*

$$K = -\frac{\mathbf{R}_{1212}}{g_{11}g_{22} - g_{12}g_{21}}. \qquad (3B.8\text{-}15)$$

Given a Riemannian n-manifold H^n and two linearly independent vectors X and Y at $p \in H^n$, there is a uniquely defined smooth 2-manifold generated by geodesics which pass through p and are tangent to the plane L determined by X_p and Y_p. The Gaussian curvature of this 2-manifold is given by

$$\mathbf{K} = - \frac{\mathbf{R}(X, Y; X, Y)}{g(X, X)\, g(Y, Y) - g^2(X, Y)} , \qquad (3B.8\text{-}16)$$

and is called the *Sectional Curvature* of H at p with respect to Section L. If any other linearly independent tangent vectors in L are used to compute K, the same real number is obtained.

If the *Sectional Curvature* is independent of L at p, then

$$\mathbf{R}(X, Y; X, Y) = - \mathbf{K}\{g(X, X)g(Y, Y) - g^2(X, Y)\}. \qquad (3B.8\text{-}17)$$

Locally, this is just,

$$\mathbf{R}_{ijk\ell} = \mathbf{K}(g_{jk}\, g_{i\ell} - g_{j\ell}\, g_{ik}) . \qquad (3B.8\text{-}18)$$

If (3B.8-17) or (3B.8-18) is true at every point of H, then for $n > 2$ the Bianchi identities can be used to show $\mathbf{K}$ = constant. This is not true for $n = 2$, however. In the case when $\mathbf{K}$ is constant, then $\mathbf{R} = n(n-1)\mathbf{K}$, so $\mathbf{R}$ is constant. However, $\mathbf{R}$ may be constant and $\mathbf{K}$ nonconstant. For example, in an Einstein n-space $n > 2$, $\mathbf{R}_{ij} = (\mathbf{R}/n)g_{ij}$, by definition and it follows that $\mathbf{R}$ is a constant. It is not generally true, however, that $\mathbf{K}$ is constant.

3B.9 Killing Fields

Suppose H^n is provided with a geodesic spray S. That is, suppose the spray curves are locally given by a linear Levi-Cività connection Γ for a Riemannian metric g. Suppose $^*\Gamma$ is another such connection for a spray $^*S$ and metric tensor g^* with the property that there is a smooth positive function $\rho : H^n \to \mathbf{R}^1$ such that

$$g^* = \rho^2 g . \qquad (3B.9\text{-}1)$$

In this case, we say g^* is *conformally related* to g and the connections must necessarily be related as,

$$^*\Gamma^i_{jk} = \Gamma^i_{jk} + \delta^i_j \rho_k + \delta^i_k \rho_j - g_{jk}\rho^i ,$$

$$\rho_i = \nabla_i \log \rho , \quad \rho^i = g^{ij}\rho_j . \qquad (3B.9\text{-}2)$$

The Riemannian curvatures of these conformally related metrics are given by

$$*R^i_{jk\ell} = R^i_{jk\ell} - \rho^i_j g_{k\ell} + \rho^i_k g_{j\ell} - \delta^i_j \rho_{k\ell} + \delta^i_k \rho_{j\ell} ,$$

$$\rho_{ij} = \nabla_i \rho_j - \rho_i \rho_j + \frac{1}{2} \rho_m \rho^m g_{ij} ,$$

$$\rho^i_k = \rho_{kj} g^{ji} . \tag{3B.9-3}$$

If Γ and $\overline{\Gamma}$ are Levi-Cività Linear Connections for Riemannian metrics g, $\overline{g}$ and $\overline{\Gamma}$ is obtained from Γ by a *projective change* as in (3B.5-8), then the Riemannian Curvature tensors are related as

$$\overline{R}^i_{jk\ell} = R^i_{jk\ell} - \delta^i_j \psi_{k\ell} + \delta^i_k \psi_{j\ell} + \delta^i_\ell (\psi_{jk} - \psi_{kj}) ,$$

$$\psi_{jk} = \nabla_j \psi_k - \psi_j \psi_k , \tag{3B.9-4}$$

for some covariant vector ψ_i. We say $\overline{g}$ is *projectively related* to g.

Now, let Ω denote any *object* defined on T*H* or TT*H* or various tensor products of these with a definite law of transformation under smooth changes of coordinates (eg. tensors, connections).

Let X be a vector field on H^n and let $\phi_t(p)$ denote the unique locally defined spray curve through p. With initial condition X (see 3B.2-1). Let $\Omega^*_{(p)} = (D\phi_t^{-1})(\Omega_{(q)})$ denote the object obtained from Ω by application of the Jacobian of ϕ_t^{-1} to the object at q, or appropriate tensor products of these Jacobians. Define the *Lie derivative relative to* X, to be the limit

$$\lim_{t\to 0} \frac{1}{t} (\Omega^*_{(p)} - \Omega_{(p)}) \equiv L_X \Omega_{(p)} . \tag{3B.9-5}$$

One can prove that for smooth function on H^n

$$L_X f = Xf = X^i \partial_i f , \tag{3B.9-6}$$

whereas, for a vector field Y

$$L_X Y = [X, Y] . \tag{3B.9-7}$$

Moreover, for a given *symmetric linear connection* ∇, one proves,

$$(L_X\nabla)^i_{jk} = \nabla_j\nabla_k X^i + \mathbf{R}^i_{k\ell j} X^\ell . \tag{3B.9-8}$$

Thus, L_X is a tensor of type $(1,2)$, yet, ∇ is not itself a tensor object, generally. In the case that,

$$L_X\nabla = 0 , \tag{3B.9-9}$$

we say X is an *affine killing vector field* on H. When,

$$\begin{cases} L_X\nabla(Y,Z) = \psi(Y)Z + \psi(Z)Y , \\ \text{or locally,} \\ L_X\Gamma^i_{jk} = \delta^i_j\psi_k + \delta^i_k\psi_j , \end{cases} \tag{3B.9-10}$$

we say X is a *projective killing field* on H. When ∇ is a Levi-Cività connection and

$$(L_X\nabla(Y,Z))^i_{jk} = \delta^i_j\rho_k + \delta^i_k\rho_j - g_{jk}\rho^i , \tag{3B.9-11}$$

we say X is a *conformal killing field on* H. Here, $\rho_k = \nabla_k\rho$ for a smooth function ρ.

If, in addition to X being an affine killing field on H, it is also true that

$$\begin{cases} L_X g = 0 , \\ \text{or locally,} \\ L_X(g_{ij}) = \nabla_i X_j + \nabla_j X_i = 0 , \end{cases} \tag{3B.9-12}$$

we say X is a *metric killing field* on H. In this case, X generates a *1-parameter local group of diffeomorphisms* ϕ_t of a neighborhood of $p \in H$, which *preserves the metric* g, and consequently, the *distance* between p and any other close point q, as given by the minimum value of the arc length of all smooth curves from p to q. If (3B.9-12) does not hold, or if the connection is only linear (and not Levi-Cività), X still generates a 1-parameter local group of diffeomorphisms which map the *affine spray curves* into themselves while *preserving the affine parameter* t. In the case of a *projective killing field,* the 1-parameter group preserves the spray curves themselves, but not their parametrization. Finally, in the *conformal* case, ϕ_t preserves angles between corresponding velocity vectors of the spray curves, but does not map sprays into sprays, generally.

PART II The Theory of Volterra-Hamilton Systems Proper

3B.10 Passive Volterra-Hamilton Sprays

A spray S on H^n is a Volterra-Hamilton spray if there is a trivializing cover of the tangent bundle TH with the property that in any particular coordinate chart (U, h) the spray equations take the local form

$$\begin{cases} \dfrac{du^i}{dt} = \xi^i, \\ \dfrac{d\xi^i}{dt} = \Gamma^i_{jk}\, \xi^j\, \dfrac{du^k}{dt} \end{cases} \tag{3B.10-1}$$

where for fixed k, these expressions are *linear* in ξ^k. If these linear functions have *constant coefficients,* we call S, *passive.* Otherwise, we say S is an *active* spray. Thus, an active Volterra-Hamilton spray is one in which the n^3 functions Γ^i_{jk} have coefficients depending on the local coordinate $u^1, \ldots, u^n$. This section will be concerned with the *existence and uniqueness of passive sprays which have an underlying Finsler metric.* That is, we exhibit a Riemannian spray which is a passive Volterra-Hamilton system and show that this example is the only one possible. If a given spray is passive and Finsler, then the Douglas Tensor D vanishes. Therefore, to show uniqueness it will suffice to consider only the Riemannian case.

Following (3B.3-1) of Part I, we write the connection as

$$\Gamma^i_{jk} = \frac{\partial G^i_j}{\partial \xi^k} = \frac{\partial G^i_k}{\partial \xi^j} = \Gamma^i_{kj}\,. \tag{3B.10-2}$$

If Γ is a Levi-Cività connection for a locally conformally flat Riemannian metric, then

$$g_{ij} = e^{2\phi}\delta_{ij}\,, \tag{3B.10-3}$$

and Γ must satisfy (3B.3-17) so that

$$\Gamma^i_{jk} = \frac{1}{2}\, g^{i\lambda}\left(\frac{\partial g_{j\lambda}}{\partial u^i} + \frac{\partial g_{i\lambda}}{\partial u^j} - \frac{\partial g_{ij}}{\partial u^\lambda}\right) = \left\{ {i \atop jk} \right\}, \tag{3B.10-4}$$

where $\left\{ {i \atop jk} \right\}$ is the traditional symbol for a Levi-Cività connection. It is easy to see that

$$\left\{ {i \atop ii} \right\} = \nabla_i \phi ,$$

$$\left\{ {i \atop ij} \right\} = \left\{ {i \atop ji} \right\} = \nabla_j \phi , \qquad i \neq j ,$$

$$\left\{ {i \atop jj} \right\} = - \nabla_i \phi , \qquad i \neq j ,$$

$$\left\{ {i \atop jk} \right\} = 0 , \qquad i \neq j \neq k . \tag{3B.10-5}$$

If we further require the Γ's to be constants, then the only possible form of (3B.10-3) is

$$g_{ij} = e^{2(\alpha_i x^i)} \delta_{ij} , \tag{3B.10-6}$$

where $\alpha_1, ..., \alpha_n$ are arbitrary constants. *Therefore,* (3B.10-6) *is the unique, locally conformally flat, Riemannian metric with constant connection coefficients* Γ.

The proof *that* (3B.10-6) *is the only possible Finsler metric with* Γ *constants* requires use of linear partial differential equations theory.

It is well known that a necessary and sufficient condition that an affine connection Γ be a Levi-Cività connection for a Riemannian metric, g_{ij}, is that the system of linear partial differential equations,

$$\nabla_k g_{ij} \equiv \frac{\partial g_{ij}}{\partial x^k} - g_{hj} \Gamma^h_{ik} - g_{hi} \Gamma^h_{jk} = 0 , \tag{3B.10-7}$$

be identically satisfied. The *integrability conditions* for (3B.10-7) are the so-called *Ricci Identities* (Eisenhart, 1949),

$$\nabla_\ell \nabla_k g_{ij} - \nabla_k \nabla_\ell g_{ij} = g_{ih} \mathbf{R}^h_{jk\ell} + g_{jh} \mathbf{R}^h_{ik\ell} . \tag{3B.10-8}$$

The system (3B.10-7) and (3B.10-8) combine to give the much simpler condition,

$$g_{hi} \mathbf{R}^h_{jk\ell} + g_{hj} \mathbf{R}^h_{ik\ell} = 0 . \tag{3B.10-9}$$

If these equations are *satisfied identically,* then (3B.10-7) is *completely integrable* and the curvature tensor $\mathbf{R}^i_{jk\ell}$ vanishes. Let us suppose that is not the case and proceed to differentiate (3B.10-9) with ∇_m. We obtain, using (3B.10-7),

$$g_{hi}\nabla_m \mathbf{R}^h_{jk\ell} + g_{hj}\nabla_m \mathbf{R}^h_{ik\ell} = 0 \,. \tag{3B.10-10}$$

Proceeding in this fashion, we obtain, via formal differentiation, the *sequence of compatible equations* in (U, h)

$$\begin{cases} g_{hi}\nabla_{m_1}\nabla_m \mathbf{R}^h_{jk\ell} + g_{hj}\nabla_{m_1}\nabla_m \mathbf{R}^h_{ik\ell} = 0 \\ \vdots \\ g_{hi}\nabla_{m_r}\cdots\nabla_{m_1}\nabla_m \mathbf{R}^h_{jk\ell} + g_{hj}\nabla_{m_r}\cdots\nabla_{m_1}\nabla_m \mathbf{R}^h_{ik\ell} = 0. \end{cases} \tag{3B.10-11}$$

Using the theory of linear partial differential equations as for example given in *Non-Riemannian Geometry* (Eisenhart, 1927, p. 14-18), we are able to conclude with P. Eisenhart that :

A necessary and sufficient condition that (3B.10-7) *admit a solution* g_{ij} *is that there exist a positive integer* N *such that the first* N *sets of compatible equations* (3B.10-9), (3B.10-10), (3B.10-11) *admit a solution* $\bar{g}_{ij}$ *which also satisfies the* $(N+1)^{th}$ *set.*

Although it may occur that $\bar{g}_{ij}$ is *not* a solution of (3B.10-7), one may be constructed by setting $g_{ij} = e^{2\phi}\delta_{ij}$ and then making use of

$$\Gamma^i_{jk} = \{^{\,i}_{jk}\}_{\bar{g}} + \delta^i_j\bar{\nabla}_k\phi + \delta^i_k\bar{\nabla}_j\phi - \bar{g}_{jk}\bar{g}^{ih}\bar{\nabla}_h\phi, \tag{3B.10-12}$$

to rewrite (3B.10-7) as,

$$\bar{\nabla}_k\phi = X_k \equiv -\frac{1}{2}\bar{g}^{ij}\left(\frac{\partial\bar{g}_{ij}}{\partial u^k} - \bar{g}_{hj}\Gamma^h_{ik} - \bar{g}_{hi}\Gamma^h_{jk}\right). \tag{3B.10-13}$$

The integrability conditions for (3B.10-13) are $\partial X_\ell/\partial u_k = \partial X_k/\partial u_\ell$ which are always satisfied in view of (3B.10-12). Finally, if $\bar{g}_{ij} = \delta_{ij}$ in (U, h), then Γ is a locally conformally flat Riemannian connection.

Now suppose g_{ij} is a solution of (3B.10-7) in (U, h). For a given fixed point p, there is always a linear coordinate change converting $g_{ij}(p)$ into δ_{ij} according to

$$g_{ij}(p) = \bar{g}_{h\alpha}(p)\,\xi^h_i(p)\,\xi^\alpha_j(p) = \delta_{h\alpha}\,\xi^h_i(p)\,\xi^\alpha_j(p) \tag{3B.10-14a}$$

and where the $\xi^h_{(i)}(p)$ is the Jacobian of this transformation or, from another point of view, it is the linear transformation in T_pU defined as

$$\frac{\partial}{\partial \bar{u}^k} = \xi^i_{(k)}(p)\frac{\partial}{\partial u^i}\,. \tag{3B.10-14b}$$

It is also the case that,

$$\bar{\nabla}_i\left(\frac{\partial}{\partial \bar{u}^j}\right) = \bar{\Gamma}^k_{ij}(p)\frac{\partial}{\partial \bar{u}^k} \tag{3B.10-15}$$

where,

$$\Gamma^\alpha_{\beta\delta}(p) = \bar{\Gamma}^i_{jk}(p)\xi^j_\beta(p)\xi^k_\delta(p)\xi^\alpha_i(p)\,, \tag{3B.10-16}$$

and where, $\xi^\alpha_i(p)\xi^j_\alpha(p) = \delta^j_i$. The equations (3B.10-16) express the fact that the connection coefficients transform as a tensor, under linear changes of coordinates. Now the last equation of (3B.10-11) becomes

$$0 = \left(\delta_{h\alpha}\bar{\nabla}_{q_r}\cdots\bar{\nabla}_{q_1}\bar{\nabla}_q\bar{\mathbf{R}}^h_{\beta\delta\gamma}(p) + \delta_{h\beta}\bar{\nabla}_{q_r}\cdots\bar{\nabla}_{q_1}\bar{\nabla}_q\bar{\mathbf{R}}^h_{\alpha\delta\gamma}(p)\right)$$
$$\times\ \xi^\alpha_i(p)\xi^\beta_j(p)\xi^{q_r}_{m_r}(p)\cdots\xi^{q_1}_{m_1}(p)\xi^q_m(p)\xi^\delta_k(p)\xi^\gamma_\ell(p)\,. \tag{3B.10.17}$$

In the case that Γ^i_{jk} are constant throughout (U,h), the $\bar{\Gamma}^i_{jk}(p)$, formed relative to the linear transformation which diagonalizes $g_{ij}(p)$ to δ_{ij}, are also constant throughout (U,h). The curvature tensor $\bar{\mathbf{R}}^h_{ijk}(p)$ is formed from $\bar{\Gamma}(p)$, and is also constant throughout (U,h). The system (3B.10-17) is equivalent to the following equations relating these systems of constants in (U,h),

$$\delta_{h\alpha}\bar{\nabla}_{q_r}\cdots\bar{\nabla}_{q_1}\bar{\nabla}_q\bar{\mathbf{R}}^h_{\beta\delta\gamma}(p)$$
$$+\ \delta_{h\beta}\bar{\nabla}_{q_r}\cdots\bar{\nabla}_{q_1}\bar{\nabla}_q\bar{\mathbf{R}}^h_{\alpha\delta\gamma}(p) = 0\,. \tag{3B.10-18}$$

In this way we arrive at a solution $\bar{g}_{ij} = \delta_{ij}$ of (3B.10-9), (3B.10-10), and (3B.10-11) in (U,h). It follows that there is a smooth ϕ in (U,h) such that

$$\Gamma^i_{jk} = \{^{\,i}_{jk}\} = \delta^i_j \nabla_k \phi + \delta^i_k \nabla_j \phi - \delta_{jk} \delta^{ih} \nabla_h \phi \tag{3B.10-19}$$

where $\{^{\,i}_{jk}\}$ is formed relative to $g_{ij} = e^{2\phi}\bar{g}_{ij} = e^{2\phi}\delta_{ij}$. This proves that (3B.10-6) is unique among all Finsler metrics with constant Γ's.

This uniqueness result for passive Riemannian sprays was preceeded by a theorem of E. Rhunau (1977). She proved that the *condition,*

$$\delta_{is}\Gamma^s_{k\ell} + \delta_{ks}\Gamma^s_{i\ell} = \gamma_i \delta_{k\ell}, \tag{3B.10-20}$$

on the constant Γ, where γ_i are also constants, *was sufficient to insure the existence of some* Riemannian metric whose Levi-Cività connection is the given Γ^i_{jk}. The relation between γ_i and α_i, as the latter occur in (3B.10-6) is,

$$\gamma_i = \frac{2}{n}\alpha_i, \qquad i = 1, \dots, n. \tag{3B.10-21}$$

The above uniqueness theorem shows that Rhunau's condition (3B.10-20) is equivalent to Γ being the Levi-Cività connection of a locally conformally flat metric, which must, in fact, be (3B.10-6). Rhunau has conjectured that (3B.10-20) was necessary as well as sufficient. Obviously, this is correct.

We shall now prove (3B.10-20) implies that $\bar{g}_{ij} = \delta_{ij}$ solves (3B.10-9), (3B.10-10) and (3B.10-11) following essentially the argument of Rhunau (1977). Proceeding by induction we first prove the anchor step. Namely, we show (3B.10-20) implies

$$\delta_{is}\mathbf{R}^s_{\ell jk} + \delta_{\ell s}\mathbf{R}^s_{ijk} = 0.$$

The left-hand side equals,

$$\Lambda_{j\ell s}\Gamma^s_{ik} - \Lambda_{k\ell s}\Gamma^s_{ij} + \Lambda_{jis}\Gamma^s_{\ell k} - \Lambda_{kis}\Gamma^s_{\ell j},$$

where

$$\Lambda_{\ell ik} \equiv \delta_{is}\Gamma^s_{\ell k} + \delta_{ks}\Gamma^s_{\ell i}.$$

Using (3B.10-20), $\Lambda_{\ell ik} = \gamma_\ell \delta_{ik}$, so that the left-hand side now becomes

$$\gamma_j(\delta_{\ell s}\Gamma^s_{ik} + \delta_{is}\Gamma^s_{\ell k}) - \gamma_k(\delta_{\ell s}\Gamma^s_{ij} + \delta_{is}\Gamma^s_{\ell j})$$
$$= \gamma_j\Lambda_{ki\ell} - \gamma_k\Lambda_{ji\ell} = 0\,,$$

which proves the *anchor step.*

The *induction hypothesis* is, for $p_n \geqslant 0$

$$\delta_{hi}\nabla_{p_n}\cdots\nabla_{p_1}\mathbf{R}^h_{jk\ell} + \delta_{hj}\nabla_{p_n}\cdots\nabla_{p_1}\mathbf{R}^h_{ik\ell} = 0\,.$$

We will show,

$$\delta_{hi}\nabla_{p_{n+1}}\nabla_{p_n}\cdots\nabla_{p_1}\mathbf{R}^h_{jk\ell} + \delta_{hj}\nabla_{p_{n+1}}\nabla_{p_n}\cdots\nabla_{p_1}\mathbf{R}^h_{ik\ell} = 0\,.$$

The left-hand side of this equation expands to

$$\delta_{is}\Gamma^s_{p_{n+1}r}\nabla_{p_n}\cdots\nabla_{p_1}\mathbf{R}^r_{\ell jk} + \delta_{\ell s}\Gamma^s_{p_{n+1}r}\nabla_{p_n}\cdots\nabla_{p_1}\mathbf{R}^r_{ijk}$$
$$- \Gamma^r_{p_{n+1}p_n}(\delta_{is}\nabla_r\nabla_{p_{n-1}}\cdots\nabla_{p_1}\mathbf{R}^s_{\ell jk} + \delta_{\ell s}\nabla_r\nabla_{p_{n-1}}\cdots\nabla_{p_1}\mathbf{R}^s_{ijk})$$
$$- \cdots - \Gamma^r_{p_{n+1}p_1}(\delta_{is}\nabla_{p_n}\cdots\nabla_{p_2}\nabla_r\mathbf{R}^s_{\ell jk} + \delta_{\ell s}\nabla_{p_n}\cdots\nabla_{p_2}\nabla_r\mathbf{R}^s_{ijk})$$
$$- \delta_{is}\Gamma^r_{p_{n+1}\ell}\nabla_{p_n}\cdots\nabla_{p_1}\mathbf{R}^s_{rjk} + \delta_{\ell s}\Gamma^r_{p_{n+1}i}\nabla_{p_n}\cdots\nabla_{p_1}\mathbf{R}^s_{rjk}$$
$$- \Gamma^r_{p_{n+1}j}(\delta_{is}\nabla_{p_n}\cdots\nabla_{p_1}\mathbf{R}^s_{\ell rk} + \delta_{\ell s}\nabla_{p_n}\cdots\nabla_{p_1}\mathbf{R}^s_{irk})$$
$$- \Gamma^r_{p_{n+1}k}(\delta_{is}\nabla_{p_n}\cdots\nabla_{p_1}\mathbf{R}^s_{\ell jr} + \delta_{\ell s}\nabla_{p_n}\cdots\nabla_{p_1}\mathbf{R}^s_{ijr})\,.$$

By the induction hypothesis, all terms with a Γ factor in front, vanish. Adding in just the right terms, which actually cancel each other out, we are left with

$$(\delta_{is}\Gamma^s_{p_{n+1}r} + \delta_{rs}\Gamma^s_{p_{n+1}i})\nabla_{p_n}\cdots\nabla_{p_1}\mathbf{R}^r_{\ell jk}$$
$$+ (\delta_{\ell s}\Gamma^s_{p_{n+1}r} + \delta_{rs}\Gamma^s_{p_{n+1}\ell})\nabla_{p_n}\cdots\nabla_{p_1}\mathbf{R}^r_{ijk}$$
$$- (\delta_{is}\nabla_{p_n}\cdots\nabla_{p_1}\mathbf{R}^s_{rjk} + \delta_{rs}\nabla_{p_n}\cdots\nabla_{p_1}\mathbf{R}^s_{ijk})\Gamma^r_{p_{n+1}\ell}$$

$$- (\delta_{\ell s} \nabla_{p_n} \cdots \nabla_{p_1} \mathbf{R}^s_{rjk} + \delta_{rs} \nabla_{p_n} \cdots \nabla_{p_1} \mathbf{R}^s_{\ell jk}) \Gamma^r_{p_{n+1} i} \, .$$

The last two parameters vanish by the induction hypothesis and we are therefore left with

$$\Lambda_{p_{n+1} ir} \nabla_{p_n} \cdots \nabla_{p_1} \mathbf{R}^r_{\ell jk} + \Lambda_{p_{n+1}} {}_{\ell r} \nabla_{p_n} \cdots \nabla_{p_1} \mathbf{R}^r_{ijk}$$

$$= \gamma_{p_{n+1}} (\delta_{ir} \nabla_{p_n} \cdots \nabla_{p_1} \mathbf{R}^r_{\ell jk} + \delta_{\ell r} \nabla_{p_n} \cdots \nabla_{p_1} \mathbf{R}^r_{ijk}) \, ,$$

which, again, vanishes by the induction hypothesis. This completes the proof.

It is interesting to note that as a consequence of the existence and uniqueness of passive Riemannian sprays, no passive spray other than (3B.10-6) can be derived from the usual variational integral (3B.3-19) for Riemannian or Finsler sprays,

$$\delta \int (L)^{1/2} \, ds = 0 \, . \qquad \text{(3B.10-22)}$$

This is especially interesting because one can define a passive spray by,

$$\Gamma^i_{jk} = \delta^i_j \alpha_k + \delta^i_k \alpha_j \, , \qquad \text{(3B.10-23)}$$

where α_i and n arbitrary constants, and yet

$$W^i_{jk\ell} = 0 \, , \qquad \text{(3B.10-24)}$$

holds for the *Weyl Projective Curvature.* Therefore, for $n \geqslant 3$, this spray may be reparameterized into the straight line spray. This also holds for the special case $n = 2$.

Now the (only) passive Finsler spray is directly computable from (3B.10-12) and has connection coefficients given by

$$\Gamma^i_{jk} = \delta^i_j \alpha_k + \delta^i_k \alpha_j - \delta_{jk} \delta^{ih} \alpha_h \, . \qquad \text{(3B.10-25)}$$

Clearly, (3B.10-25) and (3B.10-23) are not identical unless all α_i vanish. Again, this includes the $n = 2$ case, as well. Therefore, we see directly that (3B.10-23) can not be Riemannian nor Finsler.

One may ask which passive Riemannian sprays are of constant sectional curvature. This is equivalent, for $n \geqslant 3$, to (3B.10-24). This question can be easily answered from the curvature formulas for the general locally conformally flat metric (3B.10-3) and the connection coefficients (3B.10-5). First, we have

$$\begin{cases} \mathbf{R}^i_{ji\ell} = \partial_j \nabla_\ell \phi - (\nabla_j \phi)(\nabla_\ell \phi), & i \neq j \neq \ell \\ \mathbf{R}^i_{jij} = \partial_i \nabla_i \phi + \partial_j \nabla_j \phi + \sum_{m \neq i, j} (\nabla_m \phi)^2, & i \neq j, \\ \mathbf{R}^i_{jj\ell} = -\partial_i \nabla_\ell \phi + (\nabla_i \phi)(\nabla_\ell \phi), & i \neq j \neq \ell \,.^* \end{cases} \qquad (3B.10\text{-}26)$$

All other $\mathbf{R}^i_{jk\ell}$ components vanish, and there is *no summation convention employed* in (3B.10-26). From the formulas (3B.10-26), those for the *Ricci curvature* and *Riemann scaler curvature* are readily obtained. They are, respectively,

$$\begin{cases} \mathbf{R}_{j\ell} = (n-2)[-\partial_j \nabla_\ell \phi + (\nabla_j \phi)(\nabla_\ell \phi)], & j \neq \ell \\ \mathbf{R}_{jj} = (n-1)[-\partial_j \nabla_j \phi] - [(\sum_{i \neq j} \partial_i \nabla_i \phi) + (n-2) \sum_{i \neq j} (\nabla_i \phi)^2] \,, \end{cases}$$

(3B.10-27)

and

$$\mathbf{R} = -(n-1)\, e^{-2\phi} \sum_j [2\, \partial_j \nabla_j \phi + (n-2)(\nabla_j \phi)^2] \,. \qquad (3B.10\text{-}28)$$

For the case $n \geqslant 3$, formula (3B.10-28) with $\phi = \alpha_i x^i$ shows that $\mathbf{R}$ is not constant. It is well-known that if the sectional curvature $\mathbf{K}$ is constant, then $\mathbf{R}$ is a fixed constant times $\mathbf{K}$. It follows that $\mathbf{K}$ is *not constant* for $n \geqslant 3$, (Rhunau, 1977). This result appeared first in (Antonelli and Voorhees, 1975).

The case $n = 2$ is special. Here $\mathbf{R} = 0$, and from (3B.10-27), $\mathbf{R}_{ij} = 0$, for all i, j. Now every 2-dimensional Ricci Flat manifold is flat. Therefore, the two dimensional passive Riemannian spray is flat. That is, there is a change of coordinates which converts the spray equations into straight line form. This result appeared first in (Antonelli and Voorhees, 1975).

For $n > 3$, the metric tensor must satisfy *Weyl's Conformal Curvature Identities:*

* ∂_i denotes *ordinary* partial derivative here.

$$
\begin{aligned}
\mathbf{R}^i_{k\ell j} &= -\delta^i_k L_{\ell j} + \delta^i_\ell L_{kj} - \delta_{\ell j} L^i_k + \delta_{kj} L^i_\ell ,\\
L^j_k &= \frac{1}{n-2}\left(\mathbf{R}^j_k - \frac{\mathbf{R}}{2(n-1)} \delta^i_k\right),\\
L^j_k &= g^{js} L_{sk} ,\\
\mathbf{R}^j_k &= g^{js} \mathbf{R}_{sk} ,\\
\mathbf{R} &= g^{js} \mathbf{R}_{js} .
\end{aligned}
\tag{3B.10-29}
$$

We now compute the sectional curvatures of a passive Riemannian spray of dimension, $n \geqslant 3$. By a linear change of coordinates $(\alpha_1, \ldots, \alpha_n)$ becomes $(1, 0, \ldots, 0)$ and the original form of the metric tensor (3B.10-6) becomes,

$$
ds^2 = e^{2x^1}((dx^1)^2 + \ldots + (dx^n)^2) . \tag{3B.10-30}
$$

The coefficients are now simpler and we deduce that,

$$
\mathbf{R}^i_{jij} = 1, \text{ for } i \neq j \neq 1 . \tag{3B.10-31}
$$

From this we obtain,

$$
\mathbf{R}_{ijij} = \begin{cases} e^{2x^1} & i \neq j \neq 1 \\ 0 & \text{otherwise.} \end{cases} \tag{3B.10-32}
$$

If $X_r = \xi^k_{(r)} (\partial/\partial x^k)$ are n mutually orthogonal unit vectors at $p \in U$, then

$$
g_{ij}(p)\xi^i_{(r)} \xi^j_{(s)} = \delta_{rs} , \tag{3B.10-33}
$$

and $\{X_1, \ldots, X_n\}$ is an *orthonormal frame* at p.

The sectional curvatures determined by the various pairs (X_r, X_s) are given by

$$
\mathbf{K}(r, s) = -\mathbf{R}_{ijk\ell} \xi^i_{(r)} \xi^j_{(s)} \xi^k_{(r)} \xi^\ell_{(s)} . \tag{3B.10-34}
$$

Using a linear combination of X_r and X_s will not change the sign, but will generally multiply the right-hand side of (3B.10-34) by a squared constant term representing the area spanned by the new linear combinations.

If, in our case here, we choose the nonorthonormal frame $X_r = \delta_r^k(\partial/\partial x^k)$ it follows that, up to multiplication by a positive constant, (3B.10-34) gives numerically,

$$\mathbf{K}(r, s) = -\mathbf{R}_{rsrs}, \quad r \neq s, \tag{3B.10-35}$$

and from (3B.10-32)

$$\mathbf{K}(r, s) = \begin{cases} -e^{2x_1} & r, s \neq 1 \\ 0 & \text{otherwise,} \end{cases} \tag{3B.10-36}$$

again, *up to multiplication by a positive constant.* This formula has the important consequence that *solution curves of any passive Riemannian spray are unstable in the sense of Jacobi* (Laugwitz, 1965). We shall have more to say about this later.

Using (3B.10-27) and $\phi = \alpha_i x^i$ it is easy to show *that* $\mathbf{R}_{ij}$ *has only two distinct eigenvalues,* $-(n-2)|\underline{\alpha}|^2$ and zero, *where* $|\underline{\alpha}|$ *denotes the euclidean norm. Furthermore, the nonzero eigenvalue has multiplicity* $n-1$, *and the unit eigenvector corresponding to zero is* $\underline{\alpha}/|\underline{\alpha}|^2$.

If we use $\xi^k_{(1)} = \underline{\alpha}/|\underline{\alpha}|^2$ and $\xi^k_{(i)} = \delta^k_i$ for $i \neq 1$, in computing the sectional curvatures $\mathbf{K}(r,s)$ we obtain for $r \neq s$,

$$\mathbf{K}(r,s) = \begin{cases} -e^{2\phi} & r, s \neq 1 \\ 0 & \text{otherwise.} \end{cases} \tag{3B.10-37}$$

Consequently, $\mathbf{K}(1,s) = 0$ for all $s \neq 1$. From this one proves that *one of the spray curves, and only one, is actually a straight line passing through the origin in direction* $\underline{\alpha}$. *Furthermore this straight line has finite length in the subspace defined by* $x^1 < 0$. This is true merely because,

$$\int_{-\infty}^{0} e^{+x_1}\, dx_1 = 1 . \tag{3B.10-38}$$

Consequently, *this subspace of* (U, h) *is not complete as a Riemannian manifold* (geodesics are not infinitely extendible) nor as a topological space (not all Cauchy sequences converge). The reader should consult Antonelli and Voorhees, *Nonlinear growth mechanics I* (1983) for more details on the geometry of passive Riemannian sprays.

3B.11 Killing Fields for Passive Sprays

In Part I a projective killing field for a local spray is defined. Combining equations (3B.9-11) and (3B.9-13), a contravariant field ξ^k is a *projective killing field in* (U, h) if and only if there is a set of smooth functions ρ^h_i for which,

$$\rho^h_i = \nabla_i \xi^h, \tag{3B.11-1}$$

and smooth functions ψ_i , such that,

$$\nabla_j \rho^h_i = \xi^k \mathbf{R}^h_{ijk} + \delta^h_j \psi_i + \delta^h_i \psi_j . \tag{3B.11-2}$$

It is known for $n \geqslant 3$ that the *integrability conditions* for (3B.11-1) and (3B.11-2) are *identically satisfied* if and only if $W^i_{jk\ell} = 0$ (Eisenhart, 1949). In this instance, there are $n^2 + 2n$ smooth functions ψ_i , ρ^h_i, ξ^k satisfying (3B.11-1) and (3B.11-2). Indeed, $n^2 + 2n$, is the maximum dimension diffeomorphisms preserving a given spray. It follows that *the projectively flat passive spray* (3B.10-23) admits $n^2 + 2n$ projective killing fields, $\xi^k_{(q)}$, $q = 1, ..., n^2 + 2n$, or equivalently; it admits an (n^2+2n)-parameter group of projective diffeomorphisms generated by $\xi^k_{(q)}$. In the affine case, $\psi_i \equiv 0$, Rhunau showed that *an* n*-dimensional spray admits a transitive* n*-parameter abelian group of affine diffeomorphisms if and only if the affine spray is passive,* (Rhunau, 1977). The killing fields are the n contravariant fields

$$\frac{\partial}{\partial x^r} = \delta^s_{(r)} \frac{\partial}{\partial x^s} , \tag{3B.11-3}$$

and they represent translations. The *abelian* condition is just the vanishing of the Lie Bracket. *Transitivity* means that any two points of (U, h) may be mapped one into the other, by a diffeomorphism defined by motion along integral curves of the fields (3B.11-3). The proof of these facts are very easy but they are interesting because of the classical theorem of L. Bianchi that *an* n*-dimensional Riemannian manifold admits an* n*-dimensional, transitive, abelian group of metric preserving diffeomorphisms if and only if it is flat euclidean* n*-space,* $\mathbf{R}^n$ (Bianchi, 1918, p. 545). Of course, nontrivial passive Riemannian sprays of dimension greater than two are not flat. In any case, it is possible to obtain more information about metric sprays of passive type than either of the above results contain.

For example, *any passive Riemannian spray of dimension* n *admits an intransitive group of metric preserving diffeomorphisms of dimension* $n(n-1)/2$, *which is transitive on any of a cannonical family of* $(n-1)$-*dimensional invariant sub manifolds, all of which are flat euclidean.*

The proof runs as follows. First, consider the passive Riemannian metric in the form (3B.10-30) and write $y = e^{x^1}$. This converts the first fundamental form, ds^2, into

$$ds^2 = (dy)^2 + y^2[(dx^2)^2 + \dots + (dx^n)^2] . \qquad (3B.11\text{-}4)$$

According to results of Fubini and Bianchi such a metric admits a group of metric killing fields (under Lie Bracket multiplication) whose generators are the translations $\partial/\partial x^2, \dots, \partial/\partial x^r$ together with the $\binom{n-1}{2}$ fields,

$$x^r \frac{\partial}{\partial x^s} - x^s \frac{\partial}{\partial x^r}, \qquad r \neq s \neq 1 . \qquad (3B.11\text{-}5)$$

This group acts transitively on the submanifolds y = constant and these are flat. This group is the full group of euclidean motions. It is not abelian but, its maximal abelian subgroup has dimension $n-1$ and is generated by $\partial/\partial x^2, \dots, \partial/\partial x^n$.

It is clear that because Riemannian sprays of passive type are necessarily defined by locally conformally flat Riemannian metrics, any such spray admits an $\frac{1}{2}(n+1)(n+2)$ dimensional group of angle preserving diffeomorphisms. In other words, *any passive Riemannian spray admits* $\frac{1}{2}(n+1)(n+2)$ *linearly independent conformal killing fields* $\xi^k_{(q)}$, $q = 1, \dots, \frac{1}{2}(n+1)(n+2)$, *characterized by* (3B.10-29) *and* (3B.9-14).

3B.12 Active Sprays

In this section we consider a special kind of example of an active spray. In the chart (U, h) we consider a Riemannian metric of the form $g_{ij} = e^{2\phi}\delta_{ij}$ where

$$\phi = \gamma + \alpha_i x^i + \tfrac{1}{2}\beta_{ij} x^i x^j \quad \text{and}$$

$$\beta_{ij} = \beta_{ji} . \qquad (3B.12\text{-}1)$$

Generally, we take $\gamma = 0$, for convenience. Thus, ϕ may be regarded as the first few terms of the Taylor expansion of ϕ, given as a smooth function defined on U. From (3B.10-26) we have for, $i \neq j$, that

$$\mathbf{R}^i_{jij} = \beta_{ii} + \beta_{jj} + \sum_{m\neq i,j}^{n} (\nabla_m \phi)^2, \tag{3B.12-2}$$

which for, $i \neq j$, leads to

$$\mathbf{R}_{ijij} = e^{2\phi}(\beta_{ii} + \beta_{jj} + \sum_{m\neq i,j}^{n} (\nabla_m \phi)^2). \tag{3B.12-3}$$

Consequently, for $n \geqslant 2$, and $\beta_{kk} > 0$, $k = 1, \ldots, n$, $\mathbf{R}_{ijij} > 0$. This implies, by the same argument as follows (3B.10-34), that *all sectional curvatures are negative.* Now we can prove the result that, *if* (U, h) *is simply connected and has metric* $e^{2\phi}\delta_{ij}$, *with* $\phi = \gamma + \alpha_i x^i + \tfrac{1}{2}\beta_{ii}(x^i)^2$, *where* γ, α_i, β_{ii} *are positive for all* i, *then* (U, h) *is diffeomorphic to flat euclidean* n-*space.* This result follows from the celebrated Cartan-Hadamard theorem once one checks that the metric is complete (Milnor, 1968). Using Fubini's theorem on multiple integrals it is clear that the Riemannian volume element yields finite integrals

$$\int_0^{c_1} \ldots \int_0^{c_n} e^{2n\phi} dx^1 \ldots dx^n,$$

where all c_i are finite positive or finite negative. The integrals are infinite otherwise. From this it follows that any geodesic through the origin of (U, h) is infinitely extendible in both directions, and hence that the metric is complete (see 3B.10-38).

We reserve a special symbol, $\hat{H}$, to refer to the above complete metric, which preserves the nonpositive sectional curvature property of the passive metric obtained when all β_{ii} vanish. Evidently, the properties of this geodesic spray of $\hat{H}$ are well approximated by a passive spray when all β_{ii} are small (and positive).

The existence of metric killing fields for $\hat{H}^n$ is not obvious. If it were a compact manifold, then a well-known theorem of Bochner states there are no nontrivial killing fields. On the other hand, Rhunau showed the underlying space of a passive spray can not be compact (if it were, all affine diffeomorphisms would be metric preserving, a contradiction for $n \geqslant 3$). Recall, that a passive metric is not complete, so it is not obvious in this case that the underlying space is of euclidean topology, globally.

References

Antonelli, P.L., and B.H. Voorhees. 1975. Sherbrooke Notes. Sherbrooke, Quebec.

Antonelli, P.L., and B.H. Voorhees. 1983. Nonlinear growth mechanics I. Volterra-Hamilton systems. *Bull. Math. Biol.*, 45, no. 1: 103-116.

Bianchi, L. 1918. Lezioni sulla teoria dei gruppi continui finiti di transformazioni. *Spoerri,* p. 545. Pisa.

Douglas, J. 1927. The geometry of paths. *Ann. Math.*, 29.

Eisenhart, L.P. 1949. *Riemannian Geometry,* Princeton University Press.

Eisenhart, L.P. 1927. *Non-Riemannian Geometry,* Vol. VIII. Amer. Math. Soc. Colloq., New York,

Ikeda, N., and S. Watanabe. 1981. *Stochastic Differential Equations and Diffusion Processes.* Chapter V. North Holland.

Laugwitz, D. 1965. *Differential and Riemannian Geometry.* Academic Press, New York.

Milnor, J. 1968. *Morse Theory.* Annals of Math. Studies, no. 51. Princeton University Press.

Rhunau, E. 1977. *Spaces of Locally Constant Connection.* Master Thesis, University of Alberta.

Steenrod, N. 1951. *The Topology of Fibre Bundles.* Princeton University Press.

3C

A Note on Passive Volterra-Hamilton Systems

F.L. Bookstein

Consider the Antonelli-Voorhees differential system $\dot{c} = |c|^2\alpha - 2(\alpha \cdot c)c$. Now the reflection of α in c is $\alpha_r = 2(\alpha \cdot c)(c/|c|^2) - \alpha$. [For α_r is in the plane of α and c; and $|\alpha_r|^2 = |\alpha|^2 - 4((\alpha \cdot c)^2/|c|^2) + 4(\alpha \cdot c)^2(|c|^2/|c|^4) = |\alpha|^2$, while $c \cdot (\alpha - \alpha_r) = c \cdot (2\alpha - 2(\alpha \cdot c)(c/|c|^2)) = 2\alpha \cdot c - 2\alpha \cdot c = 0$.]

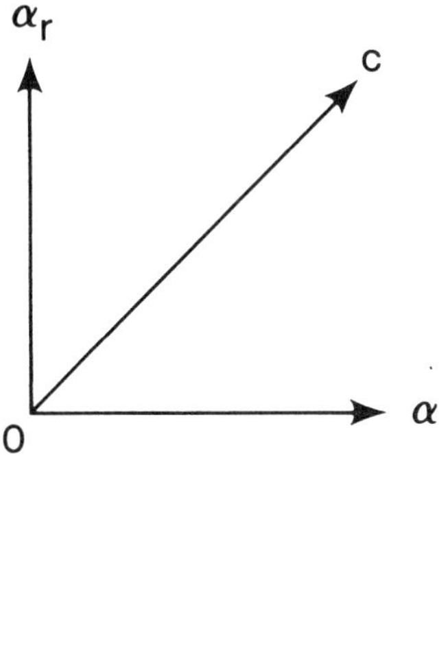

Draw a circle D through c and 0 and tangent to α at 0. Let t be the tangent to D at c. We have $\theta_3 = \theta_2$, by characterization of α_r; and $\theta_2 = \theta_1$, since tangents to a circle make equal angles with their chord. Hence $\theta_1 = \theta_3$. By Euclid, t is parallel to α_r.

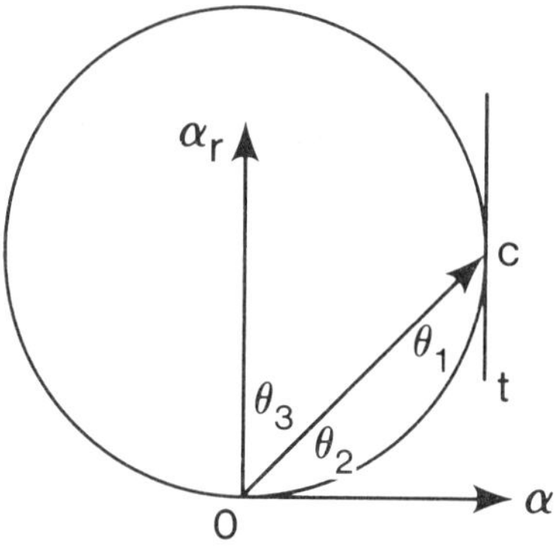

But $\dot{c}$ is proportional to α_r.

Hence D is an integral curve of the differential system.

Then in any plane through α the integral curves are a parabolic pencil of circles ...

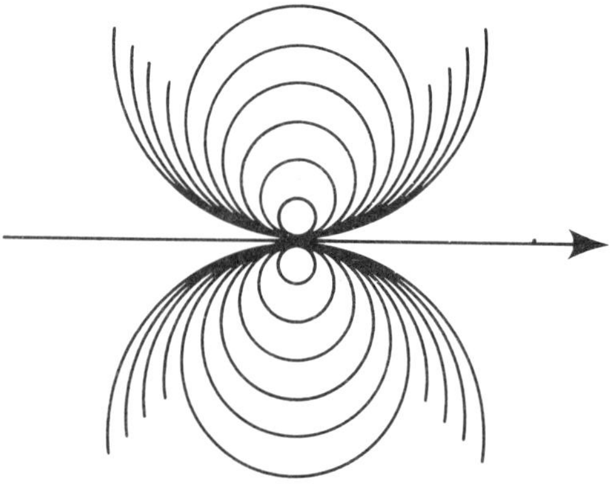

... what could be more classical?

FIELD THEORY

4

POSITIONAL INFORMATION IN IMAGINAL DISCS: A CARTESIAN COORDINATE MODEL

M.A. RUSSELL

4.1 Theory of Pattern Formation in Embryonic Development

A problem in biology which has also attracted the interest of mathematical and physical scientists is the origin of spatial patterns in embryonic development. In multicellular organisms cells are thought generally to be identical in gene content, so the origin of cellular differences in gene activity must be sought in the embryonic environment which each cell-lineage encounters during development.

It has been established by moving labelled cells from their normal embryonic locations to new ones, that the fate of a cell at differentiation is to a large degree dependent upon its position in the embryonic system. Wolpert (1969) has argued that cells acquire information about their positions from the local values of one or more continuous variables defining coordinate systems called "positional fields." Monotonic chemical gradients maintained by reaction-diffusion systems might provide a physical basis for positional fields (see Chapter 7), but in no case has the hypothesis been directly verified. It is assumed only that particular positional values somehow elicit the activation of specific sets of genes in responding cells. Thus, pattern formation is viewed as comprising two distinct steps, the global specification of positional fields by cooperative interactions among a group of cells, and the local interpretation of positional values by individual cells.

Although models of this kind often provide quite elegant explanations for experimental data, they are difficult to test directly, because positional values can be ascertained only by monitoring the patterns of cell differentiation that they are assumed to elicit. Until positional information can be measured by some independent means, the validity of the models will depend on tests for consistency of observed and expected patterns after perturbation of the developing system. Sometimes it is found that more than one model sufficient to account for a particular set of results can be devised. Additional criteria for evaluating the acceptability of a model are then required.

4.2 The System: Imaginal Discs in *Drosophila*

In *Drosophila* (a Dipteran insect), the entire exoskeleton of the adult is secreted by a single layer of epidermal cells. This takes place during the pupal period when the morphologically simple segmented larva metamorphoses to form the adult fly. The exoskeleton bears a great variety of hairs, bracts, bristles, and other sense organs. These are often products of single epidermal cells and their spatial distribution is assumed to reflect a pattern of differential gene activity in the two-dimensional array of epidermal cells which secretes them.

In the mature *Drosophila* larva, the adult epidermal cells are already present in the form of imaginal discs. These are groups of undifferentiated cells which develop as single layered invaginations of the larval epidermis into the body cavity. The epithelium of each disc therefore comprises a surface topologically similar to that of a sphere, except for a narrow hollow stalk which connects the disc to the larval body wall. The different discs are segmentally arranged in the larva, and each gives rise at metamorphosis to a different part of the adult epidermis of the corresponding adult segment. Early in the pupal period each disc turns inside out, and the edges of adjacent discs fuse to form a continuous surface replacing the larval epidermis. It is only after these morphogenetic movements have occurred that adult cuticle is secreted. Therefore, the discs can be thought of as comprising a special set of adult epidermal cells, segregated from the larval ones early in embryonic development. These cells proliferate during the larval period, and only differentiate at metamorphosis.

Although overt evidence of differential gene activity amongst the disc epidermal cells only becomes apparent at metamorphosis, processes which circumscribe the fates of individual cells have already taken place during the larval stage. When cultured in isolation, complementary fragments of an imaginal disc will differentiate complementary subsets of the inventory of pattern elements normally formed. The subset of markers differentiated by a fragment depends on the site and orientation of the cut that generates it. This implies that the cells are already differentially programmed to form specific markers, in accordance with their locations in the disc.

4.3 Positional Information in Imaginal Discs

A modified version of this experiment suggests in addition that each disc may be a single independent positional field. When fragments cut from opposite edges of the same discs are cultured together in intimate contact, the pattern elements normally only differentiated by the cells in the excised intercalary part are regenerated (Haynie and Bryant, 1976). These observations cannot adequately be explained except by postulating some kind of positional information system, for example, the one illustrated in Figure 4.3-I. To account for the results such a system must be assigned a specific property:

> that all intermediate gradient values be generated by interpolation, when normally nonadjacent values are brought into contact at the graft junction.

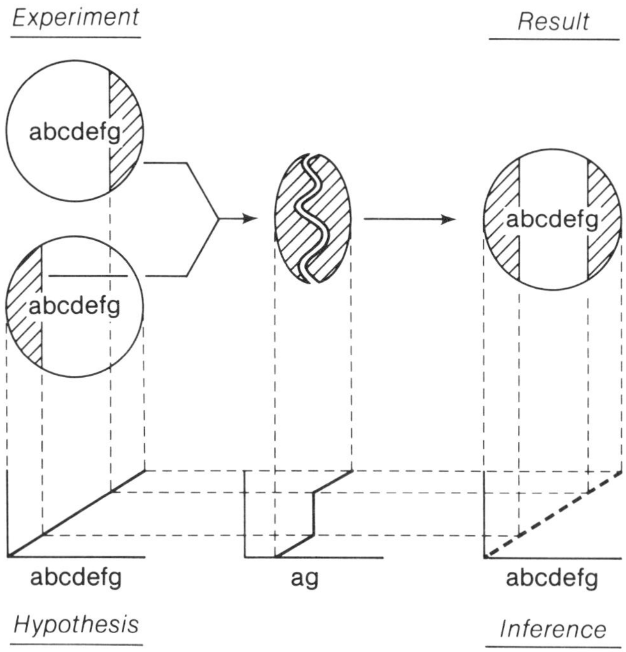

Figure 4.3-I. Hypothetical positional-information system to account for intercalary regeneration of grafted disc fragments. When fragments cut from opposite edges of two discs are mixed and cultured together, markers which fate map to the intercalary region between the cuts (represented by the letters b-f) are often regenerated. This result is susceptible to explanation in terms of a gradient of positional-information where each positional value directs the differentiation of a particular morphological marker. On the basis of this hypothesis it may be inferred that smoothing of the gradient by interpolation of missing values takes place when cells with different values are juxtaposed.

In practice, in similar systems, it is found that cell-division at the graft junction accompanies intercalary regeneration and may be necessary for it to take place. Accordingly, the new cells must be assigned positional values by interpolation between those at the two cut surfaces. It would be interesting to know if a reaction-diffusion system incorporating this feature (growth) would exhibit the required property in computer simulations.

A feature of importance is that the results described above are obtained quite generally, regardless of the positions and orientations of the two cuts generating the interacting fragments. This implies, first, that positional values must be specified uniquely at all points in a disc, with respect to a single coordinate system, i.e. a disc may be a single positional field. In addition, the result implies that the property of intercalation is isotropic in this system. This feature imposes a restriction on the models. Evidently the one-dimensional gradient of Figure 4.3-I would be insufficient.

4.3.1 Positional Information Models: Polar Coordinates

French et al. (1976) proposed a model which incorporates the required property. This was subsequently modified to take into account new experimental data and to confine the cellular interactions required to local ones between neighboring cells (Bryant et al., 1981). The following is a brief account of this most recent version of the model.

It is assumed that position is specified on a system of polar coordinates. Thus, the positional value of each cell is given by a unique combination of angular and radial values (Figure 4.3-II). In discs which develop into appendages, for example legs, the radial coordinate on the fate map transforms directly into the proximo-distal coordinate in the appendage, when the disc everts. When cells with different positional values are juxtaposed by grafting or wound healing cell-division is stimulated. Pattern transformations caused by a variety of such procedures may be explained on the assumption that positional values are assigned in the new population of daughter cells by iterative application of the following ordered sequence of rules:

1. Interpolation of the sequence of intermediates between juxtaposed values in the radial coordinate.
2. Interpolation between juxtaposed values in the angular coordinate via the shorter of the two possible routes around the circle (clockwise or anticlockwise).
3. Displacement of a cell's radial value to a value specifying a more distal level, whenever a neighboring cell with the same combination of radial and angular values is already present.

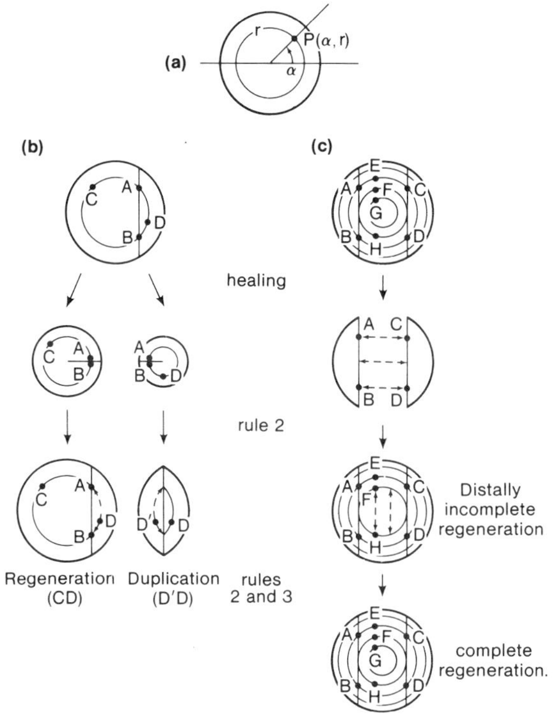

Figure 4.3-IIa. The polar coordinate model of Bryant et al. (1981). The disc is modelled as a sheet of epithelial cells within which position is specified by two variables, angular (α) and radial (r).

Figure 4.3-IIb. Regeneration and duplication of complementary disc fragments. The fragments are assumed to fold so that cut edges heal. Where cells with different positional values are thus brought into contact, cell division takes place. Positional values are assigned to new cells on the basis of three rules which operate on the initial positional values at the cut edge. Consider a cut passing through the points A & B, producing complementary fragments containing points C & D, where A B C & D all share the same radial value, r. Healing will bring into apposition the points A & B. Since these points differ only in *alpha,* Rule 2 (see Section 4.3.2) specifies that interpolation of new positional values will be via the coordinate values of D. Therefore, the smaller fragment, containing D, will duplicate (D′D) and the larger fragment, containing C, will regenerate (CD′).

Figure 4.3-IIc. An explanation of intercalary regeneration (see Figure 4.3-I) on the basis of the polar coordinate model. Suppose that grafting brings into contact points A and C, and B and D which are similar in r but differ in alpha. Rule 2 therefore specifies that intercalation of positional values via E, F & H will take place. Regeneration of G cannot be accomplished simply by application of Rule 2 since G has an r value more distal than any value in the interacting fragments. However, the operation of Rule 2 on F and H will generate a set of positional values already present in neighboring cells, so Rule 3 will come into play resulting in distalisation via G. Iteration of this process would result in complete regeneration. Note that the same rules account for the distal regeneration of a proximal disc fragment, and the duplication of a complementary distal one.

The way these rules account for the pattern transformations observed in a number of real experimental situations is illustrated in Figure 4.3-II. It will be clear that isotropic regulative behavior is a feature of this model that arises from the rules governing transformations in *alpha* and *r* . A problem with the model is the complexity of these transformational rules. It is hard to envisage a biologically plausible cellular mechanism capable of implementing such complex logic. The "shortest route" stipulation for intercalation of angular values is a case in point. The purpose of the present work is to show that a simpler and therefore more plausible explanation of the data is possible if the polar-coordinate system is replaced with a 3-dimensional rectangular coordinate description of positional information in the imaginal discs.

4.3.2 A Cartesian Coordinate System of Positional Information

French et al. (1976) argued that two parameters are necessary, and also sufficient, to define positional information in the disc epithelium. Although it is true that all points on a surface may be distinguished by just two parameters, a third is evidently required to define its curvature. Imaginal disc epithelia in fact assume changing and often highly convoluted shapes throughout their developmental history, and the shapes of the appendages they finally form can only be described by employing a third coordinate. Although the dimensionality of a positional information system need not necessarily correspond with that of the physical space in which the system exists, it must do so in order to specify the development of form. Thus if as is usually assumed, morphogenesis results from a more fundamental process of pattern formation, it may be important not to neglect the fact that the disc epithelia occupy complex curved regions of physical space. One way in which positional values might be fully defined in such a system would be to employ a 3-dimensional system of rectangular coordinates. A polar coordinate system could only describe the positions of points in a plane.

As mentioned above, the topology of the disc epithelium is essentially similar to the surface of a sphere (Figure 4.3-III). A 3-dimensional positional system could be specified in this system very simply by three orthogonally oriented gradients, maintained for example, by independent reaction-diffusion systems, based on molecular species capable of diffusing between neighboring cells in the plane of the epithelial surface. The positional value at any point on the surface would thus be specified by a unique combination of values for three morphogens. Furthermore, because of the constraints imposed by diffusion, neighboring cells would invariably be more similar in positional value than cells located at a distance. Isotropic intercalation is a natural corollary of this feature of the system as will be shown below.

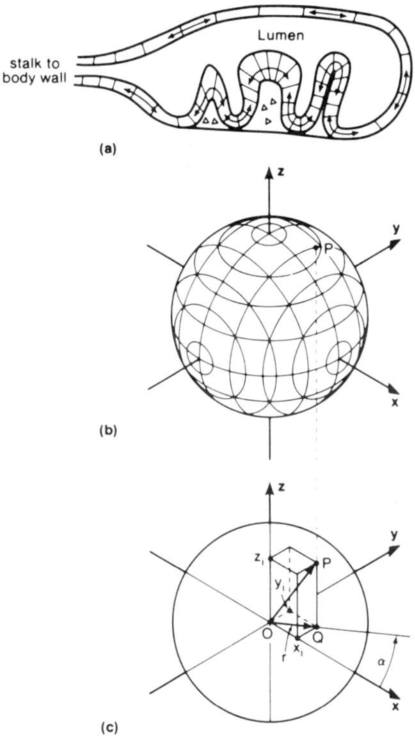

Figure 4.3-IIIa. Diagrammatic section through an imaginal disc. A disc is a single layer of epithelial cells invaginated from the body wall and connected to it by a narrow stalk. We image that positional information in the disc may be provided by chemical signals which can diffuse between adjacent cells in the plane of the epithelial surface (arrows) but not across the lumen. Considered from this viewpoint the epithelial surface of an imaginal disc is 3-dimensional and topologically similar to the surface of a sphere, which can therefore be used to model the positional information system in the disc.

Figure 4.3-IIIb. Positional information in the disc epithelium according to the 3-dimensional Cartesian coordinate model. The contours represent local concentrations of 3 chemical species such as would be established by diffusion via adjacent cells in the surface between remote high and low points. Note that although any two gradients may be locally sufficient to define position in the surface, the same two gradients are not sufficient globally (see also Figure 4.5-VIII).

Figure 4.3-IIIc. The positional information space defined by the Cartesian coordinate model. O is the origin (0,0,0). P is a point with coordinates $(x_i\ y_i\ z_i)$ on the surface of the sphere representing the set of positional values occupied by the set of cells in the disc epithelium. It is assumed that all positional values along the line-segment *OP* code for the same kind of pattern element at differentiation; i.e. it is a relation between x, y and z values and not their individual magnitudes, that codes for a particular differentiated state. Q is the projection of P on the X-Y plane showing the relationship between Cartesian and Polar coordinate models. The angle which the line OQ makes with the X-axis is alpha, and the length of the line segment OQ is r .

This kind of positional system can be represented more formally on Cartesian coordinates defining an imaginary "positional space" as illustrated in Figure 4.3-III. In this representation the set of cells comprising a disc is mapped onto a corresponding set of positional values forming a closed surface about a point with positional value (0,0,0), the origin. Thus the positional value of any given cell lies on a unique ray, emanating from the origin. If we assume that the positional values at all points along a given ray code for the same differentiated state, then it is possible to account for all of the regeneration results outlined above, (and the other examples quoted by French et al., 1976 and Bryant et al., 1981), using a single simple rule of cellular behavior, namely:

> that values in each coordinate are assigned by interpolation whenever normally nonadjacent cells are juxtaposed.

The assumption that all points along a ray code for the same differentiated state implies merely that differentiation is controlled by the *relationship* between the three coordinate values, rather than by their individual magnitudes. The way this model accounts for regeneration and duplication of disc fragments is illustrated in Figure 4.3-IV.

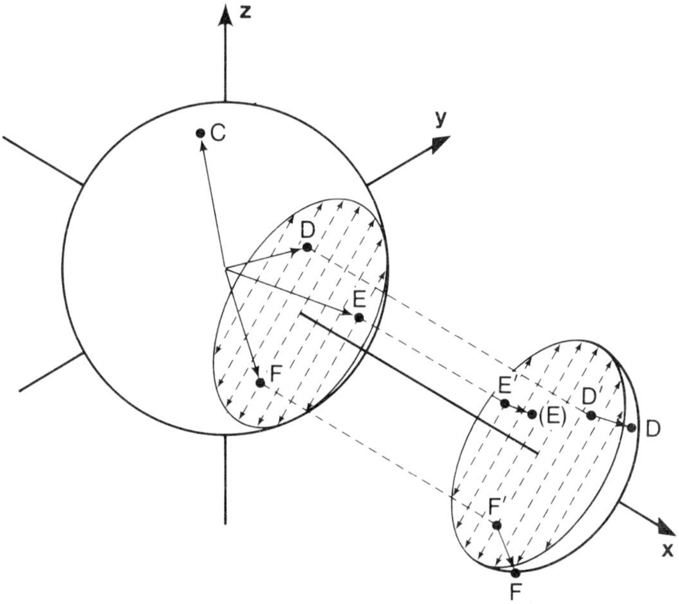

Figure 4.3-IV. Interpolation between confronted values in x, y and z after healing of any cut will generate a plane surface in both fragments as shown. Every ray that intersects the convex surface cut off also intersects both plane surfaces created by interpolation. Therefore, any fragment which includes the origin will regenerate (CD) and any which does not, will duplicate (D′D).

The relationship between this model and the one proposed by French et al. (1975) is worth exploring. It is a simple matter to transform from the Cartesian to the polar system as follows:

$$\cos\alpha = x/\sqrt{x^2+y^2}$$

$$r = \sqrt{x^2+y^2} \qquad (4.3.2\text{-}1)$$

Thus the polar-coordinate model may be viewed as a projection of the Cartesian coordinate system onto the X – Y plane. The single rule of interpolation in the Cartesian model is simply a generalization to all 3 variables of Rule 1 of the polar coordinate model (see above). By applying the transformation (4.3.2-1), it can be shown that the required behavior of alpha in the polar coordinate system (Rule 2, shortest route intercalation, see above) would arise as a result of interpolation of x and y values (see Figure 4.3-V). Finally, the special transformational rule (Rule 3, displacement of *r* to a more distal level when existing levels are occupied) necessary to account for distal regeneration in the polar coordinate system, is not required for z when the Cartesian coordinates are used, because the system is symmetrical under rotation. Thus each coordinate is equivalent and interpolation of x, y and z values will result in the isotropic intercalation behavior actually observed (Figure 4.3-VI).

At this point, we have two possible explanations for the observed behavior of disc fragments in regeneration. Both employ the same basic idea of positional information, but the Cartesian coordinate system seems preferable on grounds of parsimony, since the only property required to implement the transformational rule is simple diffusive smoothing of gradient values. However, an experimentalist may describe his results in terms of any coordinate system he decides to impose, and it may be argued that until we have independent evidence about which coordinate system the cells use, both models are equally valid. I would therefore like to introduce evidence provided by another phenomenon involved in imaginal disc development, known as compartment formation.

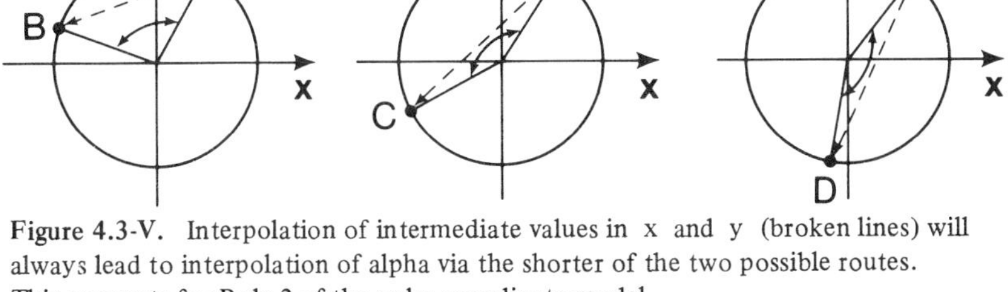

Figure 4.3-V. Interpolation of intermediate values in x and y (broken lines) will always lead to interpolation of alpha via the shorter of the two possible routes. This accounts for Rule 2 of the polar coordinate model.

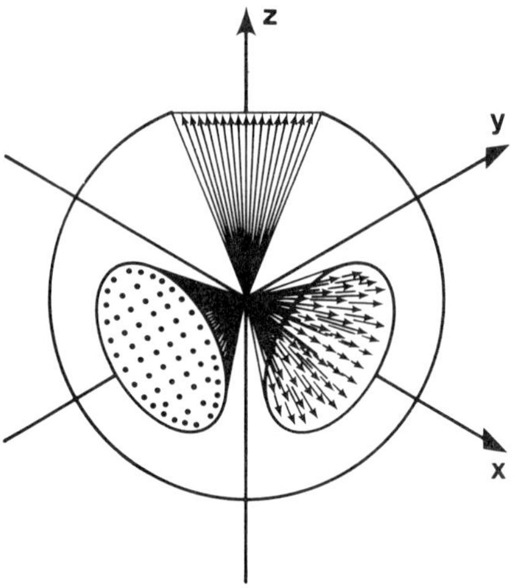

Figure 4.3-VI. According to the Cartesian coordinate model, any disc fragment which contains the origin will regenerate, and the complementary fragment will duplicate, regardless of the orientation of the cut (isotropic regulative behavior).

4.4 Compartment Formation

A method called clonal analysis has provided evidence that the way a disc-cell differentiates is not just a function of its current location, but is also influenced by its ancestry. Although a cell's fate is absolutely correlated with its location in the mature disc, it may nevertheless be causually determined by the position in the system of one of its clonal ancestors at some earlier stage of development.

Single cells can be labelled genetically at specific stages of development. Their descendents form discrete clones whose boundaries are recognizable in the adult. Unexpected restrictions in the sets of pattern elements that can be labelled by single clones have been detected. These restrictions in clonal competence are evidence for the acquisition by the labelled founder cells of inheritable "determinative" commitments which circumscribe the ultimate fates of their descendents. It was found, for example, that the clonal descendents of individual embryonic wing-disc cells could form any part of the anterior wing, any part of the posterior wing, but never parts of both. The edges of clones which fell in the middle of the wing defined an almost straight line in a region of the wing surface undistinguished by any special morphological features. Cells labelled at a later stage had acquired in addition a commitment to either dorsal or ventral wing. Still later, further restrictions separated cell-lineages committed to forming proximal or distal structures (Garcia-Bellido et al., 1973).

Thus it appears that the fate of a cell is determined by a series of developmental decisions made by its clonal ancestors. Each decision involves a choice between just two alternative pathways and is called a "compartmentalization event." Each event subdivides a population of cells into two parts. Typically, there might be 5-10 cells in a new compartment when it is first formed, and 1-2 cell divisions between successive compartmentalization events. Every cell in a given compartment has an equivalent "state of determination," which is different from that of all the cells in the alternate compartment. The commitment to one developmental pathway or another is apparently made independently by each cell on the basis of its location in the system (Crick and Lawrence, 1975). This would account at least partially for the observed final correlation between a cell's position and its ultimate differentiated state. In addition the part of a compartment that a given cell will form is also a function of its position within the compartment concerned. Figure 4.4-VII shows where the compartment boundaries fall when plotted on a simplified diagram of a wing disc. They form a geometrically simple pattern which suggests that the mode of subdivision might be based on threshold values in an underlying Cartesian coordinate system of positional information. It should also be clear that there are no equivalent features in the polar-coordinate description of position in the disc which would correspond with the pattern of compartment boundaries actually found. Therefore, these results provide a new justification for supposing that the cells use a Cartesian coordinate system to specify positional information.

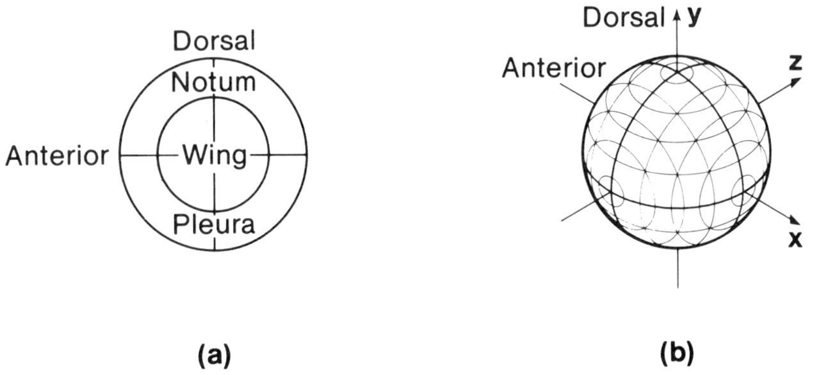

Figure 4.4-VIIa. Diagram showing the location of compartment boundaries plotted on a simplified (two dimensional) fate map of the wing disc.
Figure 4.4-VIIb. Location of the same compartment boundaries on the positional information surface representing the imaginal disc.

4.5 Evidence from Pattern Mutants

Further intriguing evidence comes from the phenotypes of some mutants which we have been studying. Most of the information in a spatial pattern of cell differentiation must be encoded genetically in the form of a program which somehow specifies and controls such parameters as cell division rates and orientations, and ultimately the temporal and spatial pattern of activity of those genes whose products are characteristic of each differentiated cell type. Most of this information must be stored in specific nucleotide sequences in DNA, so that it may be replicated and transmitted from one generation to the next.

Although the basic features of gene structure, function and control necessary for the evolution and operation of such programs are understood in principle, the logical structure of the programs is not. It may be assumed that like a computer program, an "epigenetic program" processes data, i.e. the values of the positional signals that impinge upon a cell lineage during development. Also as in a computer program, an output is generated – the spatial pattern of cell differentiation. It is characteristic of epigenetic programs that the output pattern is much more complex than the input data. The program does not merely permit expression of an existing "prepattern" (Stern, 1968): it contains coded information to generate a more complex system of relations.

Genetic information is stored in discrete heritable units called genes, which are subject to modification in various ways by mutation. Consequently, it is possible to think of mutants which modify spatial patterns of cell differentiation as defective instructions in an epigenetic program. The problem is to deduce from the pattern defect found in a mutant, the logical function of the normal gene in the epigenetic program.

Perhaps the most favorable mutants to examine from this point of view are ones which cause simple, easily characterized transformations of the normal pattern. A simplification of the normal pattern is what would be expected if information in the epigenetic program were deleted. Therefore, we have been interested in mutations which cause mirror-image duplications. In mutants of this kind, one contiguous set of pattern elements is replaced by a mirror-image copy of an adjacent set. This kind of pattern transformation is also generated rather commonly in developing systems by various extrinsic perturbations and Bateson (1971) has speculated that mirror symmetry might result from loss of positional information, which he contends will lead generally to an increase in symmetry in a positional field. Applied to the model presented above this idea generates some interesting predictions, as shown in Figure 4.5-VIII. If the information in a single (e.g. x -) coordinate is deleted, the field will become symmetric about a mirror-plane normal to the deleted coordinate. This is because for any point, $(x_i\ y_i\ z_i)$

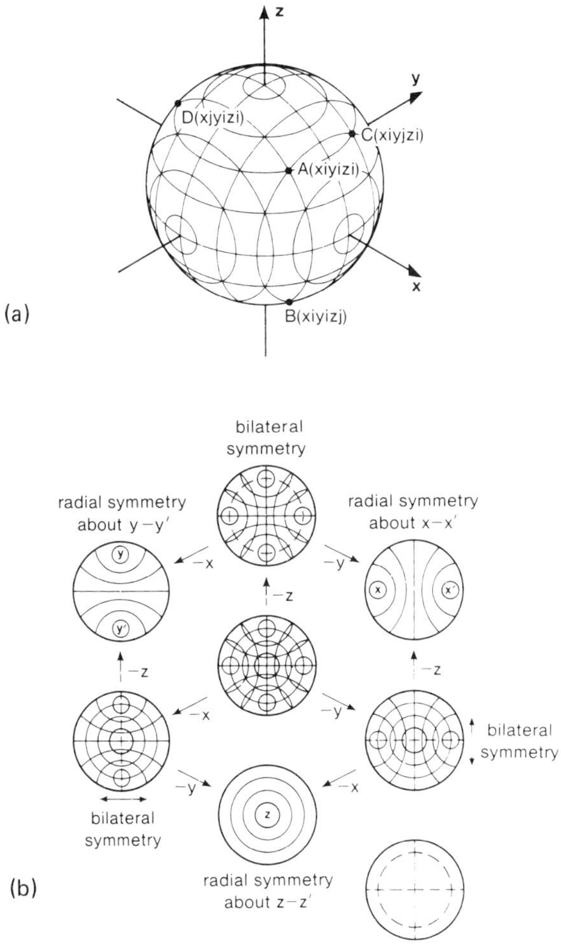

Figure 4.5-VIII. Effect of deleting information in different coordinates.
Figure 4.5-VIIIa. Each point e.g. A(x_i y_i z_i) on the spherical positional information surface, is the point of intersection of three contours in the x, y and z gradients. Each pair of contours has precisely two points of intersection. Thus xi and yi intersect at points A and B, xi and zi at points A and C, and yi and zi at A and D. Therefore, these pairs of points are distinguishable, only if information from the third gradient is available. Consequently, if information in one coordinate is deleted, the positional field becomes mirror symmetric about the plane normal to the deleted coordinate.
Figure 4.5-VIIIb. To facilitate comparison with the pattern of compartment boundaries on the fate map, plane-polar projections of the Cartesian coordinate model are presented. The diagrams show the symmetries generated by deleting individual coordinates and pairs of coordinates. The map produced by superimposing the planes of symmetry which result from the deletion of each coordinate in turn is also shown (lower right). The geometry is identical to the pattern of compartments in the wing disc (Figure 4.4-VII).

with $x_i \neq 0$ on the sphere there exists just one other $(x_j\ y_i\ z_i)$ which is distinguished from the first by a different value for x alone. Similarly deletion of information in y and z coordinates would generate symmetry normal to these two coordinates respectively.

A phenotype which corresponds to that expected if the x-coordinate were deleted is found in a mutant called *engrailed.* The posterior compartment of the wing is transformed in this mutant into a mirror-image copy of the anterior. The line of symmetry corresponds to the anterior-posterior compartment boundary (Garcia-Bellido et al., 1973). Similarly, one of the effects of another mutant called *Polycomb,* is to transform the ventral compartments of the wing into the corresponding dorsal compartments. Again the line of symmetry is the compartment boundary.

The mirror symmetry of the mutant patterns suggests strongly that each mutant might be thought of as interfering with the acquisition by the cells, of information from a single coordinate of a Cartesian positional field. Another implication would be the idea that compartment boundaries normally form along the lines of symmetry inherent in this kind of positional system. Thus the phenotypes of the mutants are consistent with our previous conclusions about the nature of the positional field in a disc.

Nevertheless, this interpretation is not without its difficulties. We have found that in *Polycomb, engrailed* double mutants, four anterior-dorsal compartment patterns oriented in mutual mirror symmetry are differentiated. Where we would have expected radial symmetry (see Figure 4.5-VIII), we see the differentiation of four copies of one part of the normal pattern. This suggests *engrailed* has no function in the anterior compartments, and *Polycomb* has no function in the dorsal compartments. *Polycomb* and *engrailed* therefore behave in one respect as if they represent *discrete* distinctions between specific compartments and their alternates, and in another respect as if they encode *continuous* variables across the entire field. A satisfactory explanation for this duality has not so far been found.

References

Bateson, G. 1971. A re-examination of Bateson's rule. *J. Genet.* 60: 230-240.

Bryant, S.V., V. French and P.J. Bryant. 1981. Distal regeneration and symmetry. *Science,* 212: 993-1002.

Crick, F.H.C. and P.A. Lawrence. 1975. Compartments and polyclones in insect development. *Science,* 189: 340-347.

French, V., P.J. Bryant and S. Bryant. 1976. Pattern regulation in epimorphic fields. *Science,* 193: 969-981.

Garcia-Bellido, A., P. Ripoll, and G. Morata. 1973. Developmental compartmentalization of the wing disc of *Drosophila. Nature New Biol.*, 245: 251-253.

Haynie, J. and P.J. Bryant. 1976. Intercalary regeneration in imaginal wing discs of *Drosophila melanogaster. Nature*, 259: 659-662.

Stern, C. 1968. *Genetic mosaics and other essays.* Harvard University Press, Cambridge, Mass.

Wolpert, L. 1969. Positional information and the spatial pattern of cellular differentiation. *J. Theoret. Biol.*, 25: 1-47.

5
DEVELOPING ORGANISMS AS SELF-ORGANIZING FIELDS

B.C. GOODWIN

5.1 Introduction

Self-organization is a property of a structurally and functionally integrated entity which is considered to be made up of, or to have, "parts." There are essentially two ways of relating the whole and the parts: either the parts are regarded as given (with or without some temporal sequence of presentation or production) and the entity is generated by their interaction; or the entity is regarded as a whole defined by certain invariant relations, the "parts" coming into being as a result of systematic transformations which preserve invariance while generating heterogeneity ("parts") within a functional and structural unity. The most extreme form of the first description is atomism, which assumes that all the information necessary for generating the entity is resident in the parts, so that spontaneous assembly occurs simply as a result of their interaction. This hypothesis, in various forms and with various modifications, is a dominant theme in contemporary models of evolutionary, developmental, behavioral, and even cognitive processes. What characterizes this conceptualization is the assumption that there are no laws of self-organization other than those governing the interaction of the parts or constituents so that the whole is reducible to these parts and their interactions. In developmental biology, the constituents are usually considered to be, ultimately, molecules, although there are theories in which cells are used as intermediate "atoms" in the analysis. Since genes are generally considered to be the determinants of which molecules are present in organisms, it follows that organisms are reducible to genes. The self-organizing process of embryogenesis is then regarded as a consequence of two activities: the operation of the "genetic program" which determines the types of molecular constituents in the organism, the sequence and spatial location of their appearance; and the interaction of these constituents according to physical and chemical laws.

Such an approach to embryogenesis means that the specific forms generated by morphogenetic processes, defining different species of organism, are irreducibly complex because there are no laws or principles which constrain the "genetic program" which is the determinant of specific form. This program is the result of random permutation and natural selection, purely contingent processes as far as organisms are concerned, the only constraint being that the organism specified by the program must be able to survive and leave offspring in some environment. In such a view, there can be no general laws of biological form. Each species is, speaking more than metaphorically, a law unto itself. This is reflected in the primacy of the species concept in neo-Darwinism and the emphasis placed upon competition and survival as the expression of the individual species' success in establishing a unique, singular relationship of order and stability within itself and with its environments. Thus, as one might expect since cognitive constructs are themselves "self-organizing," there is a clearly-defined continuity between the way organisms are conceptualized in neo-Darwinism as "survival machines" and the way in which they are considered to be generated from their molecular parts.

There is no *a priori* reason why such a description of self-organization should not be valid, and indeed it is clear that there are special cases in both the animate and the inanimate realms where atomistic explanations appear to be appropriate. As regards structure or form, these are the instances where a crystallization or "self-assembly" type of process leads to a unique morphology. However, even in inorganic chemisty one encounters many instances of polymorphism in which the same substance can crystallize into different forms, familiar examples being graphite and diamond, or rhombic and monoclinic sulphur. Thus, in general, composition does not determine form. A similar polymorphism is observed in biological structures at different levels of organization, the molecular (Oosawa et al., 1966), the cellular (Sonneborn, 1970), the tissue (Saunders et al., 1957) and so on. Much of embryology consists in fact of generating a variety of "abnormal" forms out of cells of identical genotype: for example, the induction of supernumerary limbs in an amphibian by a simple manipulation of tissue in the embryonic limb bud involving no addition or deletion of cells, but simply a change of relative position, and no change in the external environment. Furthermore, organisms containing specific mutant genes (e.g. homoeotics) may or may not express them; while organisms with wild-type genes can show the "mutant" phenotype (spontaneous homoeotic transformation). Hence there is no one-to-one relationship between genotype and phenotype, always assuming a constant external environment, so that genes are not the specific determinants of morphology. That is to say, the form of an organism is not determined by its genome, with the consequence that self-assembly theories together with a genetic program are inadequate to provide a generative theory of biological self-organization (see Webster and Goodwin, 1982, for a more detailed argument leading to this conclusion).

We must now consider whether or not there is empirical evidence relating to general organizational principles, or laws of form in biology, manifesting as regularities of morphology over large taxonomic groups. We have seen above that neo-Darwinism, which takes the view that "the chief part of the organization of every being is simply due to its inheritance" (Darwin, 1859; inheritance meaning, in contemporary usage, the genetic program), provides no basis for understanding any such regularities since biological form in this theory is determined by contingency, not by law. However, if ordering constraints do exist in the biological realm, then this must be taken into account of in any theory of biological self-organization. This leads us to the work of the pre-Darwinian rational morphologists, who were animated by a belief in the possibility of a rational, intelligible ordering or classification of organisms which would provide an insight into the laws of organic creation (i.e. generative rules). This tradition reached its peak in the work and insights of the great comparative morphologists of the late eighteenth and early nineteenth centuries such as Geoffroy St. Hilaire, Cuvier, Reichert, and Owen, who searched for and discovered empirical regularities of organismic structure. These regularities appeared as invariant structural relations or "typical forms" which were seen to define that which is common to a variety of particular realizations of the same type. Owen's demonstration of the structural homologies which exist between the great variety of vertebrate limbs, leading to the concept of the pentadactyl limbs as the typical form, is characteristic of this work. Each specific member of the invariant set, such as the limb of the horse, of the bat, of the frog, etc., can then be seen as equivalent to every other member under a transformation, so that a common plan is revealed which unifies the diversity of manifest forms. It is, in fact, straight-forward to demonstrate the simple proposition that tetrapod limb morphogenesis may be understood in terms of some basic generative principles capable of producing a great variety of limb forms which are all transformable one into the other under modifications of the limb generating process (Goodwin and Trainor, 1982). This is analogous to the realization that the different forms of motion shown by bodies under the action of a central attracting force, obeying Newton's laws, all belong to the same invariant set known as the conic sections; and indeed the rational morphologists were inspired by the same vision as Newton, which was the Enlightenment Ideal of a mathematical natural science. Their conviction was, and they provided good evidence for the belief, that the morphological complexity of organisms is not irreducibility complex but that there exist rational principles or laws of form which render the diversity intelligible.

Despite the fact that this tradition was largely eclipsed by Darwinism, which adopted the diametrically opposite view that organismic form is determined not by rational law but by historical accident, by contingency, a few rather isolated and sometimes misunderstood biologists have pursued this approach further. Among these the embryologist Hans Driesch stands out and his work is very relevant to the view of self-organization which will be developed below. He introduced the field concept into embryology as a result of his demonstration that relative position in the whole embryo is an important determinant of cell fate. He used the concepts of wholeness, self-regulation, and transformation to define the properties of tissues which respond to a variety of disturbances (e.g., removal or addition of cells, or spatial reordering of parts) by a reorganization such that the normal form is generated. Examples of such fields are the amphibian embryo from fertilization up to about the gastrula stage, the limb and eye primordia, and a variety of other tissues domains which define secondary fields. Within such domains, relative position is a primary determinant of cell fate and the parts which emerge during individuation and differentiation come into being as a result of local and global ordering principles, generating a structural and functional unity. Driesch assumed, like the rational morphologists, that there are organizing rules which operate within organisms to constrain or limit the forms which can be generated (Driesch, 1929) but, again like his predecessors, he failed to give them any mathematical formulation. The problem to be addressed now is what type of mathematical description may be appropriate for these organizing principles, for which there is clear biological evidence.

5.2 Organisms as Fields

The proposition which emerges from the above analysis is that living entities are wholes or *structures* defined by internal relations which remain invariant under certain categories of transformation, the latter limiting the possible generative processes which can result in organisms of specific form (species). Organisms are not, in this view, generated as a result of the interaction of "atomic" constituents, whatever these may be construed to be. Heterogeneity ("parts") arises as a result of systematic transformations of the organized whole, which may be described as the manifestation of states selected from a potential set which satisfies a primary property of invariance characteristic of organisms. Thus the organism is not so much a self-organizing system which generates an ordered state from disordered or less ordered parts; it is more a self-organized entity which can undergo transformations preserving this state. The problems faced by this conceptualization are those of making explicit the nature of the invariant internal relationships which define the whole; the type of transformation which it can undergo; and the relationship between whole and part which confers upon it the properties of generation (reproduction) and regeneration.

Following Driesch's insight that developing organisms have field properties, we may proceed to the question of what type of field and how it may be characterized mathematically. A very extensive body of experimental work in developmental biology since Driesch's pioneering studies has led recently to the observation that an appropriate arithmetic description of the spatial smoothness characteristic of developmental fields is a simple spatial averaging or intercalculation rule applied to field values (French, Bryant, and Bryant, 1976). This states simply that the field value at any point within the boundaries of a developmental field is the arithmetic mean of the values at equidistant neighboring points. Mathematically, this leads to the most general field equation used in physics, namely Laplace's equation. The question then naturally arises whether one can use solutions of this and related equations, known as harmonic functions, to describe developmental fields and hence biological form. Preliminary essays in this direction have been published (Goodwin, 1980; Goodwin and Trainor, 1980; Goodwin and Trainor, 1982). This approach will now be illustrated by an analysis of the earliest stage of amphibian embryogenesis, following the treatment of Goodwin and Trainor (1980), and then certain conclusions regarding the problem of self-organization in biology will be drawn.

5.3 A Field Description of the Typical Cleavage Process

The first five stages of the typical cleavage pattern is described by classical investigation as shown in Figure 5.3-I, starting from the two-cell stage after the first division of the egg. From the 32-cell stage, cell divisions continue to show an alternation between vertical and horizontal cleavage planes,

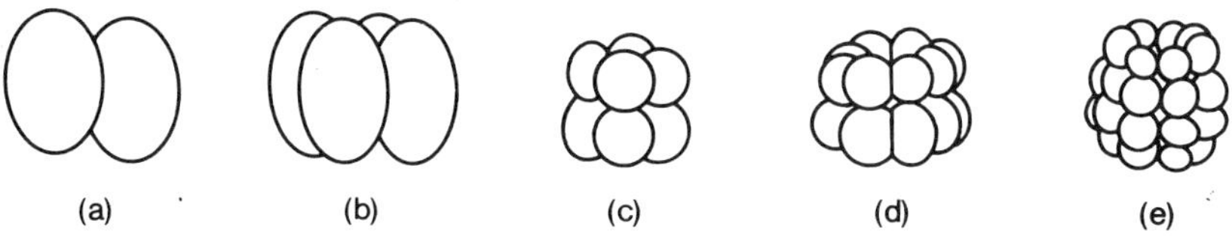

Figure 5.3-I. The holoblastic radial cleavage pattern.
(a) 2 cells, (b) 4 cells, (c) 8 cells, (d) 16 cells, (e) 32 cells.

but there is at some stage a loss of spatial and temporal order (synchrony) which differs between species. Since our interest is in the geometry of the cleavage planes, we project the typical pattern onto the original spherical egg to get the schematic sequence in Figure 5.3-II, which illustrates the first seven cleavages up to the 128-cell morula. This is immediately suggestive of a sequence of harmonic functions on the sphere, the cleavage lines corresponding to nodal lines of spherical harmonics. Accordingly, we develop an "eigenfunction" description of cleavage on the basis of a minimization principle wherein the eigenstates of a morphogenetic surface field describe the successive stages of the cleavage process in the early embryo. The cleavage process is then seen as a series of transformations to successively higher characteristic states of the morphogenetic field as metabolism proceeds. An analogy to this may be found in the electron density distributions of the hydrogen atom, in which the transitions to successively higher energy states results from the action of some external optical pumping field. A characteristic biological feature is that the cleavage transformations are "pumped" or induced internally, the system being self-generating.

5.4 A Variational Principle for Cleavage Planes

Proceeding with the analysis at a fairly abstract level, let us now introduce a field function $u(\theta,\phi)$ over the surface of the sphere and adopt the convention that its nodal lines represent lines of least resistance to a furrowing process preliminary to the development of cleavage planes. This function may be taken to be some kind of order-disorder parameter relating to the organization of microfilaments in the cell surface. The basic stability of the typical cleavage pattern suggests the use of a minimization principle on this field function. An appropriate surface density function, which in a physical problem would be the energy density, is

$$E(\theta,\phi) = A\left[\left(\frac{\partial u}{\partial \theta}\right)^2 + \frac{1}{\sin^2\theta}\left(\frac{\partial u}{\partial \phi}\right)^2 + \beta u^2\right] \tag{5.4-1}$$

where the constants A and β incorporate relevant physiological units. Then suppose that the characteristic cleavage planes correspond to a minimum of the integral of this density function over the surface energy E,

$$\delta \int_0^{2\pi}\int_0^{\pi} E(\theta,\phi)\sin\theta\, d\theta\, d\phi = 0 \tag{5.4-2}$$

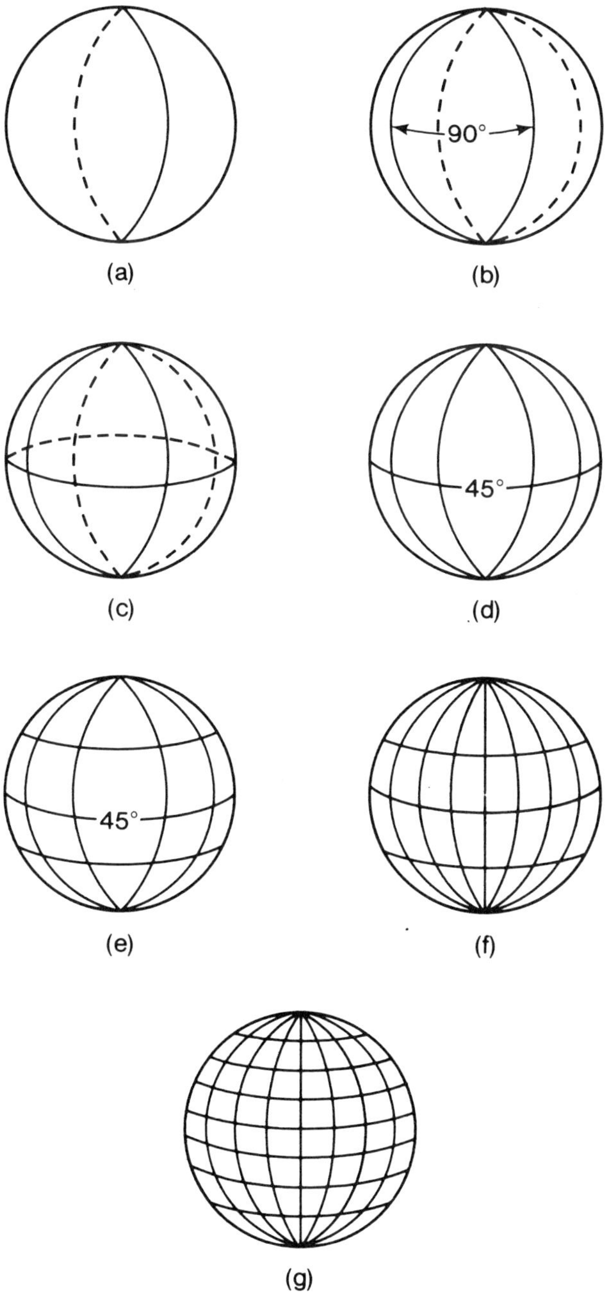

Figure 5.3-II. The geometry of typical cleavage planes up to the 128-cell stage. In (f) and (g) only, the silhouette forms one of the 8 longitudinal sections. (a) 2 cells, (b) 4 cells, (c) 8 cells, (d) 16 cells, (e) 32 cells, (f) 64 cells, (g) 128 cells.

subject to a conservation law on u^2

$$\int_0^{2\pi} \int_0^{\pi} u^2(\theta,\phi) \sin\theta \, d\theta \, d\phi = 1 \tag{5.4-3}$$

which amounts to a normalization condition on the field variable u. Equations (5.4-2) and (5.4-3) require satisfaction of the Euler-Lagrange equation (see Trainor and Wise, 1979)

$$\frac{1}{\sin\theta} \frac{\partial}{\partial\theta} \left(\sin\theta \frac{\partial u}{\partial\theta}\right) + \frac{1}{\sin^2\theta} \frac{\partial^2 u}{\partial\phi^2} - \alpha u = 0 \tag{5.4-4}$$

where α incorporates the parameter β and an undetermined multiplier of equation (5.4-3).

The usual conditions on u, that it be finite, single-valued and continuous over the sphere restricts the possible solution u to a characteristic (eigenfunction) set, viz. the spherical harmonics (real part taken):

$$u(\theta,\phi) \rightarrow Y_{\ell m}(\theta,\phi) = \sqrt{2}\, N_{\ell m} P_\ell^m(\cos\theta) \cos m\phi \tag{5.4-5}$$

where ℓ takes on the integral values 0, 1, 2, etc., and for given ℓ, the m values are integers ranging from $-\ell$ to $+\ell$. The parameter α is restricted to the corresponding characteristic values (eigenvalues) $\ell(\ell + 1)$. In equation (5.4-5), the P_ℓ^m are associated Legendre polynomials (Hobson, 1955) and the $N_{\ell m}$ are the normalization constants given by

$$N_{\ell m} = \left[\frac{(2\ell+1)(\ell - |m|)!}{4\pi(\ell + |m|)!}\right]^{1/2} . \tag{5.4-6}$$

It is easy to calculate the surface "energy" corresponding to each characteristic cleavage state. The result is:

$$E_\ell = A\{\ell(\ell+1) + \beta\} . \tag{5.4-7}$$

However, not every characteristic state (5.4-5) is realized in the cleavage process since the mitotic apparatus imposes a biological constraint (somewhat analogous to superselection rules in physics: Wick, Wightman and Wigner, 1952) that the number of cells is doubled in each cleavage stage.

According to the ideas set out above, the nodal lines of the characteristic function $Y_{\ell m}(\theta,\phi)$ on the sphere are in correspondence with the furrow lines of a characteristic cleavage state, except that the set of characteristic states is limited by the requirement that the number of cells is given by 2^p where p is the number of cell divisions. It is easily shown that the number of cells in a state characterized by $Y_{\ell m}$ is $2m(\ell - m+1)$ unless $m = 0$ in which case the number is $\ell + 1$. (It is sufficient to choose the real part of $Y_{\ell m}$, i.e. the cos $m\phi$ solutions, so that we need consider only $m \geqslant 0$.) Hence the biological constraint requires that

$$
\begin{aligned}
2^p &= 2m(\ell - m + 1) \quad \text{if} \quad m \neq 0, \quad \text{or} \\
&= \ell + 1 \quad \text{if} \quad m = 0. \qquad (5.4\text{-}8)
\end{aligned}
$$

In general this equation, for a given characteristic cleavage state corresponding to p divisions, is satisfied by more than one set of (ℓ, m) values. It is natural to suppose that the choice is made primarily on the basis of lowest ℓ value, since according to equation (5.4-7), this minimizes the "energy." The choice is then nearly unique, except for a two-fold degeneracy every second division. It is assumed that the animal-vegetal polarity of the embryo defines a secondary polar field weaker than the primary field which removes this degeneracy in favor of the highest m value for a given ℓ, in much the same way as a magnetic field removes the $(2\ell+1)$-fold degeneracy of magnetic states in the hydrogen atom.

Table 5.4-I shows the correspondence between cleavage states and characteristic functions $Y_{\ell m} = 2N_{\ell m} P_\ell^m \cos m\phi$ by listing the number of cells to be expected from each set of (ℓ, m) values up to $\ell = 7$. The appropriate (ℓ, m) pair is then selected out uniquely by the conditions expressed in equations (5.4-8), together with the minimization condition (5.4-7), except for the two-fold degeneracies at the first, second, fourth, etc., cell divisions. As remarked above a unique correspondence is achieved by assuming that a weaker polar field selects (1,1) over (1,0), (2,2) over (2,1) and (5,4) over (5,2), that is, it favors highest m value for a given ℓ.

In Table 5.4-I the selected states for the 6th and 7th cleavages corresponding to the 64 and 128 cell stages have been included, without listing all of the rejected (ℓ, m) values. Figure 5.3-II illustrates the nodal lines for the successive stages in a typical cleavage process. (Note that in the model used here the furrowing process corresponds to an accumulative conjunction of surface harmonic patterns. In a variant of the model (Goodwin and LaCroix, 1982), the pattern is not cumulative and each stage is described by a single harmonic pattern of nodal lines.)

5.5 Fields and Self-Organization

This example clarifies some of the abstract concepts introduced in the

Table 5.4-I Correspondence between $Y_{\ell m}(\theta, \phi)$ and cell number, defined by $\ell + 1$ if $m = 0$, otherwise by $2m(\ell - m + 1)$.

ℓ value	m value	cell number	(ℓ, m) pair selected
1	0	2	(1, 1)
	1	2	
2	0	3	(2, 2)
	1	4	
	2	4	
3	0	4	(3, 2)
	1	6	
	2	8	
	3	6	
4	0	5	
	1	8	
	2	12	
	3	12	
	4	8	
5	0	6	(5, 4)
	1	10	
	2	16	
	3	18	
	4	16	
	5	10	
6	0	7	
	1	12	
	2	20	
	3	24	
	4	24	
	5	20	
	6	12	
7	0	8	(7, 4)
	1	14	
	2	24	
	3	30	
	4	32	
	5	30	
	6	24	
	7	14	
11	8	64	(11, 8)
15	8	128	(15, 8)

discussion on organisms as fields, and it is now of some interest to elaborate on these and their relationship to the concept of self-organization in biology. The first point to emerge is the primacy of the organism as the fundamental biological entity, replacing the usual definition of the cell as the unit of life. This follows from the fact that the field is the self-organizing entity, and this is coextensive in the above description with the organism. The orderly geometry of the cleavage planes is a reflection of this organization at the level of the developing embryo. This does not mean that the parts, which in this case are cells, have no properties of their own. On the contrary, the constraint of binary division, a cellular property, is a major source of the distinctive geometry of the cleavage process, since this defines one of the three selection rules which determine the harmonic functions allowable as descriptions of this process. What thus becomes apparent is that the spatial organization of the whole derives from principles relating to global field behavior together with constraints coming from the properties of the entities which are generated as parts. Such a description avoids both atomistic reduction and a holistic description which identifies the whole with some (conceptually or materially) isolable essence. The genome is such an isolable essence, traceable historically to its conceptual roots in idealistic holism via Weismann's conceptualization of the organism as separable into germ plasm (essence) and somatoplasm (expression of the essence; see Webster and Goodwin, 1982). A field description of developing organisms sees spatial organization as the expression of distributed influence, global order being constrained to give specific morphology as a result of autonomous properties of parts. This defines a "decentered structure."

The ambiguity between the concepts of "cell" and "organism" can be resolved in terms of the above description. The fertilized egg is both a cell and a developing organism. It is an organism insofar as it is a totality describable by a field; it is a cell insofar as it embodies the specific constraints (e.g., binary division) characteristic of such an entity. As cleavage proceeds, the organism continues to be identified with the whole field (the embryo), while cells are identified as parts which play specific roles within a transforming context. After gastrulation, more complex parts such as neural plate, limb fields, eye fields, etc., come into existence. These consist of aggregates of cells so that an integrated hierarchy of parts emerges within the context of the organism as the global field which continues to impose organizational constraints upon the whole, the parts imposing reciprocal constraints so that further specific form emerges.

This type of description allows one to make comparisons between developmental processes in a cellular (or unicellular) and in multicellular organisms and to understand them in the same terms, which is the classical view. If one adopts a position such as that of Wolpert (1971), that development is to be understood in terms of the responses of cells (really, of their genomes:

Wolpert and Lewis, 1975) to "positional information" established over multicellular domains, then such comparisons become problematical, as discussed by Frankel (1974). The view of organisms as fields overcomes this difficulty and leads to the suggestion of some unexpected homologies between the morphology of ciliate protozoa and of amphibian gastrulae (Goodwin, 1980). This is very much in the tradition of rational morphology, since this makes clear that one must seek homologies at the level of "deep structure," i.e. at the level of generative principles such as those defining field properties, not in terms of "surface structure" such as whether or not an organism is partitioned into cells. Again, this is not to deny that surface structure imposes constraints which affect manifest form; it is simply that in comparing field effects at the level of the whole organism, these are secondary.

A very important problem in embryogenesis relates to the means whereby new or emergent aspects of morphogenesis are initiated at specific times in the process. For example, the relatively simple recursive process of cleavage in amphibian embryos is followed by the dramatic phenomenon of gastrulation, which transforms the hollow blastula into the three-layered late gastrula. The transition from cleavage to gastrulation has been described (Goodwin, 1980) as the expression of a cortical or surface field which is radially symmetric in the unfertilized egg but develops bilateral symmetry, normally as a result of sperm entry. The description of such a surface field in terms of harmonic functions reveals that with bilateral symmetry there appears a special point on the surface, a saddle point, which is identified with the future dorsal lip of the blastopore. However, it is assumed that the influence of this singularity on the morphogenetic process cannot become significant and be expressed until cleavage transforms the initially solid sphere of the egg into a hollow spherical shell, the blastula. Then the saddle point can make its presence felt as a point on the blastula where surface polarity is absent and a radial influence can manifest as the force causing bottle cell formation and the initiation of invagination. Thus cleavage is seen as a process which establishes a necessary condition for the emergence of a new phase of morphogenesis resulting from the interaction between a singularity in the surface or cortical field and the residual radial component of the cleavage field, which is described by solid harmonics.

5.6 Generation and Regeneration

There is another basic property of organisms relating the whole and the part, and this is the capacity of a part to generate or to regenerate the whole. So far, in discussing embryogenesis, the view has been developed that the fertilized egg is a whole which undergoes transformations resulting in the appearance of parts which have distinctive properties but are not, within the context of the whole organism, autonomous in the sense that atomistic theories would have them be. However, there was a time when the egg was a cell within the ovary, itself a part of the parent organism. The oocyte during its maturation develops the capacity to develop into a new whole. Such a transformation can be achieved also by parts (multicellular fragments) of hydroids such as *Hydra,* or parts (noncellular fragments) of ciliate protozoa such as *Stentor,* or a cell of a carrot, any of which can regenerate the whole organism. The capacity of plants to propagate from leaves and stems, of insects and urodeles to regenerate limbs from stumps, and of higher organisms to regenerate skin and liver, are other manifestations of this same property of parts to transform into wholes. It is evident that different organisms vary widely in their regenerative capacities, but it is true of all organisms that from particular parts, wholes can be produced. This defines reproduction or generation, as well as regeneration, and it is one of fundamental self-organizing properties which living creatures display. What does a field description of organisms have to say about such behavior?

There is an interesting property of harmonic functions which is suggestive of precisely this capacity. If such a function is described over any part of its domain of definition (e.g., part of the surface of a sphere), then the function can be recovered uniquely over the whole domain by analytic continuation. Thus, in a particular sense, the part contains the whole. This gives us a kind of existence theorem for the generative and regenerative properties of organisms, defined as fields describable by harmonic functions. It is just this type of property that needs to be embodied in a description of biological self-organization, although the specific property of harmonic functions described here is neither necessary nor sufficient to account for the actualities of generation and regeneration in the living realm.

5.7 Structuralist Biology

It may well have become apparent to the reader before this point that the general context within which this essay has been constructed is that of contemporary Structuralism, as defined by Levi-Strauss (1968) and Piaget (1968), and developed in another, more extended analysis by Webster and Goodwin (1982). A field as used above is an example of a *structure* in that it belongs to an invariant set (the harmonic functions) defined by internal relations (those defining the field equation over a domain), each member of which is a transformation of the other (by a change of boundary values).

The particular field functions which have been used in this treatment are, in a general sense, not as important as the structuralist principles which inform the analysis. For these principles emphasize the necessity to identify that which is specific to a particular area of study, such as biology, before attempting to develop a theory to "explain" the phenomena. This is why, in approaching the problem of self-organization in embryogenesis, it was necessary first to identify any empirical regularities emerging from the study of biological form which might suggest the existence of principles of organization or invariant relationships in organisms. The evidence points clearly in this direction so that organisms, and hence embryos, are indeed structures in the technical sense: i.e., entities with the defining characteristics of wholeness and self-organization, capable of undergoing transformations which preserve these deep properties while changing manifest form. This is just a more elaborated description of a view which has been clearly articulated by those who insisted that the primary problem of biology is that of organization and form, not of composition or heredity, the latter finding their place within the context of the former (cf. Russell, 1930; Needham, 1936). The use of the field concept and the more specific description of biological form in terms of harmonic functions simply makes more explicit the implications of this view. Gene products can in certain instances select specific form, such as left- or right-handed spiralling in the third cleavage planes and, consequently, in the shells of the snail, *Limnaea*. However, in general we have seen from the field description of cleavage that the constraints determining specific form arise from other organizational levels than simple molecular composition, these being such processes as binary cleavage, animal-vegetal polarity, and a minimum "energy" condition. This emphasizes once again the primacy of organizational principles in biological process and the inadequacy of any theory based upon genes and molecules.

We may say that the role played by the genes is to specify the potential molecular composition of an organism and to define some temporal sequences in which molecular components are made. These constitute constraints which impose some limitations on the forms which organisms can assume, and in certain instances may actually determine higher-order form; but in general the relationship between "genotype" and "phenotype" is one of causal necessity, not sufficiency, since gene products act within the context of fields (electrical, visco-elastic, etc.) which generate morphology. A linguistic analogy would be that genes essentially determine the set of words out of which a text can be constructed. Words are clearly insufficient to define a text, which embodies higher-order syntactical, semantic, and contextual constraints or rules. These are rules of organization which limit the set of allowed arrangements of the words within the text. Such organizational principles are described in this essay as field constraints, which limit the allowed range of biological forms. The generative principles of organismic morphogenesis then consist of organizational constraints or rules (laws of form) common to all organisms in the form of fields (thus defining biological universals) together with specific constraints characteristic of individual species (which then define particulars). Genes specify some of the latter but none of the former.

One of the most significant aspects of the structuralist approach is the deliberate avoidance of any *a priori* material reduction of the organism to parts such as cells, molecules, or genes. Once the problem of biological (self-) organization is clearly and explicitly described, and an appropriate description is available, then it *may* be possible to carry out a relevant material reduction. Certainly it will be possible to achieve an *abstract* reduction to laws and rules, such as those which have emerged from the analysis and description of cleavage given above. This description was itself inspired by a paper which was, in my opinion, a landmark in the discovery of general rules of morphogenesis expressed in terms which make no reference to composition or inheritance and are of a purely relational nature (French, Bryant, and Bryant, 1976). To start with the assumption that one knows what the basic parts of the organism are is to make a strategic error at the outset, which can lead one badly astray in seeking at once the most economical and the most rigorous analytical treatment of the problem. The description of organisms as fields which embody self-organizing properties and can undergo transformations preserving invariant relationships is, despite its limitations, a step towards a biological science of form and organization which relates part to whole in a manner which preserves organismic unity in the diversity of manifest morphology.

Acknowledgements

I am indebted to my colleague, G.C. Webster, for the intellectual stimulus which has led to many of the ideas developed in this essay.

References

Darwin, C. 1859. *On the Origin of Species.* Reprint of the first edition, 1950. Watts & Co., London.

Driesch, H. 1929. *The Science and Philosophy of the Organism.* Black, London.

Frankel, J. 1974. Positional information in unicellular organisms. *J. Theoret. Biol.*, 47: 439-481.

French, V., P.J. Bryant, and S.V. Bryant. 1976. Pattern regulation in epimorphic fields. *Science,* 193: 969-981.
Goodwin, B.C. 1980. Pattern formation and its regeneration in the protozoa. In G.W. Gooday, D. Lloyd, and A.P.J. Trinci, eds., *The Eukaryotic Microbial Cell.* Symp. Soc. Gen. Microbial, 30: 377-404.
Goodwin, B.C., and N. LaCroix. 1984. A further study of the holoblastic cleavage field. *J. Theoret. Biol.,* 109: 41-58.
Goodwin, B.C., and L.E.H. Trainor. 1980. A field description of the cleavage process in embryogenesis. *J. Theoret. Biol.,* 85: 757-770.
Goodwin, B.C., and L.E.H. Trainor. 1982. The ontogeny and phylogeny of the pentadactyl limb. Brit. Soc. Dev. Biol., Symp. on Development and Evolution, p. 75.
Hobson, E.W. 1955. *The Theory of Spherical and Ellipsoidal Harmonics.* Chelsea Publishing Company, New York.
Levi-Strauss, C. 1968. *Structural Anthropology.* London.
Needham, J. 1936. *Order and Life.* Yale University Press.
Oosawa, F., H. Kasai, S. Hatano, and S. Asakura. 1966. Polymerization of actin and flagellin. In G.E.W. Wolstenholme and M. O'Connor, eds., *Principles of Biomolecular Organization,* pp. 273-286. Little, Brown and Co., Boston.
Piaget, J. 1968. *Structuralism.* Routledge and Kegan Paul, London.
Saunders, J.W., J.M. Cairns, and M.T. Gasseling. 1957. The role of the apical ridge of ectoderm in the differentiation of the morphological structure and inductive specificity of limb parts in the chick. *J. Morphol.,* 101: 57-87.
Trainor, L.E.H., and M.B. Wise. 1979. *From Physical Concept to Mathematical Structure.* University of Toronto Press.
Webster, G.C., and B.C. Goodwin. 1982. The origin of species: a structuralist view. *J. Soc. Biol. Struct.,* 5: 15-47.
Wick, G., A. Wightman, and E. Wigner. 1952. *Phys. Rev.,* 88: 101.
Wolpert, L. 1971. Positional information and pattern formation. *Curr. Top. Dev. Biol.,* 6: 183-224.
Wolpert, L., and J. Lewis. 1975. Towards a theory of development. *Fed. Proc.,* 34: 14-20.

6

REMARKS ON EMERGENCE IN PHYSICS AND BIOLOGY

L.E.H. TRAINOR

6.1 Introduction

In Chapter 5, Goodwin has analyzed the problem of self-organizing systems in developmental biology and has made the case for a field point of view which expresses an irreducible degree of wholeness or globality in developing systems rather than an essentially reductionist view of material parts (e.g. molecules, or cells) responding to a genetic program through the laws of physics and chemistry. Goodwin terms the latter view neo-Darwinism, since it envisions the development of a species largely as the development of the genetic program which derives from random permutations and natural selection. Goodwin envisions some structural laws of growth and form, embedded in the laws of physics and chemistry, which complement the laws of molecular interactions but which also transcend them and are not directly emergent from them. Since both the modern structuralists and the neo-Darwinians subscribe to the laws of physics and chemistry, it is useful to examine the laws of physics to see whether these divergent viewpoints are two sides of one and the same coin or whether they represent fundamental points of departure in an approach to understanding in developmental biology, and, perhaps, in other areas of study relating to living organisms including even the elusive questions of cognition and consciousness.

In what follows we do not take sides in this issue, but simply examine, within well-established paradigms in modern physics, whether a case could be made by analogy for a distinction between what I will term *irreducible* and *reducible* forms of emergence, the former arguing in favor of an irreducible degree of wholeness in biological systems in which the emergent properties are primitively embedded, the latter arguing for emergence as arising out of interactions at a more fragmented or reduced level.

Classical physics in the sense of Newton, and perhaps also of Maxwell, was highly reductionistic, particularly when seen from an atomistic viewpoint: Matter is composed of small interacting particles, and large scale behavior is derivable, at least in principle, from studying large aggregates of such particles. Emergence is the appearance of unexpected structures, or the appearance of unexpected behaviors of large structures. Crystallization of liquids is an example of emergence—a new and unexpected behavior emerging at a macroscopic level which does not appear when only a few particles are considered, even though no new forces are invoked at the macroscopic level. Emergence of this kind is clearly *reducible* since it arises naturally, if surprisingly, from the assumption of pair-wise interaction between "elementary particles." The phenomenon may transcend the intuition, but not the microscopic dynamics underlying the phenomenon.

6.2 Global Aspects of Quantum Physics

We now address the question whether physical systems exhibit a more basic emergence, which is *a priori irreducible,* i.e. it cannot be reduced to or derived from the basic interactions between elementary particles. Curiously enough they do. Physics, which is supposed often to be highly reductionistic, and is, is also highly global in approach. A dramatic example is the emergent property of superconductivity. This phenomenon is not reducible in understanding, nor in mechanism, to the interaction of "elementary particles"; to achieve this emergent phenomenon one has to impose at the outset a remarkably global property on the system, namely the Pauli principle, which requires amongst other things that every electron is correlated with every other electron by the requirement of antisymmetrization of the wave function. This property is essential to the phenomenon of superconductivity and is a compelling and necessary property of the system which is *a priori* global or holistic. Indeed, not only superconductivity but nearly every quantum phenomenon is supremely affected by the so-called Fermi and Bose conditions. Phenomena such as superconductivity, superfluidity and the Josephson effect are particularly striking examples because the quantum effects are manifested on a macroscopic scale where the drama of their existence is transparent.

In quantum physics one speaks of both statistical and dynamical correlations in the behavior of many-particle systems. Dynamical correlations between elementary particles arise out of the forces between these particles; the motion of any one particle affects its interactions with other particles and thus influences the motions of the other particles. One says their dynamical motions are "correlated." Such microscopic correlations can have macroscopic or emergent effects with measurable properties, such as the latent heat of vaporization in the transition of a liquid to its vapor. Emergent behavior of this kind is *reducible* because it arises out of microscopic behavior and can be best understood by reduction of the macroscopic problem to its microscopic support.

Statistical correlations are another matter. The Pauli principle and its extension into so-called Fermi-Dirac statistics asserts that all spin one-half particles of a particular kind in the universe (e.g. electrons, protons, or neutrons) have an *a priori* correlation which does not arise out of the dynamics. In the case of these so-called Fermi particles or fermions this correlation is such as to give them an anti-social behavior, which is expressed in the antisymmetrization of the wave function (see below). The Fermi-Dirac correlations express a very global property of elementary particles which cannot be reduced to properties of individual particles or their two- or three-body interactions. Observed behavior of physical systems expresses aspects of both kinds of correlation, the dynamical/reducible type and the statistical/global type. Thus emergence of ferromagnetism, for example, cannot be said to be either a reducible or a global phenomenon, but arises out of both aspects in a physical symbiosis.

Identical particles of integer spin (e.g. photons or pi-mesons), the so-called bosons, obey an *a priori* Bose-Einstein statistics which requires symmetrization of the wave function for all such particles in the universe, as opposed to antisymmetrization for fermions. This requirement endows these particles with an obsessive social behavior—they all prefer to be in the same quantum state—which manifests itself in a variety of macroscopic phenomena such as helium superfluidity associated with a modified Bose-Einstein condensation. The modifications arise out of dynamical correlations, so that again the observed, emergent behavior arises out of both dynamical (reducible) and statistical (global), interparticle correlations.

It is sometimes said that the symmetrization (or antisymmetrization, as the case may be) is only important when the particles are sufficiently close so that their wave functions overlap. This is small consolation in cases, e.g. like conduction electrons in a metal, where the electrons are nonlocalized so that the global property of antisymmetrization must be expressed at the outset. Moreover, particles which are very distant may at some future time come together, even momentarily as in a scattering experiment, so that their persistent symmetrization or antisymmetrization must already be manifest in the so-called asymptotic states when the particles are far apart with nonoverlapping wave functions (Trainor and Wu, 1953).

6.3 Field Expression of Quantum Globality

How does the quantum theory come to terms with what we have described as its *a priori* globality? Historically, the formalism developed in stages with attention at first directed at the idealized one-particle problem, then at the two-particle problem (i.e. two "identical" particles), then at the many-particle problem. For an idealized one-particle system, the problem of antisymmetrization of fermion wave functions does not arise. For the idealized two-particle problem (e.g. the hydrogen molecules or the helium atom) antisymmetrization is imposed as an axiom on any solutions obtained to the purely dynamical problem. Let $\psi(x_1, x_2, t)$ be a solution of the purely dynamical problem where x stands for position and whatever internal variables such as spin. The antisymmetrization principle then states that ψ is not a solution realized in nature, but only its antisymmetrized version:

$$\phi_A(x_1, x_2, t) = \frac{1}{\sqrt{2}} \left[\psi(x_1, x_2, t) - \psi(x_2, x_1, t)\right]$$

where we have assumed ψ normalized to unit probability and the $1/\sqrt{2}$ factor renormalizes the linear combination. Since probabilities and other measurable quantities correspond to quadratic expressions of the wave function, such as

$$|\phi_A|^2 = \frac{1}{2} \{ |\psi(x_1, x_2, t)|^2 + |\psi(x_2, x_1, t)|^2 - \psi^*(x_1, x_2, t)\psi(x_2, x_1, t) - \text{c.c.}\}$$

where c.c. stands for complex conjugate, it is clear that interference effects arise and particles lose their identification as individuals to some degree. Pauli (1947) has discussed the complex nature of the probabilities associated with these many-particle, antisymmetrized wave functions.

In this and other conventional examples, the global aspect of many electron systems is expressed as the antisymmetrization of the ψ-field, i.e. it is expressed through a *field function.* The matrix formulation of quantum mechanics might seem to avoid use of a field function (Kaempffer, 1965), but this is illusory since the Hilbert space vectors have again to be interpreted as fields in order to accord with what is experimentally measured.

Further developments in quantum mechanics lead to a more explicit recognition of the value of the field concept in expressing the global nature of the identical particle problem. In these developments one went beyond first quantization of the states of a single particle to so-called second quantization of the fields describing many particle systems (Jordan and Klein,

1927; Jordan and Wigner 1928). In second quantized formalism, all electrons are simultaneously recognized and placed on an equal footing at the outset. Quantum field theory in all of its aspects, including such modern developments as grand unified theories, quantum chromodynamics, etc. use the field concept in an essential way to express the global constraints imposed on system behavior by the antisymmetrization or symmetrization procedures required by fermion, respectively boson systems.

A degree of globality in quantum theory also exists in aspects other than the quantum statistical aspects discussed above. These are less transparent in the formalism and more controversial in interpretation, and will only be mentioned here. The famous Einstein, Rosen and Podolsky paradox (Einstein et al., 1935) and its codification in Bell's theorem (Bell, 1966) clarifies an essential nonlocality in the relationship between particles which is imposed by conservation laws, but which are additional to the nonlocalities imposed by statistics as presented in the preceding paragraphs. The summary position to which most physicists agree is that the apparent validity of quantum mechanics means that the physical world has some very strange aspects when viewed with our normal perceptual biases. This situation and its possible relationship to the so-called explicate and implicate orders of perception and explanation have been discussed extensively by Bohm (1980).

6.4 Discussion

The remarkable features of the quantum world discussed in the previous section, no doubt carry over into the biological realm, but are not necessarily of great consequence there in the direct sense of that word, though that question is still very much open. Our discussion here was merely to illustrate possibility by possibly unrelated example. Goodwin and Webster (1982), and Goodwin in Chapter 5, are addressing the question whether one can, indeed, keeping in view questions of complexity, arrive at an understanding of biological systems, and development specifically, in terms of a molecular approach only. The point is not to deny the importance of the molecular approach, including DNA and its role in inheritance, but to question whether explanation can be achieved in this way. One could go further and ask whether explanation can be achieved without appeal to some as yet unknown biological analogues of the antisymmetrization principle in quantum field theory.

The existence or nonexistence of such analogues does not gainsay the utility of the developmental field approach. (An analysis of this approach and its relation to physical analogues has been discussed by Trainor (1982)).

Even if biological phenomena can, in principle, be derived as emergents from a vast, complex and hierarchical scheme arising out of molecular dynamics, including quantum effects, the concept of a developmental field is still a strong and useful one, even a necessary one. One can raise, however, the interesting question whether the developmental fields are expressing something else within the domain of physical laws, but which cannot be reduced even in principle to molecular dynamics.

References

Bell, J.S. 1966. On the problem of hidden variables in quantum mechanics. *Rev. Mod. Phys.,* 38: 447.

Bohm, D. 1980. *Wholeness and theImplicate Order.* Routledge and Kegan Paul, London.

Einstein, A., N. Rosen, and B. Podolsky. 1935. Can quantum-mechanical description of physical reality be considered complete? *Phys. Rev.,* 47: 777.

Jordan, P., and O. Klein. 1927. Zum mehrkörper problem der quanten theorie. *Zeits fur Phys.,* 45: 751.

Jordan, P., and E.P. Wigner. 1928. Über das Paulische aquivalenzverbot. *Zeits fur Phys.,* 47: 631.

Kaempffer, F. 1965. *Concepts in Quantum Mechanics.* Blaisdell, Waltham, MA.

Pauli, W. 1947. *Die allgemeinen Prinzipien der Wellenmechanik.* Edwards, Ann Arbor. Reprinted from *Handbuch der Physik,* 2 Aufl., Band 24, 1 Teil.

Trainor, L.E.H., and T.Y. Wu. 1953. Symmetry requirements in electron scattering by an atom. *Phys. Rev.,* 89: 273.

Trainor, L.E.H. 1982. A field approach to pattern formation in living systems. *Physics in Canada,* 38, No. 5: 117.

Webster, G.C., and B.C. Goodwin. 1982. The origin of species: a structuralist approach. *J. Soc. Biol. Struct.,* 5: 15.

7

PATTERN FORMATION AND MORPHOGENETIC FIELDS

N.D. KAZARINOFF

7.1 Introduction

Three papers on mathematical biology, "Control of sequential compartment formation in *Drosophila*" by S. Kauffman, R. Shymko, and K. Trabert (1978), "Field description of the cleavage process in embryogenesis" by B.C. Goodwin and L.E.H. Trainor (1980), and "Pattern formation and its regeneration in the protozoa" by B.C. Goodwin (1980) appear to be closely related. This essay will discuss, from a mathematician's viewpoint, their findings. This essay contains some speculation and conjecture while leaving many large and significant questions unanswered. It is my hope that this review and discussion of the ideas presented in their papers will contribute to the advance of theories of developmental biology.

Kauffman et al. (1978) present a reaction-diffusion model of the development of the imaginal wing disk in *Drosophila melanogaster.* Goodwin and Trainor (1980) examine the beginning stages of embryogenesis using eigenfunctions of the Laplacian operator on the boundary of a solid ball and a selection rule. Goodwin (1980) uses harmonic functions and the averaging property of the Laplacian to formulate the outline of a morphogenetic field theory for both protozoa and metazoa. Goodwin (1980, p. 379) is critical of the reaction-diffusion approach. His objection that the qualitative character of reaction-diffusion generated spatial patterns depends upon the size of the space domain considered, I believe is based upon some misunderstanding.

My goal in this essay is to reconcile the two approaches, reaction-diffusion and that of the morphogenetic field, at least partially, and to advocate that in any mathematically based theory of developmental biology inclusion of time as an independent variable is vital.

7.2 The Imaginal Wing Disk of *Drosophila Melanogaster*

Kauffman et al. (1978) treat the imaginal wing disk of *Drosophila melanogaster* as if it were an elliptical disk. It is really not a disk at all. Rather it is a 2-sphere (a continuous distortion of $\{x^2 + y^2 + z^2 = 1\}$) that is composed of relatively many cells in its upper hemisphere (this hemisphere is a disk) and many fewer cells in its lower hemisphere. It is this fundamental topological difference between domains that leads to the striking differences between the eigenfunctions used in Goodwin and Trainor (1980) and Kauffman et al. (1978). I shall show that if Kauffman et al. (1978) had treated this imaginal wing disk as a 2-sphere, then their eigenfunctions would be *qualitatively* the same as those appearing in Goodwin and Trainor (1980). I write qualitatively because eigenfunctions on the surface of a family of ellipsoids $\{x^2/a^2 + y^2/b^2 + z^2/c^2 = 1\}$ vary smoothly as a, b, and c vary smoothly from 1, 1, and 1 (a, b, and c remaining positive).

Before continuing with my review and discussion of Kauffman et al. (1978), I want to digress. The fruit fly *Drosophila melanogaster* has been and is being intensively studied by biologists, for example, by M.A. Russell at the University of Alberta in Edmonton. One means of study of the morphogenesis of this fly is through its many mutant forms. Some of them can be distinguished with the aid of a microscope, preferably, a teaching microscope. They are truly wonderous to behold, especially the color and the pattern of hairs on the fly. The biologist uses gynadromorphs and mitotic recombination (induced by radiation) to discover the sequential formation of lines of clonal restriction in the wings of the fruit fly. Moreover, using microsurgery on the embryo after it has undergone sufficient development, the biologist can remove particular groups of cells and replace them with other groups of cells from other positions in the same embryo or from other embryos at different stages of development. The biologist thus conducts experiments testing theories and giving rise to their modification.

The model introduced by Kauffman et al. (1978) is based upon experiments that revealed nodal or switching lines on the imaginal wing disk of *Drosophila melanogaster*. Cell fate and function differs from one side of such a curve to the other. The reaction-diffusion model of Kauffman et al. (1978) is:

$$X_t = D_1 \Delta X - AX + BY^p/(1 + Y^p)$$

$$Y_t = D_2 \Delta Y - CX + D(Y^p + b)/(1 + Y^p)\,, \qquad (7.2\text{-}1)$$

where A, B, C, D, D_1, D_2, b, and p are positive numbers, and where

$$\Delta = \partial^2/\partial x^2 + \partial^2/\partial y^2\,,$$

$\vec{x} = (x, y)$ lies in an elliptical domain in $\mathbf{R}^2$ and $t \geqslant 0$ is time.* The dependent variables X and Y are thought of by Kauffman et al. (1978) to be the concentrations, depending on $(\vec{x}, t)$, of biochemical morphogens. Because these morphogens have not been identified experimentally, how they might influence the fate of cells, should they exist, is also unknown.

If we rescale the time and X by

$$t \to k_1 t \text{ and } X \to k_2 X ,$$

where

$$k_1 = 1/A \text{ and } k_2 = A/C ,$$

then (7.2-1) becomes

$$\begin{aligned} X_t &= d_1 \Delta X - X + EY^p/(1 + Y^p) \\ Y_t &= d_2 \Delta Y - X + F(Y^p + b)/(1 + Y^p), \end{aligned} \qquad (7.2\text{-}2)$$

where

$$E = BC/A^2, \quad F = D/A, \quad d_i = k_1 D_i \quad (i = 1, 2).$$

For the reaction-diffusion system (7.2-1) there is a unique constant equilibrium solution on the first quadrant of $\mathbf{R}^2$, provided

$$E > F \quad \text{or} \quad BC > AD, \qquad (7.2\text{-}3)$$

namely,

$$\begin{aligned} X_e &= EFb(E - F)/(E + F(b - 1)) \\ Y_e &= EFb/(E - F). \end{aligned} \qquad (7.2\text{-}4)$$

Thus, if we let

$$u = X - X_e , \quad v = Y - Y_e , \qquad (7.2\text{-}5)$$

then (7.2-2) becomes

*X_t, Y_t denote partial derivative with respect to the time t.

$$u_t = d_1 \Delta u - u + \delta v + f(v)$$

$$v_t = d_2 \Delta v - u + \gamma v + g(v) . \tag{7.2-6}$$

Here

$$\delta = pEY_e^{p-1}(1+Y_e^p)^{-2}, \quad \gamma = pFY_e^{p-1}(1-b)(1+Y_e^p)^{-2},$$

$$f(v) = E\left[\frac{(v+Y_e)^p}{1+(v+Y_e)^p} - \frac{Y_e^p}{1+Y_e^p}\right] - \delta v, \text{ and}$$

$$g(v) = F\left[\frac{(v+Y_e)^p + b}{1+(v+Y_e)^p} - \frac{Y_e^p + b}{(1+Y_e^p)}\right] - \gamma v . \tag{7.2-7}$$

We consider (7.2-6) for $\vec{x} \in \Omega$, where

$$\Omega : \left\{ (x, y) \mid x^2/a' + y^2/b' < 1, \ a', b' > 0 \right\}$$

and we impose the Neumann boundary conditions $\partial u/\partial n = \partial v/\partial n = 0$ on $\partial\Omega$.

Now suppose

$$\binom{u}{v} = e^{\lambda t} \binom{\phi}{\psi} \tag{7.2-8}$$

is a solution of the system (7.2-6) linearized about (0, 0). Then

$$\lambda \binom{\phi}{\psi} = \begin{pmatrix} d_1 \Delta \phi - \phi + \delta \psi \\ d_2 \Delta \psi - \phi + \gamma \psi \end{pmatrix}. \tag{7.2-9}$$

If on the elliptical disk Ω, $-\Delta$ (with Neumann boundary conditions) has eigenvalues $\upsilon_{m,n}^2$ with corresponding eigenfunctions $\mu_{m,n}$, then the eigenvalues $\lambda_{\pm(m,n)}$ of (7.2-9) satisfy

$$\det\begin{pmatrix} -1-\upsilon_{m,n}^2 d_1-\lambda & \delta \\ -1 & \gamma-\upsilon_{m,n}^2-\lambda \end{pmatrix}=0\,,\quad \begin{pmatrix} m=0,1,2,\ldots \\ n=1,2,3,\ldots \end{pmatrix}, \tag{7.2-10}$$

where I have used

$$\binom{\phi}{\psi}=\binom{a_{m,n}}{b_{m,n}}\mu_{m,n}\,. \tag{7.2-11}$$

Note that for each eigenvalue $\upsilon_{m,n}^2$ of the Laplacian $(m \geqslant 0, n \geqslant 1)$ there correspond two eigenvalues $\lambda_{\pm(m,n)}$ of the linear system (7.2-9). The eigenfunctions $\mu_{m,n}$ are zero on nodal curves described in Kauffman et al. (1978). This is, of course, classical, physical mathematics. Kauffman et al. (1978) interpret these nodal curves as switching curves for the behavior of cells in the imaginal wing disk. The "concentrations" u, v of the "normalized" morphogens change sign as one crosses these curves. Kauffman et al. (1978) point out evidence for persistence of these curves, in distorted shape, in the wing of an adult fly. From the point of view of Goodwin (1980) and Goodwin and Trainor (1980) these eigenfunctions represent morphogenetic fields that are *whole* organism determined and that are not necessarily genetically controlled, at least in protozoa.

There exist countably many $\upsilon_{m,n}$, and hence there exist infinitely many nodal curves. But in the imaginal wing disk there exist only finitely many switching curves. This dilemma is unresolved in Goodwin (1980), Goodwin and Trainor (1980), and Kauffman et al. (1978). Moreover, there exist many nodal lines, even for small m and n, that are not observed in the imaginal wing disk. Goodwin and Trainor (1980) overcome this difficulty by introducing selection rules that exclude certain eigenstates.

I now take a mathematically dubious step. I suppose that the bifurcation parameter γ is controlled in some (unknown) way and that $\gamma = \gamma(t)$ increases with t from 0 to some $N > 0$ as the cells of the imaginal wing disk subdivide. As we shall presently see, there are finitely many critical values of γ at which Hopf Bifurcation occurs for the system (7.2-6) with Neumann boundary conditions. I assume that in between these finitely many values $\gamma(t)$ may increase rapidly, while in a neighborhood of each of them $\gamma(t)$ is almost constant. The first Hopf Bifurcation occurs at $\gamma = 1$ with $m = 0$, $n = 1$. Subsequent Hopf Bifurcations occur for general (m, n) provided

$$\gamma = 1 + \upsilon_{m,n}^2(d_1 + d_2) \tag{7.2-12a}$$

and

$$\delta > (1 + \upsilon^2_{m,n} d_1)(\gamma - \upsilon^2_{m,n} d_2) = (1 + \upsilon^2_{m,n} d_1)^2. \tag{7.2-12b}$$

The first eigenvalue $\upsilon^2_{0,1}$ for $-\Delta$ with Neumann boundary conditions is 0; the corresponding eigenfunction is 1. The conditions for a Hopf Bifurcation at $\gamma = 1$ reduce to

$$\delta - \gamma = \delta - 1 > 0 \qquad (m = 0, n = 1) \tag{7.2-13a}$$

$$\min_{\substack{(m,n)\\(m,n) \neq (0,1)}} [\delta - 1 + \upsilon^2_{m,n}(d_2 - d_1) + \upsilon^4_{m,n} d_1 d_2] = \varepsilon > 0. \tag{7.2-13b}$$

Note that γ is positive only if $b < 1$. Further the condition $\delta > 1$ restricts the parameters in the model through (7.2-7), (7.2-4), (7.2-3), and (7.3-2). The bifurcating periodic solutions at this first Hopf Bifurcation are described by the formulae

$$\binom{u}{v} = zq + \bar{z}\bar{q} + O(\varepsilon^3),$$

where

$$q = \frac{1}{\delta}\begin{pmatrix} \delta \\ 1 + i\omega_0 \end{pmatrix}, \qquad \omega_0 = (\delta - 1)^{1/2},$$

$$z = \varepsilon e^{i\omega_0 t} + O(\varepsilon^3), \text{ and } \varepsilon = [(\gamma - 1)/\mu_2],$$

and where μ_2 can be computed by the recipe given in Chapter 5 of Hassard et al. (1980).

Since the eigenvalues $\upsilon^2_{m,n}$ increase monotonically for fixed n or m, the condition (7.2-12b) is satisfied for only finitely many pairs (m,n). Thus there exist only finitely many values of $\gamma > 1$ for which a Hopf Bifurcation occurs. It follows that if the zero sets of the corresponding bifurcated solutions are interpreted as nodal lines (switching curves), then the total number of nodal lines is finite, which agrees with observation. Of course, each of these subsequent Hopf Bifurcations is a bifurcation to unstable periodic solutions. This is so because at

$$\gamma = 1 + \upsilon^2_{m,n}(d_1 + d_2) \qquad (m, n \geqslant 1) \tag{7.2-14}$$

the eigenvalues $\lambda_{\pm(i,j)}$ that correspond to $0 \leqslant i < m$, $1 \leqslant j < n$ all have positive real parts since the $\upsilon^2_{i,j}$ are each smaller than $\upsilon^2_{m,n}$.

As γ increases from any value where a Hopf Bifurcation to unstable periodic solutions occurs, the corresponding periodic solutions of (7.2-1) change. In particular, the shapes and positions of their zero sets (nodal lines) change as the nonlinear terms in (7.2-1) gain a dominant influence. These changes in the nodal lines may, if the nonlinear terms are correctly chosen, perhaps correspond to changes in the actual switching lines observed in the developing wing disk of the fly.

I observed next that similar conclusions hold for a bifurcation analysis of (7.2-1) but with Dirichlet boundary conditions ($u = v = 0$ on $\partial\Omega$), although the mathematical analysis is much more difficult (see Chapter 5 of Hassard et al. (1980) for an example). Further, I have carried out the analysis on the 2-sphere $\{x^2 + y^2 + z^2\} = a^2$. The conclusions are similar. The eigenfunctions are the spherical harmonics restricted to the 2-sphere. Also, the existence of such Hopf Bifurcations only depends on the linear part of the reaction-diffusion system studied. Consequently the bifurcations to the periodic solutions exist for a wide variety of nonlinear terms in (7.2-6) corresponding to different biochemical kinetics in various organisms. The region Ω may be homeomorphically (continuously and 1–1) distorted as well.* By continuity, for the reaction-diffusion system considered on the distorted domain the qualitative structure of the bifurcations will be the same; such changes in Ω will not cause changes in either the number or topological character of the nodal curves.

M. Russell theorizes that three biochemical morphogens exist that switch on or off particular genes in *Drosophila melanogaster,* with the switching occurring at the nodal curves. I have not carried out a bifurcation analysis for any three-dependent-variable reaction-diffusion model; but, in principle, through an analysis of a three-dependent-variable reaction-diffusion system on a two-sphere or a solid ball, one could demonstrate the existence of a finite number of Hopf Bifurcations and give an interpretation much like that provided above. If Russell is correct, once a gene is turned off or on, it has had its determinative effect on the morphogenesis of the fly; what happens to the concentrations of the three mutagens subsequently is of no consequence for that gene, or combination of genes, and the same is true for subsequent changes in the switching curves.

I call attention to two defects in my interpretation. I have no biological mechanism for controlling the growth of the bifurcation parameter γ in time; and the subsequent bifurcations introduce nodal lines that do not correspond to reality. The selection rules of Goodwin and Trainor (1980), explained in the next section, may remove the latter defect.

*A homeomorphism is a continuous 1-1, onto map with a continuous inverse.

Finally, I remark that there exist other possible types of bifurcation, such as secondary bifurcations to torii or bifurcations from a single steady state to multiple steady states. For example, for a single reaction-diffusion equation there cannot exist a Hopf Bifurcation, but there can exist bifurcation from a steady state to multiple steady states by virtue of a real eigenvalue passing through zero (see Hassard et al., 1980, pp. 78-85). This type of bifurcation too may have some relevance to developmental biology.

7.3 Modelling the Initial Stages of Embryogenesis

In 1980, Goodwin and Trainor considered the initial stage of embryogenesis, the first several subdivisions of a holoblastic egg in their paper, "Field description of the cleavage process in embryogenesis." The original holoblastic egg is topologically a solid ball; but after several cell divisions, the egg becomes a hollow ball of cells, the blastula, which is topologically a (2-sphere) $\times$ (an interval). Thus a topological change in the embryo is observed. This is even the case upon the very first cell division. Two tangential 3-balls, the stage after the first cleavage, are topologically distinct from a single 3-ball.* I do not see that Goodwin and Trainor (1980) account for these topological changes. Perhaps, for the biologist it is not important to explain them.

Goodwin and Trainor (1980) call attention to a "regularity of the first succession of cleavage planes in different species." They suggest that the evolving spatial organization of an embryo is, at its initial stages, controlled by spatially defined functions whose selection is given by minimizing a generalized function analogous to a surface energy, which gives the basic symmetries of the cleavage process. They write, "we make no attempt to interpret the field descriptors in terms of biological variables such as membrane permeabilities, ion fluxes, segregated cytoplasms, microfilament orientations, etc. A closer approach to this level of analysis, consistent with the present treatment, will be given in a subsequent publication which will deal in some detail with mitotic spindle orientation at successive cleavages, its relation to cytoplasmic and cortical fields, and the nature of the influences which give different cleavage patterns in different species. Our objective in the present paper is to show how some quite general considerations relevant to the physical description of cleavage lead to a field analysis of the process."

Goodwin and Trainor (1980) draw an analogy between their model and the quantum mechanical model of the hydrogen atom. The hydrogen atom undergoes transitions to successively higher energy states if it is suitably pumped with energy. The cleavage process, in its initial stages, is interpreted

* A 3-ball is any homeomorph of the standard unit radius solid ball.

as a series of transformations to successively higher energies, characteristic states of the morphogenetic field as metabolism proceeds. This appears a bit far out to me—though I grant the charm and simplicity of the concept.

The surface energy function chosen by Goodwin and Trainor (1980) is defined on

$$S^2 = \{(r, \theta, \phi) \mid r = 1\}$$

by

$$E(\theta, \phi) = A\left[u_\theta^2 + \frac{1}{\sin^2\theta} u_\phi^2 + \beta u^2\right] \qquad (A, \beta > 0)\,. \qquad (7.3\text{-}1)$$

They assume a conservation law—perhaps, better termed a normalization—

$$\int_0^{2\pi}\int_0^{\pi} u^2 \sin\theta \, d\theta \, d\phi = 1\,.$$

Then the Euler-Lagrange equation obtained by minimizing the total surface energy

$$\int_0^{2\pi}\int_0^{\pi} E(\theta, \phi) \sin\theta \, d\theta \, d\phi$$

is

$$\frac{1}{\sin\theta}(\sin\theta \, u_\theta)_\theta + \frac{1}{\sin^2\theta} u - \alpha u = 0 \qquad (7.3\text{-}2)$$

for some $\alpha \geqslant 0$, dependent on β, together with periodic boundary conditions.* I am unsatisfied with the *surface* energy field (author's italics) introduced by Goodwin and Trainor (1980). As the egg subdivides, the cells are in contact, at least biochemical contact. The *difference* in surface energy as between free surfaces and surfaces in contact would appear relevant to a theory that explains cleavage. The egg is not a hydrogen atom, nor is it a biological analog of a hydrogen atom.

The eigenfunctions of (7.3-2) are, of course, the spherical harmonics

$$\mu_{n,m} = 2^{1/2} N_{n,m} P_n^m(\cos\theta) \cos m\phi$$

$$(n = 0, 1, 2, \ldots;\ -n \leqslant m \leqslant +n,\ m \text{ an integer}).$$

*Subscript θ in (7.3-2) indicates partial derivative with respect to θ.

They correspond to the eigenvalues $\alpha_n = n(n+1)$. Here

$$N_{n,m} = \left[\frac{(2n+1)}{4\pi} \cdot \frac{(n-|m|)!}{(n+|m|)!} \right]^{1/2} .$$

The "surface energy" corresponding to $\mu_{n,m}$ is

$$E_n = A[n(n+1)+\beta] .$$

To select the viable energy states of the developing embryo, Goodwin and Trainor (1980) introduce selection rules, again in analogy with quantum mechanics. Since at each cell division, the number of cells doubles, Goodwin and Trainor (1980) introduce the 2^p-selection rule that requires

$$2m(n-m+1) = 2^p \text{ for some } p, \text{if } m \neq 0,$$

$$n = 2^p - 1 \text{ for some } p, \text{ if } m = 0.$$

This selection rule is not quite sufficient. There is a two-fold degeneracy every second cell division. Goodwin and Trainor (1980) assume that "a secondary polar field, weaker than the primary field, removes this degeneracy in favor of the highest m value for a given n, much in the same way that a magnetic field removes the $(2n+1)$-fold degeneracy of magnetic states in the hydrogen atom." (The rejected state $m = 0$, $n = 1$ may, I conjecture, be interpreted to represent the birth of the primary morphogenetic field.)

In their last section Goodwin and Trainor (1980) present discussion, which I quote here in part: "The principles and rules which are used in this paper to generate the typical holoblastic cleavage pattern define what may be referred to as a morphogenetic law or laws of form which, in terms of the viewpoint adopted, reveal aspects of the intrinsic properties of the living state just as states of the hydrogen atom are an expression of its intrinsic organization. Thus cleavage is regarded as a spontaneous expression of spatial ordering principles inherent in organisms, described and interpreted as fields.
This gives an example of how the structuralist description of organisms (Webster and Goodwin, 1980) can be usefully applied to a particular morphogenetic process. The fields used in the analysis are structures describing the state of the whole organism; they undergo systematic transformations which conform to constraints defined by selection rules; and the whole process is in broad aspect selfregulating."

It is interesting to me that the eigenfunctions of both approaches (Kauffman et al., 1978; and Goodwin and Trainor, 1980) are eigenfunctions of the

Laplacian. If Kauffman et al. (1978) had modelled the imaginal wing disk as a 2-sphere, which seems to be reasonable, then the results would have been the same as those of Goodwin and Trainor (1980), except for the important selection principles introduced by Goodwin and Trainor (1980) and except for the improvements due to the inclusion of nonlinear terms by Kauffman et al. (1978). I suggest that a combination of the morphogenetic field-energy potential concept and the reaction-diffusion approach is possible. It appears to me that, whatever may be the agents that cause development to proceed as it does, whether or not they be gene regulated or whole organism determined, that the mathematics is insensitive to this, that it accommodated either viewpoint. Also, I strongly believe that inclusion of time as a variable is essential if one is to model any developmental process in biology. This is a strong argument in favor of the reaction-diffusion analysis outlined in Section 7.2. At least, my hope is that it is a step in a fruitful direction.

7.4 Pattern Formation in Protozoa

Goodwin (1980) presents a remarkable combination of biology and mathematics. In his concluding section he writes: "The theory of pattern formation described in this chapter is based upon the idea that organismic morphology arises from the existence of a basic organizing constraint in living organisms, the rule of spatial averaging described above (p. 379). This rule generates Laplace's equation, so that *biological form must be expressed by solutions of this equation known as harmonic functions.* The particular solutions which describe the morphology of any species are determined by the boundary conditions that characterize the morphogenetic field of the species, and *these are specified by genes, by other determinants such as cortical and cytoplasmic states, and by organelle structure*" (author's italics).

Goodwin describes the form of the ciliate protozoa *Tetrahymena pyriformis* by a pair of conjugate harmonic functions on $\{ x^2 + y^2 + z^2 = a^2 \}$:

$$U(\theta, \phi) = \alpha \, \ell n \cot(\theta/2) + \beta\phi ,$$

$$V(\theta, \phi) = \beta \, \ell n \cot(\theta/2) - \alpha\phi ,$$

where θ, ϕ are the usual angular spherical coordinates. The level line of U and V on the sphere are spirals joining the north and south poles of the sphere. In some species *(Amphiletus claparedel)* the cilia are located on curves

$$-V(\theta, \phi) = \alpha\phi = \text{constant},$$

which are meridians on the sphere; in others the cilia are located on curves

$$U(\theta, \phi) = \alpha \ln \cot(\theta/2) = \text{constant},$$

which are circles of latitude on the sphere, for example in *Didinium nasutum.*

Goodwin (1980) calls U and V conjugate fields. To account for further features of ciliate protozoa, Goodwin generalizes these conjugate fields. For example, *Tetrahymena pyriformis* has a mouth, which Goodwin describes in terms of a solution of Laplace's equation on the sphere that has a saddle point at the mouth

$$\theta = 2 \arctan(b/\alpha), \quad \phi = 0$$

if $\beta = 0$, namely

$$U(\theta, \phi) = \alpha \ln \cot(\theta/2) + \beta\phi - b[\cot(\theta/2)] \cos \phi .$$

The position of this mouth in various species corresponds to various (α,β,b).

One's first question upon reading this description is, if one is a mathematician, which boundary conditions determine U and V? Goodwin (1980) writes: "The restriction of the basic pattern to the cell surface, together with the fact that the anterior (or north) pole is different from the posterior (or south) pole, is sufficient to select from among those functions satisfying Laplace's equation (more generally, Poisson's equation) on the sphere a unique one which describes the basic polar organization of the polar field. .... The more general case in which the field lines or ciliary meridians have a spiral orientation is defined by a function

$$U(\theta, \phi) = \alpha \ln \cot(\theta/2) + \beta\phi ,$$

a right-handed spiral being generated when β is negative and a left-handed spiral when β is positive. Thus the spiral is the general solution as observed, while a nonspiral polarity corresponds to the special case $\beta = 0$."

It is true that $k_1 \cot \theta/2 + k_2$ and $k_3 + k_4$ are the most general single variable solutions of $U = 0$ on the sphere, the k_i being constants. What Goodwin (1980) appears to have done is to choose the *only* such functions available, rather than to select among them. Perhaps this makes good biological sense, namely, that on a sphere basic pattern organization is governed by two morphogenetic fields, each dependent on a single coordinate.

The relationship between Goodwin's theory and reaction-diffusion based models should be considered. Goodwin (1980, p. 379) writes: This property of regulation, i.e. pattern invariance irrespective of size, I take to be a basic feature of regeneration which any general model of pattern formation must be able to account for. Models based upon chemical reaction and diffusion, so-called reaction-diffusion processes ... have difficulty with this phenomenon because the wave lengths of the standing waves of chemicals which underly all spatial patterning in such descriptions are fixed by the diffusion constants of the 'morphogens'." I do not understand this criticism. For example, the radius of the egg or blastula is reflected in the energy states obtained by Goodwin and Trainor (1980) via their constant A in (7.3-1). The qualitative features of patterns generated by reaction-diffusion on a domain Ω are not changed as the size and even the shape of Ω are changed, provided Ω is changed homeomorphically or diffeomorphically. For example, the eigenfunctions for the boundary-value problem

$$y'' = -\lambda^2 y, \quad y(0) = y(L) = 0$$

are

$$A_n \sin[(2n+1)\pi x/(2L)] .$$

Each of these has amplitude independent of L, and the number of zeros each had depends only on n and not upon the interval length L. That the eigenfunctions in general depend upon the diffusion constants and are scaled by domain size is correct. But the pattern they generate is invariant with respect to domain size. Additional dependence of the eigenfunctions of reaction-diffusion systems upon domain size is evident in the time dependent factor $\exp(\lambda_n t)$. This is natural and helpful: something must control time rates in the developmental process.

Now, if Goodwin (1980) believes that some morphogenetic fields may be determined by genes, as he states in the first quotation in this section, then Goodwin does not exclude and, indeed, may permit a "morphogen-reaction-diffusion" approach that has the advantage of including time as an essential ingredient. In his article Goodwin (1980) writes rather loosely of time. Nowhere is time included in his mathematics. Yet, Goodwin (1980, p. 399) states: "There is now a specific temporal order in the appearance of the harmonics. This implies particular constraints on the processes which are beyond the scope of the present analysis. Indeed, the whole question of the time-dependence of developmental processes, including cell-division, is of fundamental significance and requires extensive investigation."

Acknowledgements

I await further developments. I thank Professors P. Antonelli, P. van den Driessche, and L. Salvadori for their encouragement, and opportunity to write this essay during visits to Edmonton, Alberta, Victoria, British Columbia, and Povo, Italy in 1980-81.

References

Goodwin, B.C. 1980. Pattern formation and its regeneration in the protozoa. In G.W. Gooday, D. Lloyd, and A.P.J. Trinci, eds., *The Eukaryotic Microbial Cell,* Society for General Microbiology Symposium 30, pp. 377-404. Cambridge University Press, Cambridge.

Goodwin, B.C. and L.E.H. Trainor. 1980. Field description of the cleavage process in embryogenesis. *J. Theoretical Biology,* 85: 757-770.

Hassard, B., N.D. Kazarinoff, and Y.H. Wan. 1980. *Theory and Application of the Hopf Bifurcation,* Lecture Notes of the London Math. Soc., Vol. 41, pp. vi + 311 + Microfiche. Cambridge University Press, Cambridge.

Kauffman, S., R. Shymko, and K. Trabert. 1978. Control of sequential compartment formation in *Drosophila. Science,* 199: 259-270.

8

TRANSFORMATIONS OF QUADRILATERALS, TENSOR FIELDS, AND MORPHOGENESIS

F.L. BOOKSTEIN

8.1 Introduction

Morphogenetic explanations often invoke geometrical metaphors. The spatial reference may be subtle, as in the classic notion of induction, or explicit, as with polarity or the recent construct of positional information. There is a growing literature, for instance, which argues at length the relation between experimental evidence and a model of polar coordinates for encoding position around a limb. In the course of regeneration, it is argued, the animal intercalates missing structures in the sequence which, modeled as a change of polar coordinate, is shortest. Other uses of geometry have a more conventional topology, as in the attempts to derive biological compartment boundaries from the resonance geometry of diffusion-reaction partial differential equations.

In the study of shape change, this sort of speculation, whatever its apparent fit to some experimental facts, is in my view an oversimplification of the logical and parametric structure of geometric objects. The "coordinates" of a point in the plane ought to be seen not as a pair of quantities but as a pair of curves through the point specifying sets of other points with which it is, literally, *co-ordinated,* sharing a coordinate value. (For an extended discussion of this, see Bookstein, 1981.) The manner in which these curves are interconnected in the large is crucial to their effectiveness in explaining biological order. For instance, coordinates which appear radial with respect to one polar center may serve at the same time as azimuths about another center at some distance, as in Figure 8.1-Ia; or neighboring isopleths of one coordinate axis, Figure 8.1-Ib, may prove to be opposite ends of the same coordinate curve. (This pattern plays a major role in root meristem growth in plants;

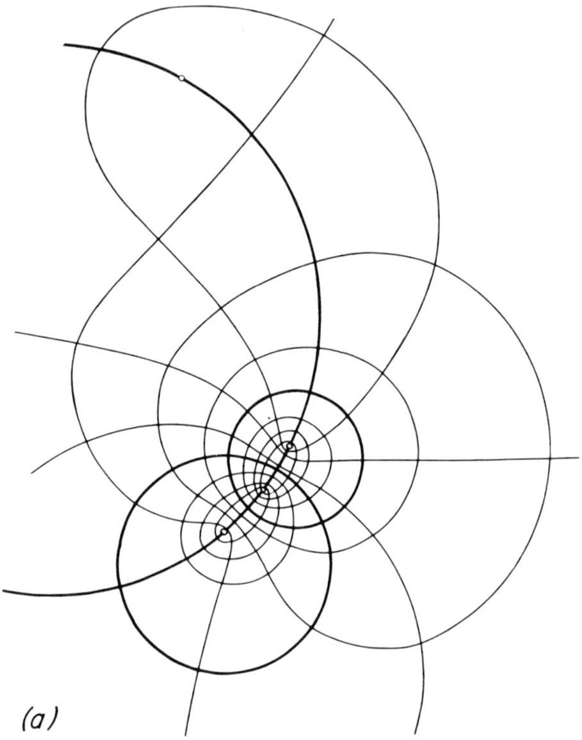

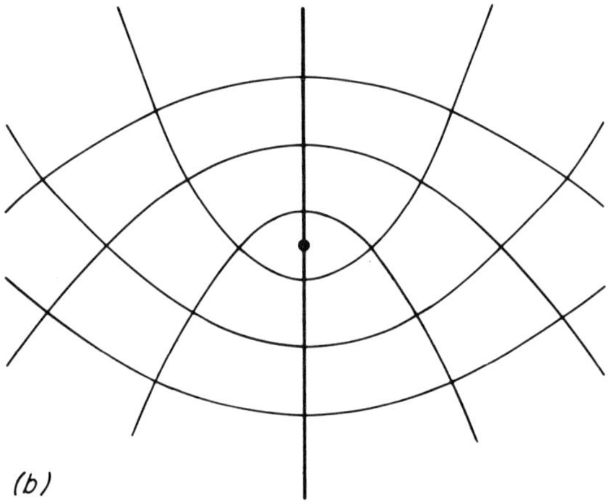

Figure 8.1-I. Two garden-variety oddities of orthogonal coordinate systems. (See text.)
Figure 8.1-Ia. A bicircular quartic coordinate system. (From Bôcher, 1894).
Figure 8.1-Ib. Parabolic coordinates.

see Schüepp, 1966). The vector fields of morphogenesis, presumed gradients of an underlying "morphogen" (scalar field or positional coordinate), may have singularities of a structure rich in two dimensions, even richer in three. This has been amply shown in Winfree (1980).

But scalars and vectors are only the zeroth and first levels of a hierarchy of complexity for the *tensor fields* which represent the interactions of multiple geometric parameters at a point. This essay explores implications for morphogenesis of the next level in this hierarchy, the *second-order symmetric tensor* representing two positive strains or other rates in two perpendicular directions. Just as scalars are indicated graphically by decimal numbers or contour lines, and vector fields by little line-elements, a symmetric tensor field is diagrammed as a little perpendicular cross at every point. Coordinate systems based on these tensors have two generic types of singularity quite different from those arising in the study of vector fields. Section 8.4 of this essay will derive the two typical forms of these singularities, and Section 8.5 will speculate on their role in morphogenetic explanation. These sections are preceded by some algebraic preliminaries which, beginning with the interplay of tensor and vector descriptions for deformations of triangles, ultimately call our attention to certain transformations of quadrilaterals that demonstrate the tensor singularities in which we are interested.

8.2 The Symmetric Tensor Field as a Coordinate Grid for Deformation

In the biophysical study of embryology several mechanisms are considered that might be described by symmetric tensors: changes in cell-cell contacts, relations between layers of cells, programs for orientation of cell division. When a tensor theory of morphogenesis finally confronts data, any of these might serve as actual encoded morphogen over a region. In this essay I will rely on the simplest embodiment of a symmetric tensor field, its role in describing deformation. Point by point, biological form-change may be described by two rates of linear extension in two directions at 90°.

Such a representation begins with the *homology map,* the "Cartesian transformation" formalized by D'Arcy Thompson (1961). A biological homology is a spatial or ontogenetic correspondence among definable structures or "parts"–separate bones, nerves, muscles, etc. In the context of mathematical morphology it becomes a homology *map,* a smooth geometrical *deformation* not of parts to parts but of points to points. For any choice of point or curve upon or inside any particular form, the homology map associates well-defined and biologically acceptable counterparts, the *homologues* of the point or curve, on all the other geometric forms in the data set.

A homology map may be drawn, after the fashion of Thompson, as the distortion of square graph paper into a more general configuration. The map relating a typical pair of four-cornered regions might look as in Figure 8.2-IIa. Of course, because data are supplied only at the corners, what is displayed must represent an interpolation formula, realistic, perhaps, but arbitrary. That shown here is the bilinear map described in Section 8.4; many others are possible (Bookstein, 1978).

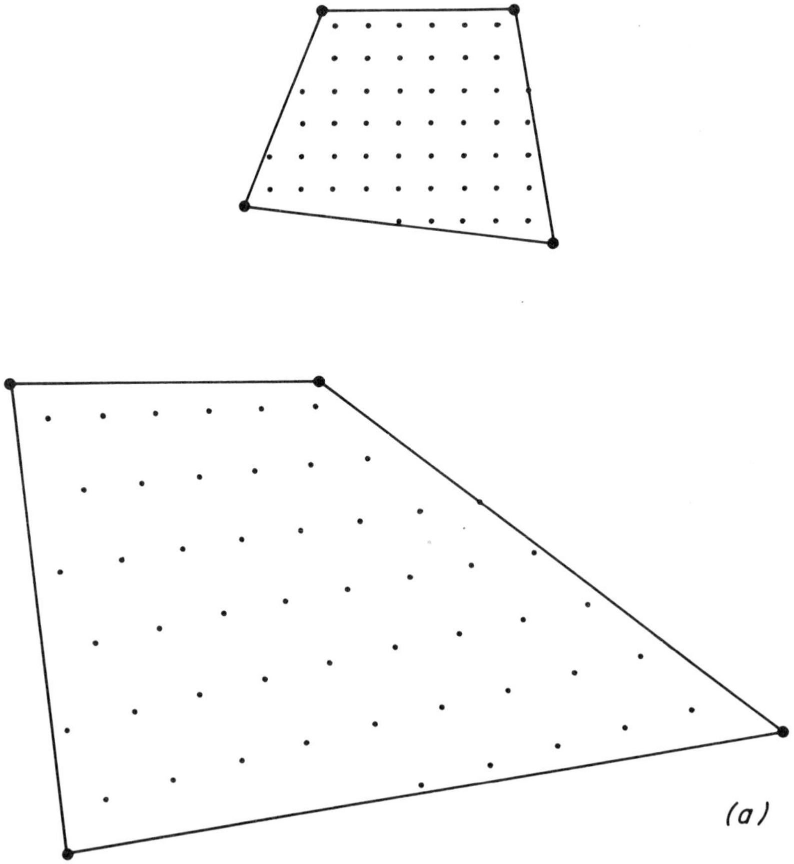

Figure 8.2-II. Representation of a homology map by a symmetric tensor field.
Figure 8.2-IIa. The "Cartesian grid," after Thompson (1961), smoothly interpolating the correspondence of corners around the boundary and throughout the interiors of the forms.

At almost every point interior to either quadrilateral, Figure 8.2-IIb, there is one cross of directions that is orthogonal in both forms according to the interpolated homology map. In the neighborhood of this cross, the transformation consists in independent expansion or contraction of each arm of the cross by separate ratios without change of angle between the arms. In one of these directions, rate of change of length is greatest, and in the other, least, of all directions across the triangle. These directions, the *principal axes* of the deformation, are at 90° in both forms. The rates of change of length along them, computed as dimensionless ratios, are called the *principal strains* or *principal extensions* at the point. The larger will be denoted d_1, the *major* principal strain; the smaller, d_2, the *minor*. The axes and principal strains together make up the *principal cross*, a visualization of the *symmetric strain tensor* familiar from continuum mechanics and engineering.

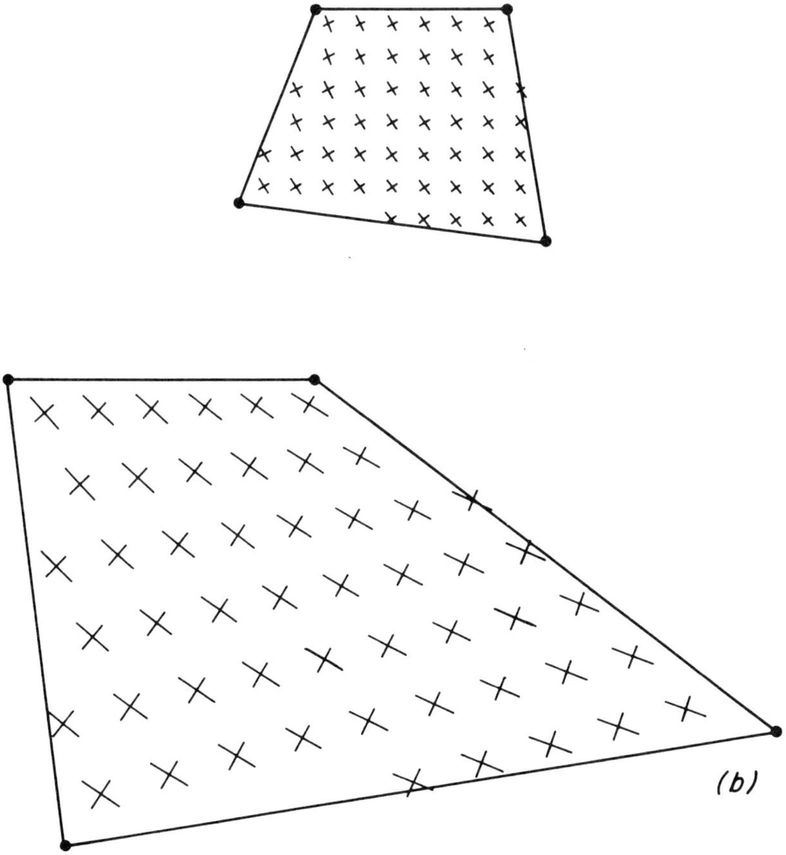

Figure 8.2-IIb. The tensor field for the mapping: its symmetrized affine derivative, point by point. The length of each cross arm is drawn proportional to the principal strain along it.

Viewed in this way, locally the deformation has a *size component* quantified by $d_1 + d_2$: the rate of change of area, unrelated to direction. Complementary to this is a *shape component,* or *anisotropy,* measured as $d_1 - d_2$, the difference of the principal strains. This quantity expresses change over the deformation in the ratio between measured distances along the two principal axes.

During the course of any change, there appear translations and rotations between larger parts of the form. The deformation model attributes these to the net influence of the local strains, their magnitudes and directions, summated across the form. The integration of these principal strains into extended curves, Figure 8.2-IIc, assembles the depiction of the tensor field into a recognizable pattern: a pair of curvilinear coordinate systems, each bearing a constant angle of 90° in *both* forms.

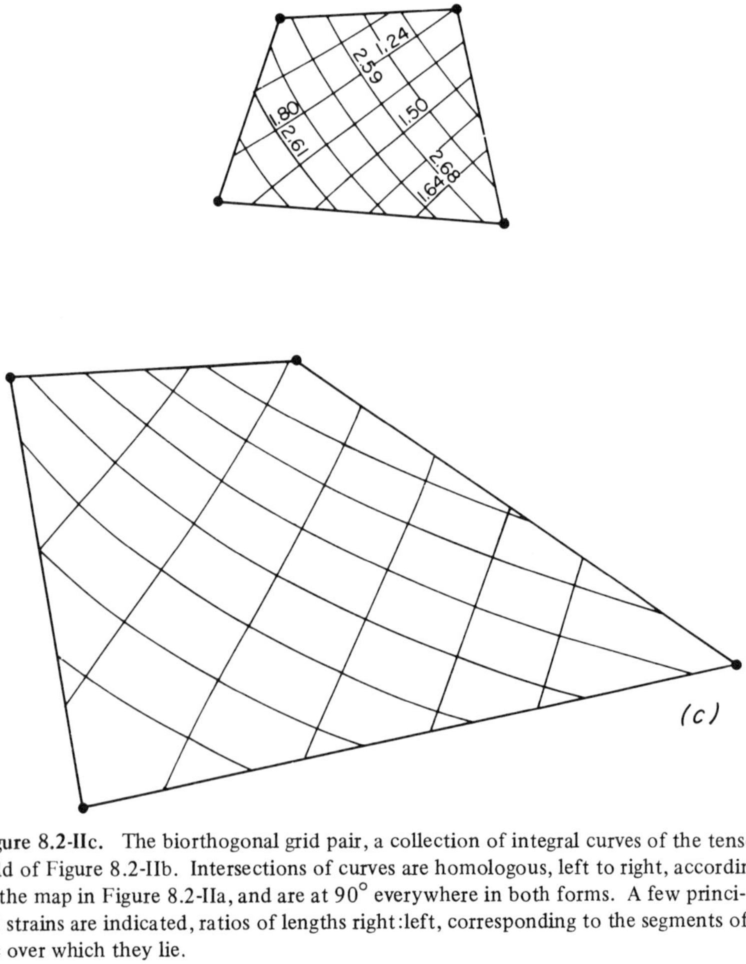

Figure 8.2-IIc. The biorthogonal grid pair, a collection of integral curves of the tensor field of Figure 8.2-IIb. Intersections of curves are homologous, left to right, according to the map in Figure 8.2-IIa, and are at 90° everywhere in both forms. A few principal strains are indicated, ratios of lengths right:left, corresponding to the segments of arc over which they lie.

These make up the *biorthogonal grid pair* representing the mapping. Upper and lower grids correspond, intersection for intersection, according to the interpolating map we are using in Figure 8.2-IIa. The selection of these curves is arbitrary, but their orientation is not: each curve is precisely parallel to one arm or the other of the principal cross at every point through which it passes. These curves are at 90° in both forms wherever they intersect; they constitute a coordinate system customized for this particular shape transformation. A relative rate of extension may be read from the grids as the ratio of the two homologous lengths (left and right) cut off along one curve by successive transects with the perpendicular system. The reader interested in further study of these coordinate systems and their application to empirically observed changes should consult Bookstein, 1978, or Bookstein et al., in preparation.

This essay takes the grids to represent real morphogens rather than, as Thompson averred, mere expression of a system of "forces" at a distance. In other words, the cell, organism, or tissue is assumed to generate its change of form actively, not passively, within the region under study. (For a related model, see Jacobson and Gordon, 1976.)

Shape Change of a Single Triangle

For the parametric treatment of deformation grids, we require some algebraic maneuvers relating to the apparent relative displacements of corners of the form over the course of a deformation. Explanations are simplest if we begin with a single cross representing a strain homogeneous in its little region. We may refer this to a set of three vertices, Figure 8.2-IIIa, because a change in shape of any configuration of three homologous points can be modeled as a homogeneous deformation of the interior of the triangle they define. The symmetric tensor representing this deformation everywhere in the interior may be visualized directly by its effect on a circle, Figure 8.2-IIIb.

By changing the scale of one triangle or the other, we are free to superimpose the pair of triangles on any pair of vertices, for instance, A and B. In doing so, Figure 8.2-IIIc, we have of course altered the size component of the deformation relating the triangles; but the directions of the principal axes and (if the size difference is small) the anisotropy of the principal strains are left nearly unchanged. We use these rescaled registrations to extract the principal directions and extensions for small changes of form by the geometric construction shown in Figure 8.2-IIId. Let the earlier and later positions C, C' of the third vertex be separated by the distance δ, and let h be the distance of vertex C from the baseline AB in the starting form. Then the strains in the principal directions are approximately

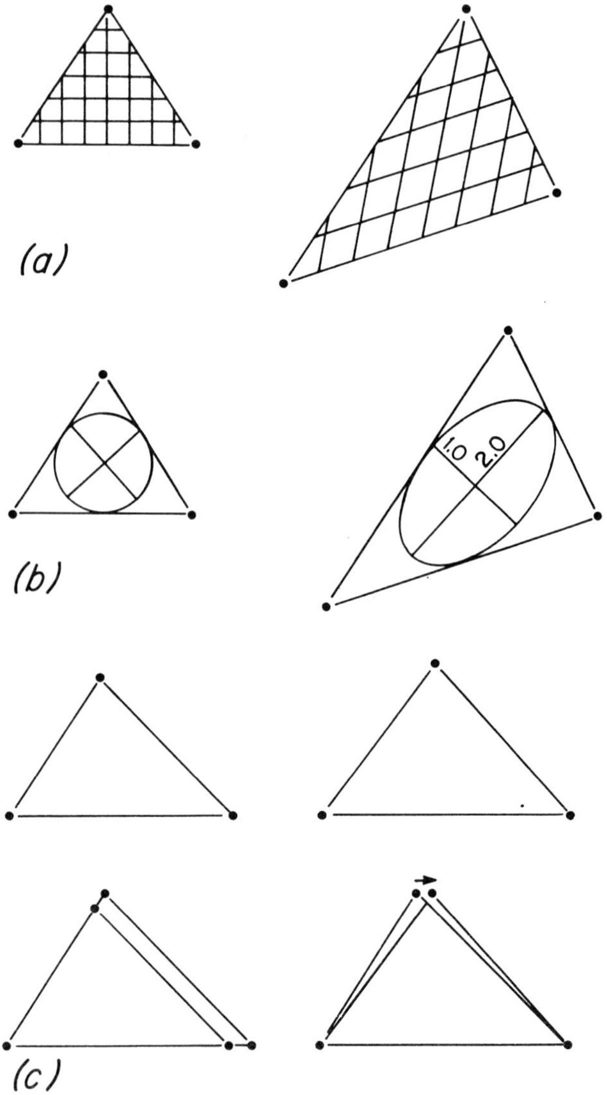

Figure 8.2-III. Expression of the deformation tensor by a displacement with respect to a baseline.
Figure 8.2-IIIa. A homogeneous deformation, after Thompson.
Figure 8.2-IIIb. The tensor representing this deformation, its axes the principal axes of the ellipse into which a circle is deformed.
Figure 8.2-IIIc. Superposition of two triangles upon a common baseline. A change of scale is required.

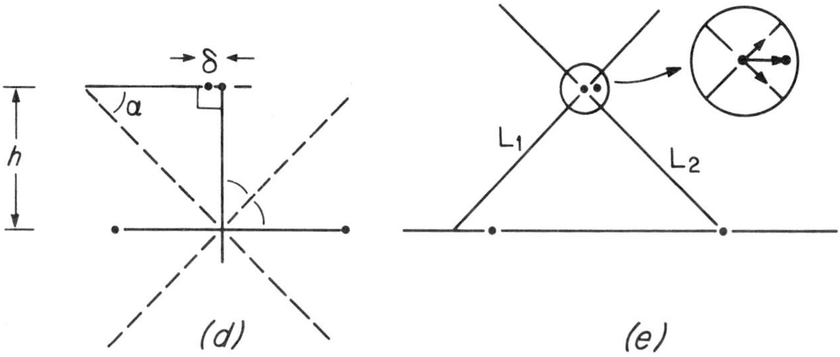

Figure 8.2-IIId. Construction of the deformation tensor (up to a scale factor) from the displacement vector.
Figure 8.2-IIIe. Construction of the displacement vector from the tensor.

$$d_1 = \frac{\delta}{h} \cos^2 \alpha ,$$

$$d_2 = -\frac{\delta}{h} \sin^2 \alpha ,$$

and so

$$d_1 - d_2 = \frac{\delta}{h} (\cos^2 \alpha + \sin^2 \alpha) = \frac{\delta}{h} ,$$

where α is the angle between the apparent displacement CC' and one principal axis, as drawn. We may also derive the relation

$$\alpha = 45° - \frac{\beta}{2} ,$$

where β is the angle between the path CC' and the fixed edge AB. This construction is explained in detail in Bookstein, 1984.

The apparent displacement of point C is made up of two vector components (Figure 8.2-IIIe): extension at a rate d_1' of the length L_1 along principal axis 1, and extension at a rate d_2' of the length L_2 along principal axis 2, where d_1', d_2' are the major and minor principal strains as scaled, that is, after the rate of extension along the baseline AB is subtracted. As long as we do not change the baseline, the ratio $L_1 : L_2$ is the same for all choices of the point C. If the deformation is uniform, then, it will displace any other point D to its image D′ by a vector DD′ parallel to CC′. The vector DD′ will be lengthened or shortened with respect to CC′ according to the distance from D to the same baseline. (The position of D along line AB is irrelevant, as whenever a linear transformation leaves two points of a line fixed it leaves all points of that line fixed.)

Interpreting the locations of points A, B, C as complex numbers, we can write this construction as the replacement of point C by [C − A/B − A]. (Cyclic permutation of A, B, C, referring point A, for instance, to the baseline BC rather than C to AB, replaces this ratio by one of the associated cross-ratios

$$\left[\frac{B-C}{A-C}\right] = 1 - \frac{1}{[C-A/B-A]}\,, \quad \left[\frac{A-B}{C-B}\right] = \frac{1}{1-[C-A/B-A]}\,.$$

We use this formalism to explore the consequence of perturbing two vertices for the displacement reported at a third. Suppose that landmarks A, B of Δ ABC have shifted by vectors $[x_A, y_A]$, $[x_B, y_B]$. Even if vertex C of Δ ABC is fixed in this coordinate system, Figure 8.2-IVa, we can interpret the shape component of the deformation of Δ ABC as a displacement imputed to C after registering on A and B as if they had been fixed instead. To simplify our algebra, let the reference coordinate system place point A at [0, 0] before its shift, and point B at [1, 0]. Point A has thus been shifted to point $A' = [x_A, y_A]$, and B to $B' = [1 + x_B, y_B]$. Assume C is fixed at the point [r, s] in this coordinate system; we wish to compute the displacement it undergoes when we register A′ at A and B′ at B. We will assume that the displacements AA′, BB′ are both small, so that we can ignore all second and higher powers of the x's and y's.

We have

$$\Delta C \sim \frac{\partial C}{\partial A}\Delta A + \frac{\partial C}{\partial B}\Delta B = \frac{[(C-B)\Delta A - (C-A)\Delta B\,]}{(B-A)^2}\,.$$

Reverting to a vector notation, Figure 8.2-IVb, the contribution of $[x_A, y_A]$, the variation at A, is scaled and rotated by the transformation taking [1, 0] to C − B = [r − 1, s]; the contribution of $[x_B, y_B]$, the variation at B, is scaled and rotated by the transformation taking [1, 0] to A − C = [−r, −s]. For small changes, the factor B − A is almost constant and can be dropped.

Then we have, for the coordinates of the normalized point C,

$$x_C = x_A(r-1) + y_A(-s) + x_B(-r) + y_B(s),$$
$$y_C = x_A(s) + y_A(r-1) + x_B(-s) + y_B(-r). \qquad (8.2\text{-}1)$$

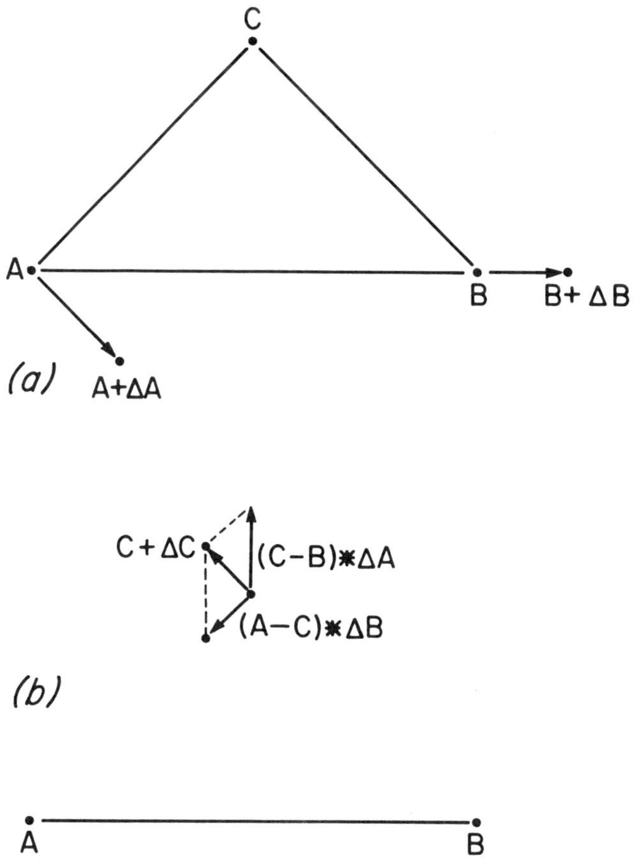

Figure 8.2-IV. Propagation of displacement.
Figure 8.2-IVa. Points A and B are shifted in a coordinate system holding C fixed.
Figure 8.2-IVb. The effective shift of C when we register on the shifted A and B by rotation and rescaling. In the figure, the asterisk represents complex multiplication.

8.3 Vector Components of Shape Change for a Quadrilateral

This section shows how to generalize from triangles to quadrilaterals the interpretation of shape change through displacement: Instead of displacing vertices one at a time, we take them two at a time.

Superposition on the Diagonal

It is convenient to begin with a starting configuation of landmarks that is exactly square. The diagonal of a square divides it into two triangles whose vertices are endpoints of the other diagonal. The shape change of the configuration of four points in a square may be represented by the simultaneous displacement of *both* these vertices when the other diagonal is fixed at both ends, as in Figure 8.3-Va. The pair of vectors E and G together represent the shape change with reference to the particular diagonal we have selected.

We have specified one diagonal as "fixed," responsible for the displacements of the remaining two vertices. We might as well have reversed the roles of the two diagonals, Figure 8.3-Vb. Equations (8.2-1) permit us to compute the vectors F, H which would have resulted had we done so. For the upper triangle, we use $[r, s] = [½, ½]$; for the lower, $[½, -½]$. The components $[x_F, y_F]$, $[x_H, y_H]$ of F and H are thereby set forth in terms of the components $[x_E, y_E]$, $[x_G, y_G]$ of E and G:

$$x_F = -\frac{x_E}{2} - \frac{y_E}{2} - \frac{x_G}{2} + \frac{y_G}{2},$$

$$y_F = \frac{x_E}{2} - \frac{y_E}{2} - \frac{x_G}{2} - \frac{y_G}{2},$$

$$x_H = -\frac{x_E}{2} + \frac{y_E}{2} - \frac{x_G}{2} - \frac{y_G}{2},$$

$$y_H = -\frac{x_E}{2} - \frac{y_E}{2} + \frac{x_G}{2} - \frac{y_G}{2}. \qquad (8.3\text{-}1)$$

In this way we represent the same shape change twice over, using eight coordinates (four two-vectors E, F, G, H) instead of the mere four coordinates mathematically required.

Some Identities

Manipulation of these equations leads to several interesting observations. We note first that, by virtue of cancellation of signs,

$$x_F + x_H = -(x_E + x_G),$$
$$y_F + y_H = -(y_E + y_G).$$

In other words,

$$E + F + G + H = 0.$$

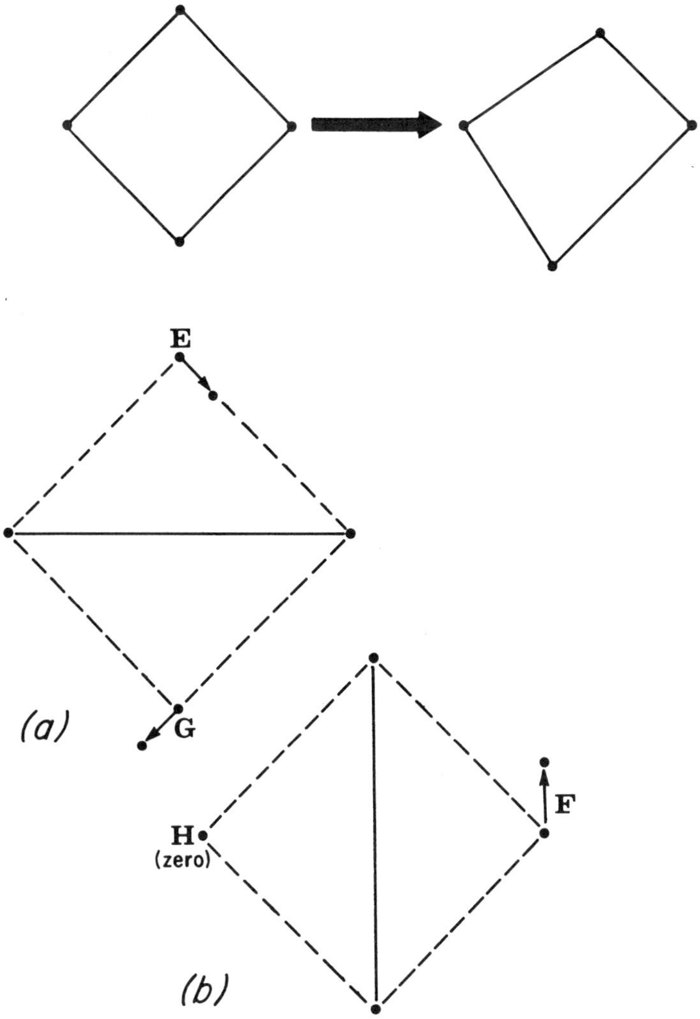

Figure 8.3-V. Displacements of corners of quadrilaterals.
Figure 8.3-Va. Diagonal FH fixed, corners E and G displaced.
Figure 8.3-Vb. Diagonal EG fixed, corners F and H displaced.

That is, the vector sum of the four imputed displacements, two each corresponding to the fixing of each of the two diagonals, is exactly zero. The diagram of four vectors thus bears no net shift component (Figure 8.3-Vc).

We may also verify, by a different cancellation, that

$$x_F - x_H = y_G - y_E,$$

$$y_F - y_H = x_E - x_G.$$

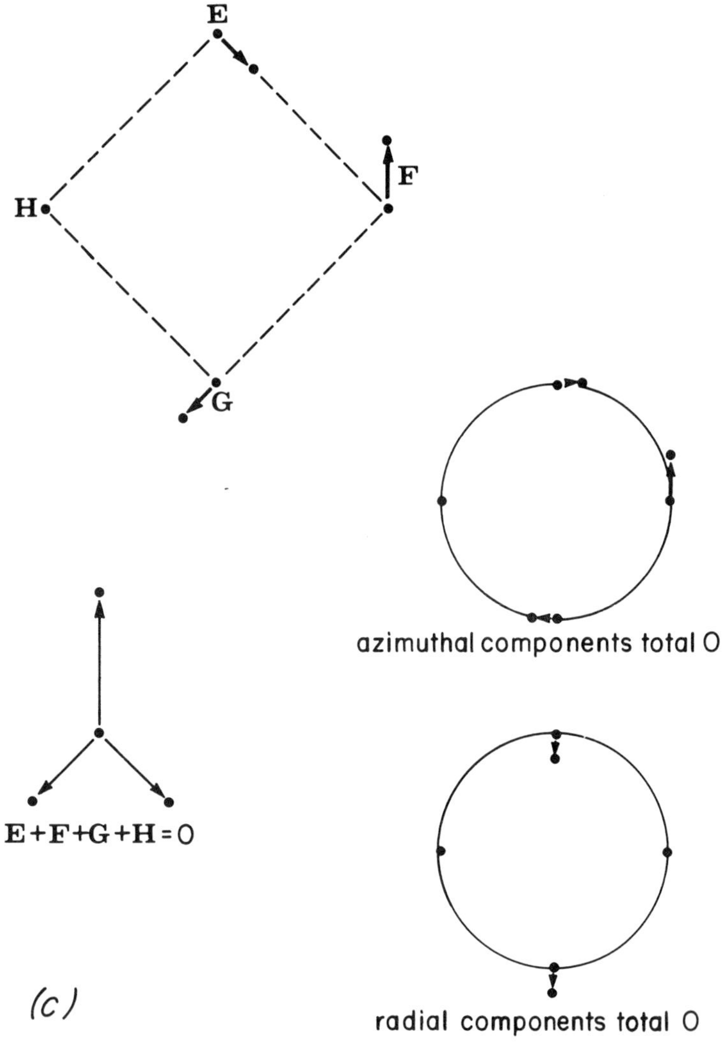

Figure 8.3-Vc. Identities involving the four corner displacements.

This may be rewritten in terms of Cartesian coordinates at 45° to the set we are presently using:

$$(y_E + x_E) + (x_F - y_F) + (-y_G - x_G) + (-x_H + y_H) = 0\,,$$

$$(y_E - x_E) + (x_F + y_F) + (-y_G + x_G) + (-x_H - y_H) = 0\,.$$

That is, the "net component" of change representing expansion or contraction about the center of the square is exactly zero, and the "net component" representing rotation about the center is likewise exactly zero.

Two Components

There are two other symmetries to be extracted from this coupled set of vector equations. Suppose, for instance, that $E = G = [x_{EG}, y_{EG}]$, as in Figure 8.3-VIa. By substituting in equations (8.3-1), we find

$$x_F = x_H = -x_{EG}\,,$$

$$y_F = y_H = -y_{EG}\,.$$

That is, if two opposite vertices are identically translated with respect to the other diagonal, then the endpoints of that diagonal show an equal and opposite translation with respect to the diagonal joining the first pair. The situation is therefore a motion of either diagonal with respect to the other without change of angle. We will call this case *pure translation* or the *purely inhomogeneous transformation*; it will occupy our attention throughout the remainder of this essay.

Instead of specifying $E = G$, we might instead consider the case $E = -G$: opposite corners of the square displayed equally and oppositely after registration upon the other pair of vertices. Substituting $x_E = -x_G$, $y_E = -y_G$ in equations (8.3-1), we obtain

$$x_F = -y_E\,,$$

$$y_F = x_E\,,$$

$$x_H = y_E\,,$$

$$y_H = -x_E\,.$$

Hence F is a 90° clockwise rotation of E (or 90° counterclockwise rotation of G), and H, which equals $-F$, is a 90° clockwise rotation of G (or 90° counterclockwise rotation of E), Figure 8.3-VIb.

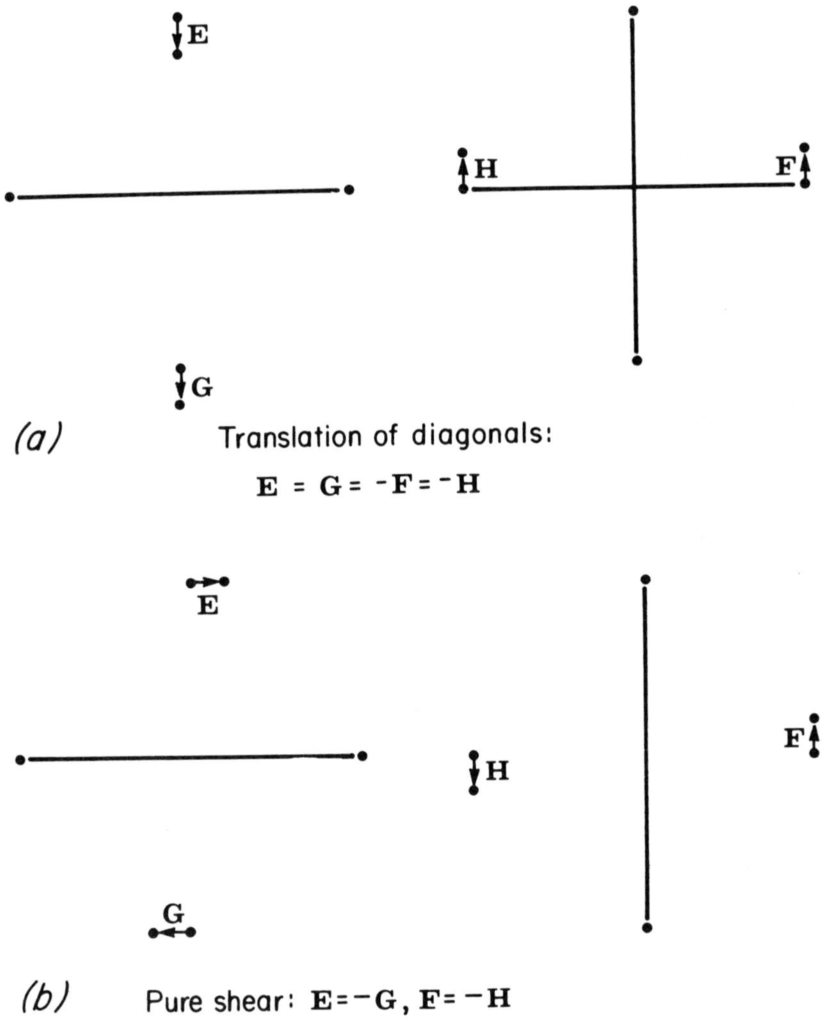

Figure 8.3-VI. The two components of shape displacement for squares.
Figure 8.3-VIa. The purely inhomogeneous transformation, translation without rotation of one diagonal with respect to the other.
Figure 8.3-VIb. Pure homogeneous shear.

In the displacements representing these deformations, the origins of vectors H and F (or E and G) are at equal and opposite distances from their common diagonal of reference, and the displacements imputed to them by the deformation are likewise equal and opposite. It follows (recall Figure 8.2-IIIe) that the deformations of the square with $E + G = F + H = 0$ represent *pure* (or *uniform,* or *homogeneous*) *shears,* those transformations which are the same on both sides of (either) diagonal. The characterizations of F and H as rotated versions of E or G merely express the effect of a 90° change in baseline direction without change of baseline length.

Now for any vectors E, G whatever, it is the case that

$$E = \tfrac{1}{2}(E + G) + \tfrac{1}{2}(E - G),$$
$$G = \tfrac{1}{2}(E + G) - \tfrac{1}{2}(E - G).$$

In this way we decompose *any* small deformation of the square into the composite of two deformations, one a pure shear and the other a pure translation. If one corner is displaced by E and other by G, the pure shear component of the pair displaces one corner by $\tfrac{1}{2}(E - G)$, the opposite corner by the opposite translation; the pure translation component displaces each of the pair of opposite corners by $\tfrac{1}{2}(E + G)$. By the identities relating E, G and F, H that were established above, the magnitudes of these two components are invariant under change of the choice of diagonal.

The General Starting Quadrilateral

Up to change of size, we have reduced all deformations of a square to four specific components: pure shear, pure translation, pure shift (which is identically zero), and pure rotation/expansion (also identically zero). The deformation of the general quadrilateral manifests the same four components. Let us, for instance, compute the picture of a pure shear for the quadrilateral of Figure 8.3-VII (top). Suppose the northwest corner is displaced by a vector E in relation to the northeast-southwest diagonal. The vector expressing the effect of this same shear upon the southeast corner is antiparallel to E with a length given by the ratio of the signed distances of these two corners to the other diagonal. There results the vector G as drawn. Were we to register this same shear upon the northwest-southeast diagonal instead, we would have rotated the baseline clockwise by 60° (the angle between the diagonals), and thus rotated the apparent displacement vector by 60° counterclockwise with respect to the baseline. The vectors F, H representing displacements of the other two corners are aligned in this new direction, and their lengths are proportional to the distance of *these* corners from their baseline diagonal. We may verify all this by noting that the two quadrilaterals drawn in dashes in Figure 8.3-VII are exactly similar.

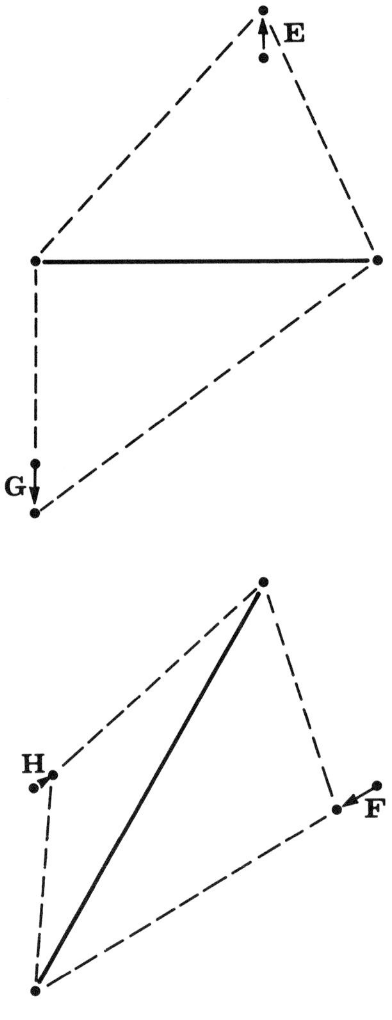

Figure 8.3-VII. Sketch of a pure shear component for a general starting quadrilateral.

8.4 The Purely Inhomogeneous Transformation as a Mapping

The discussion of triangular configurations in Section 8.2 showed how to pass between two modes of description: change as a deformation, described by a (symmetric) tensor, and change as a relative displacement of landmarks, described by a (baseline-dependent) vector. This equivalence was extended to the homogeneous component of quadrilateral shape transformations. Up to scale change, for instance, a homogeneous shear of the square might be interpreted as equal and opposite vectors of displacement at the endpoints of just one diagonal.

The case of the purely inhomogeneous component, Figure 8.3-VIa, is rather less familiar. Section 8.3 described it only in terms of a displacement: translation of the diagonals with respect to each other, without rotation. For explanation of biological form-change—morphogenetic theory or morphometrics—we must pass to the transformational point of view and model this pure translation of diagonals as generated by a smooth deformation relating the interiors.

Consider, then, the task of modeling the mapping of the square into the kite, Figure 8.4-VIIIa. I will refer to the left end of the square, the one apparently increasing in size, as the *positive pole*, and the right end, decreasing in size, as the *negative pole*. This diagonal will be called the *polar axis*; the other, the *nonpolar axis*. Smooth maps of the quadrilateral to the kite ought to express the aspect of "translation" in a manner analogous to the little vectors E, G of Figure 8.3-VIa: relative extension or compression of length along the polar axis. Lengths measured along the vertical diagonal will be, on the average, unchanged.

The deformations to be considered are, like the square and kite forms themselves, symmetric around the polar axis. Everywhere on this axis of symmetry, the principal cross describing the deformation by its derivative must be aligned with horizontal and vertical, the polar axis itself and its perpendicular. At the right, the horizontal compression is the minor principal strain, dominated by the vertical quasi-stasis; at the left, the horizontal expansion is the major principal strain. This major strain has thus rotated by 90° as we pass 180° around the form (and by 180° as we pass 360° around the form—in analogy to the winding number of a vector field, this feature implies a singularity inside).

The rotation of this principal direction by 90° may be clockwise or counterclockwise around the top of the form, as sketched in Figure 8.4-VIIIb. In one case, which will be typified by the projection mapping, the directions of greater principal strain at the top and bottom corners of the quadrilateral pass approximately through the negative pole of the form. In the other case, typified by the bilinear mapping, the positive dilatation lingers upon the positive pole of the form.

Analysis of maps from either class involves a single parameter a characterizing the magnitude of the relative translation of endpoints embodying the deformation. In terms of the square starting form, a may be imagined a convenient multiple of the shift in either endpoint of either diagonal, scaled to the edge-length of the square. For both the classes of maps used as standard forms below, there is only one distortion and one biorthogonal grid for the entire family of transformations. The starting square cuts out a greater- or lesser-sized area, depending on the value of a, from this single grid.

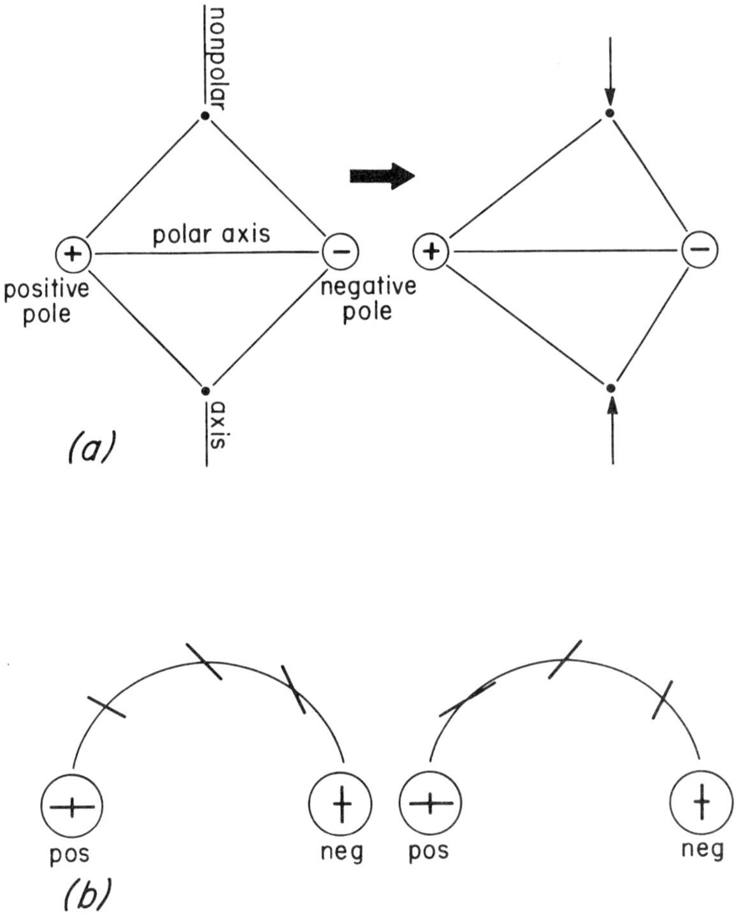

Figure 8.4-VIII. Mapping the square into the kite.
Figure 8.4-VIIIa. Symmetries, poles, and axes mentioned in the text.
Figure 8.4-VIIIb. There are two possible senses of rotation of the major principal strain over the path from the positive to the negative pole.

Projection

The reader will recall that projections, a familiar class of mappings, take all straight lines of the plane into straight lines, are linear in the extended plane of homogeneous coordinates, etc. The equations for projecting the square onto the kite have a satisfactory symmetry when the square is positioned and scaled as in Figure 8.4-IXa, with diagonals of length 2a centered at the point [1, 0] and aligned horizontally and vertically. The kite shares the vertical (nonpolar) diagonal through [1, a] and [1, -a] — these points are

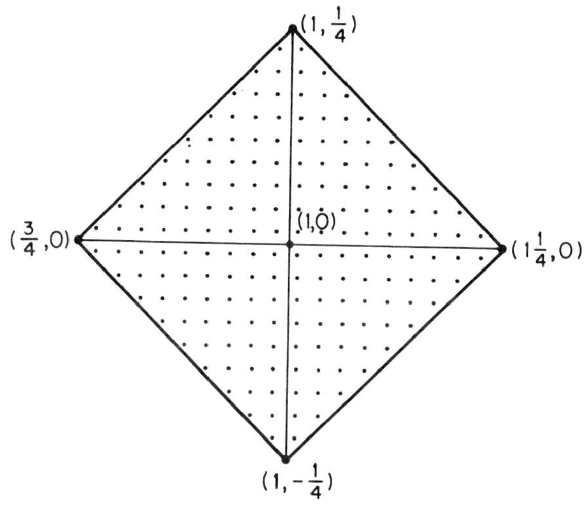

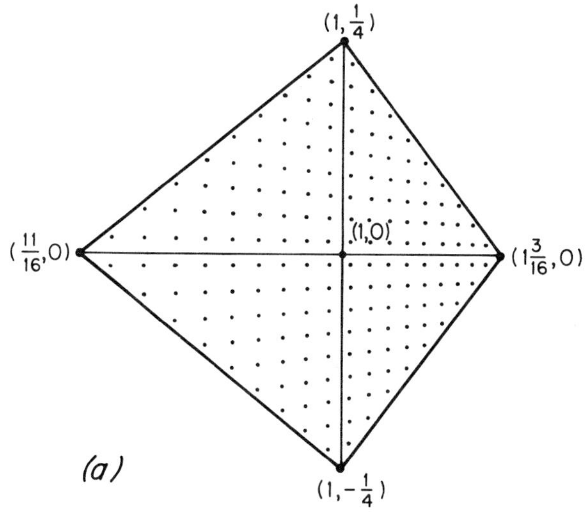

Figure 8.4-IX. Projection of the square onto the kite.
Figure 8.4-IXa. Placement of forms for $a = .25$, and Cartesian grid of the mapping.

fixed by the transformation—but its horizontal diagonal is shifted by the distance a^2, the fraction $a/2$ of diagonal length, at either end.

The (unique) projection which maps the four corners of this square onto the corresponding four corners of the kite has the equation

$$[x, y] \rightarrow \left[\frac{a^2 - 1}{x} + 2 - a^2,\ \frac{y}{x}\right] \tag{8.4-1}$$

–one easily verifies that this mapping, a linear fractional transformation, leaves $[1, \pm a]$ fixed and takes $[1 \pm a, 0]$ to $[1 \pm a - a^2, 0]$. As drawn in Figure 8.4-IXa, the map looks like an example of familiar Renaissance perspective.

The affine derivative of this projection at any point $[x, y]$ is given by the matrix

$$1/x^2 \begin{bmatrix} 1-a^2 & 0 \\ -y & x \end{bmatrix}$$

(operating from the left on column vectors). At the positive pole $[1-a, 0]$ this matrix becomes $\mathrm{diag}[\frac{1+a}{1-a}, \frac{1}{1-a}]$ –expansion in all directions, but primarily horizontally. At the negative pole $[1+a, 0]$ it becomes $\mathrm{diag}[\frac{1-a}{1+a}, \frac{1}{1+a}]$ –compression in all directions, but primarily horizontally. The map thus exemplifies the qualitative problem which interests us, the rotation of the dominant principal strain by $90°$ as we pass $180°$ around the form, from the positive to the negative pole.

For applications to morphogenetic explanation we will find most interesting the behavior of this mapping near its *conformal point,* the point at which the derivative is mere combination of rotation and change of scale. At such points, even though the principal strain values are well-defined, they are equal; the principal axes do not exist, so that a crucial geometric parameter is undefined. A plane mapping has a conformal point wherever its affine derivative takes the form $[\begin{smallmatrix} r & -s \\ s & r \end{smallmatrix}]$, skew-symmetric with diagonal terms equal. For the projection, this is the case only at the point $[1-a^2, 0]$.

The projection (8.4-1) of square into kite was derived by a combination of translation, reflection, and scaling from the canonical involutory projection $[x, y, z] \rightarrow [z, y, x]$, or, in Cartesian coordinates,

$$[x, y] \rightarrow \left[\frac{1}{x}, \frac{y}{x}\right].$$

In this version, $[1, 0]$ is an *anticonformal point* at which angles are changed in sense but not in magnitude; the conformal point is at $[-1, 0]$.

It may be verified by direct substitution that this map preserves the grid of confocal conics

$$\frac{x^2}{b} + \frac{y^2}{1+b} = 1$$

shown in Figure 8.4-IXb. This grid of conics is at $90°$; therefore it must itself be the biorthogonal grid pair (*both* coordinate meshes!) for the mapping. The action of the map is simply to interchange, for all $k > 1$, the ellipse through [k, 0] with the hyperbola through [1/k, 0]. The points on the y-axis of the ellipse go to the points at infinity on the matched hyperbola.

The line $x = 1$, the nonpolar axis, is pointwise fixed by the transformation; in particular, the center [1, 0] of the square is left fixed. The extension ratio along the polar axis is graded as x^{-2}. Nowhere except on the polar axis itself does either arm of any biorthogonal cross point to the conformal point. In particular, at the ends of the nonpolar axis, the principal strains are nearly at 45° to their orientations at the positive or negative poles. As in the left sketch of Figure 8.4-VIIIb, the direction corresponding to the more positive strain points approximately to the negative pole, and vice versa.

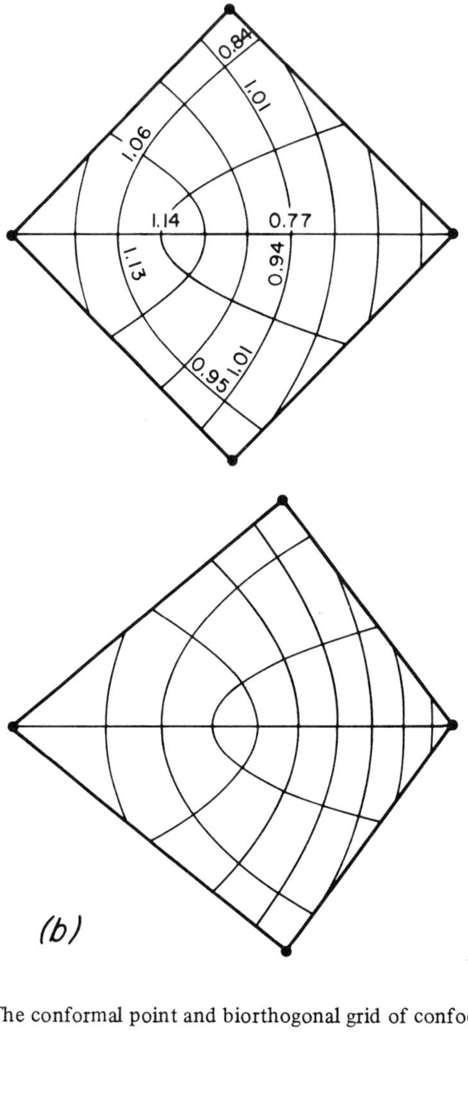

Figure 8.4-IXb. The conformal point and biorthogonal grid of confocal conics about it.

It can be shown that the general projection of quadrilateral onto quadrilateral has exactly the same biorthogonal grid pair as this highly symmetrical example. In other words, up to rotation and change of scale, every projection of one quadrilateral onto another can be observed as the effect of this specific involution on some configuration of four starting corners. Every projection, whether the configuration of displacement it represents is nearly pure shear, pure translation, or a combination, has a single conformal point and another (biologically irrelevant) anticonformal point; and its grids are confocal conics about that pair of points. (Because projection leaves straight lines straight, all finite angles measured at the conformal point are unchanged, and all measured at the anticonformal point are merely reversed in sense; in this role the point has been useful to photogrammetrists for some time.) If the conformal point lies within the interiors of the forms, the grids have the aspect of this focus. If it lies outside, then everywhere one direction of gradient dominates, and the grid may be seen as a smooth warping of a rectangle, Figure 8.2-IIc. These cases are located on the canonical confocal plot as in Figure 8.4-IXc.

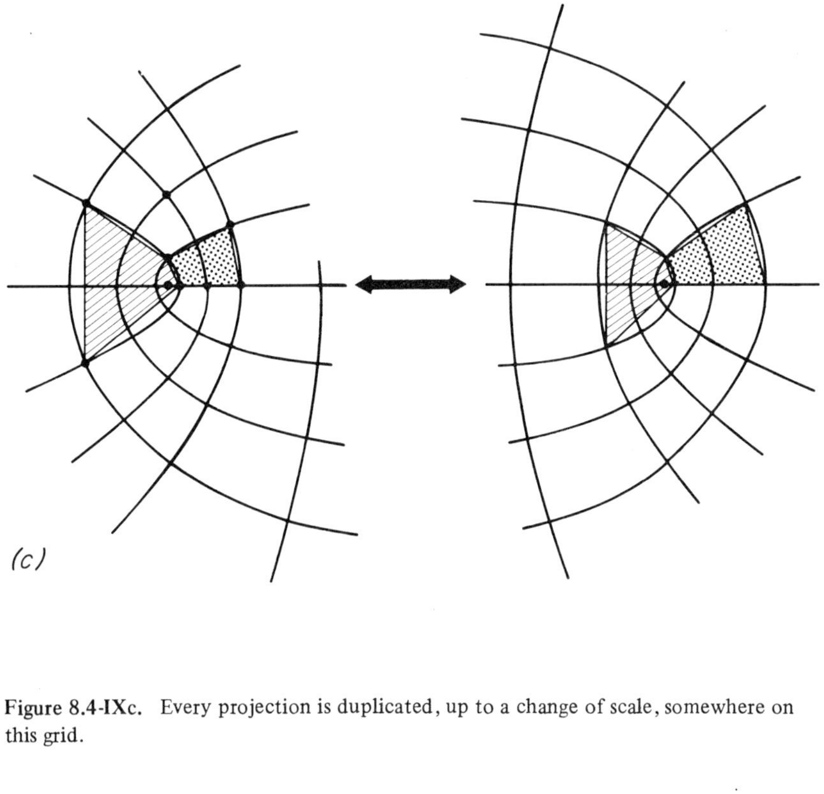

Figure 8.4-IXc. Every projection is duplicated, up to a change of scale, somewhere on this grid.

The Bilinear Mapping

Projection takes straight lines into straight lines but is highly nonlinear in its treatment of length along most lines. The homogeneous shear we seek to escape likewise takes straight lines to straight lines, and furthermore is linear along every line of the plane—which is why it lacks interest for the morphogeneticist. A compromise between these purposes is needed: a mapping that is linear on the edges of the quadrilateral, but that nevertheless allows for regional differences in directionality, in particular, for conformal points where the affine derivative is isotropic.

One class of maps satisfying these requirements is the *bilinear* family. While analytically almost as simple as the projection, they are nevertheless fairly unfamiliar to the applied geometer, and so I shall explain their algebraic and geometric properties in some detail.

The fundamental appearance of the bilinear mapping relating a pair of quadrilaterals is as in Figure 8.4-Xa. While projection deals with the quadrilateral as a set of four points in any order, bilinear mapping requires that one select four edges from the six possibilities. The transformation will be linear on these four edges, but not along the other two, those serving as the diagonals of the construction. Choose any point [x, y] interior to this quadrilateral, and consider the set of all straight lines through it. Each line divides each edge of the quadrilateral in some ratio. We are interested in the lines through [x, y] which divide one pair of opposite edges in the *same* ratio. For the general quadrilateral (no edges parallel) this is a quadratic criterion which results in a single pair of lines through [x, y], lying on the point as drawn. For squares, the lines we seek are just the lines through [x, y] parallel to the sides.

The image of the point [x, y] under bilinear mapping is the intersection in the other form of the lines that divide the homologous edges in the same pair of ratios. In Figure 8.4-Xa, for instance, line AA′ divides edges P_1P_2 and P_3P_4 in the ratio 1:2, and line BB′ divides edges P_1P_3 and P_2P_4 in the ratio 1:3. We find the points which divide the edges of the opposite form in the same ratio: C (resp. C′) divides Q_1Q_2 (resp. Q_3Q_4) in the ratio 1:2 and D (resp. D′) divides Q_1Q_3 (resp. Q_2Q_4) in the ratio 1:3. Then the point [x, y] at the intersection of AA′ and BB′ is mapped to the point at the intersection of CC′ and DD′.

The analytic geometry of this construction becomes clearer when the quadrilateral $P_1P_2P_3P_4$ that we selected (that is, the set of four edges chosen out of six) is circumscribed about a parabola, Figure 8.4-Xb. The quadrilateral $Q_1Q_2Q_3Q_4$ may likewise be circumscribed about a parabola of its own. Because all parabolas are similar, up to a change of scale we may treat the polygon of P's and the polygon of Q's as circumscribed about the same parabola.

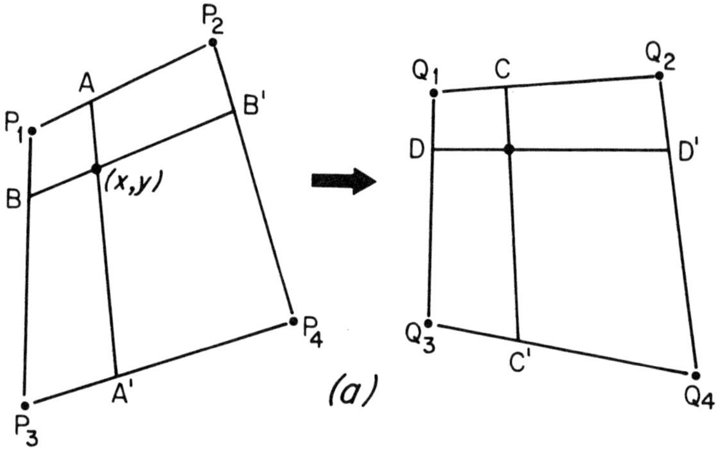

Figure 8.4-X. The bilinear mapping.
Figure 8.4-Xa. Through the point [x, y] pass two lines dividing opposite edges of the quadrilateral in equal ratios. The image of [x, y] is the intersection in the other quadrilateral of joins of the points dividing *their* edges in the same ratios.

It is an old theorem that for any three tangents to a parabola, every other tangent is cut by the three in the same affine ratio. It follows that the lines we seek through any point [x, y] are simply the unique pair of tangents to the parabola through [x, y]. If we assign each edge of the quadrilateral a coordinate by its intersection with any fixed tangent to the parabola (for example, the vertex tangent, Figure 8.4-Xc), then the bilinear map is linear, separately, on the coordinates of the two tangents through [x, y].

For starting forms which are exactly square, a different characterization of this map is available, one which is of some use in computer graphics: the bilinear map is the simplest *blending function* for the square. Suppose corners [0, 0], [1, 0], [0, 1], [1, 1] of a square are mapped into points Q_{00}, Q_{10}, Q_{01}, Q_{11} respectively. The vertical line through a point [x, y] inside the square is the connection of the point $xQ_{11} + (1-x)Q_{01}$ on the top edge with the point $xQ_{10} + (1-x)Q_{00}$ on the bottom edge. The horizontal line through [x, y] intersects this join at the point a fraction y of the way from bottom to top, the point

$$(1-y)[xQ_{10} + (1-x)Q_{00}] + y[xQ_{11} + (1-x)Q_{01}] .$$

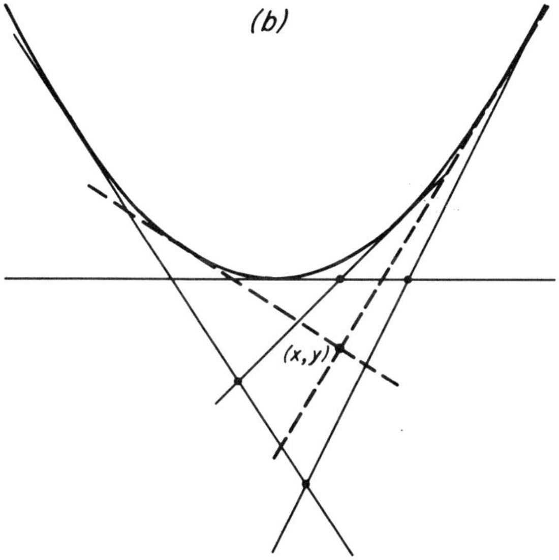

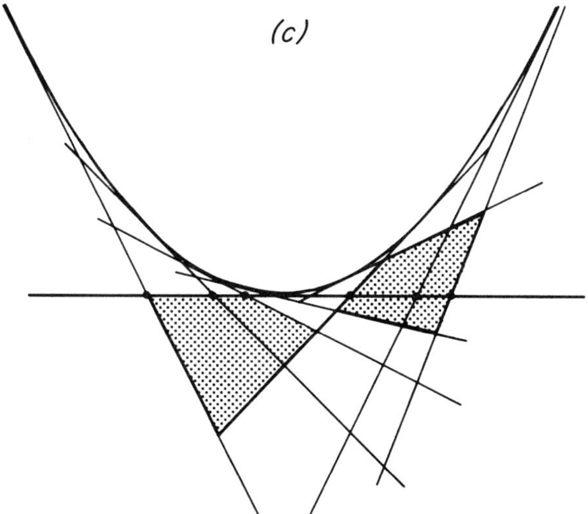

Figure 8.4-Xb. Quadrilaterals in general position can be circumscribed about a parabola. The lines through [x, y] in Figure 8.4-Xa are the two tangents through [x, y] to this parabola.

Figure 8.4-Xc. The bilinear map is linear, separately, on the coordinates of the intersections of these tangents with any fixed tangent to the parabola.

Expanding, we see that the map sends $[x, y]$ to the point

$$(1-x)(1-y)Q_{00} + x(1-y)Q_{10} + (1-x)yQ_{01} + xyQ_{11}$$

–a weighted average of the four image points corresponding to the four corners of the square, each weight given by the product of the distances from $[x, y]$ to the pair of grid lines through the diagonally opposite corner of the square.

Our particular concern is the bilinear map corresponding to the purely inhomogeneous transformation (pure translation), square onto kite. The algebra of this map is simplest under a standardization somewhat different from that of Figure 8.4-IXa, which simplified the algebra of projection. Regardless of the parameter a of the kite, place the corners of the starting square at $[\pm 1, \pm 1]$. The image quadrilateral for transformations having shift parameter a shifts the main diagonal (the polar axis) by the vector $[-2a, -2a]$, a fraction a of its length. Hence, while the ends of the nonpolar axis, $[-1, 1]$ and $[1, -1]$, are left fixed, the point $[-1, -1]$ is mapped to $[-1-2a, -1-2a]$, and likewise $[1, 1]$ to $[1-2a, 1-2a]$. The bilinear mapping for this pair of quadrilaterals is shown in Figure 8.4-XIa; its equation is

$$[x, y] \rightarrow [x, y] - a[1+xy, 1+xy] \ . \tag{8.4-2}$$

The derivative of this map may be found by direct calculation or by bilinear mixing of its derivatives at the four corners; it may be written in matrix form as

$$\begin{bmatrix} 1-ay & -ax \\ -ay & 1-ax \end{bmatrix} .$$

The bilinear map is potentially biologically meaningful wherever the determinant $1 - a(x+y)$ of this matrix is positive. Whenever a is less than 0.5, this will be the case throughout the square. For $a = 0.5$, the two edges $Q_{11}Q_{01}$, $Q_{11}Q_{10}$ of the kite lie on the same line (see Figure 8.4-XIe), so that the image Q_{11} of the point $[1, 1]$ is undefined. The projection map likewise had a line of singularities, the real axis; while for projection areas near the line "blow up," the bilinear map squashes them flat.

At $[-1, -1]$ the derivative of the bilinear map is $\left[\begin{smallmatrix} 1+a & a \\ a & 1+a \end{smallmatrix}\right]$, representing an *increase* in size in all directions except the nonpolar axis; then the polar axis bears the major principal strain. At $[1, 1]$ the derivative $\left[\begin{smallmatrix} 1-a & -a \\ -a & 1-a \end{smallmatrix}\right]$ represents a *decrease* of size in all directions except the nonpolar axis; the polar axis here bears the minor principal strain. Thus the qualitative behavior of this map, like that of the standardized projection, suits the import of the purely inhomogeneous transformation: the major principal strain rotates by 90° as we travel halfway around the starting square.

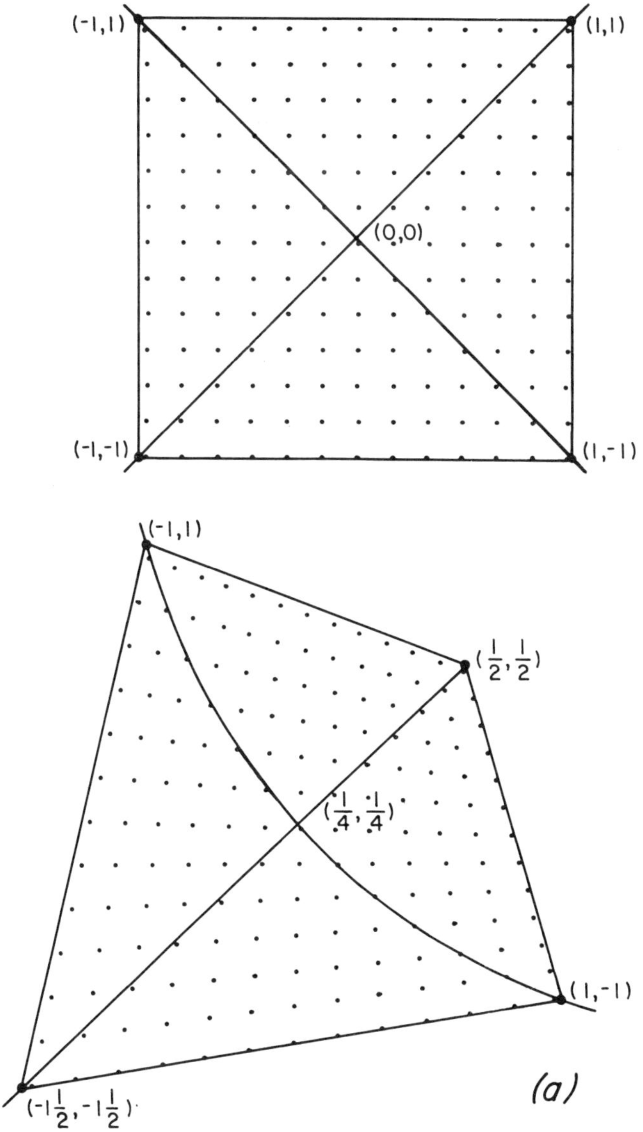

Figure 8.4-XI. Bilinear mapping of the square onto the kite.
Figure 8.4-XIa. Placement of forms for a = .25, and Cartesian grid of the mapping.

Notice that along the polar axis, axis of symmetry of the scene, the bilinear mapping bears a different nonlinearity from that of the projection. The image of the point $[x, x]$ is $[x - a - ax^2, x - a - ax^2]$; the derivative of this map is linearly graded from $1 + 2a$ to $1 - 2a$ along the polar axis, and the derivative perpendicular to that axis is constant at 1. The image of the center $[0, 0]$ of the square is not the intersection of the diagonals of the kite, but rather the point $[-a, -a]$ displaced from $[0, 0]$ by half as much as the endpoints of the major diagonal are displaced from *their* starting loci. Notice, too, that the minor diagonal is not left straight under this mapping; it is mapped (cf. Figure 8.4-XIa) into the arc of a parabola.

The representation of this mapping by the symmetric tensor field corresponding to its derivative is shown in Figure 8.4-XIb. As we pass over the upper half of the square from $[1, 1]$ to $[-1, -1]$, the major principal strain for the bilinear map rotates in a sense opposite to that for the projection. At the ends of the nonpolar axis, the principal strains are again at about $45°$ to those at the endpoints of the polar axis; but the greater principal strain points toward the positive, not the negative, pole. This is associated, naturally, with the linearity the bilinear map enforces on each edge of the quadrilateral separately.

Grids Near the Conformal Point of the Bilinear Map

There is a conformal point located where $1 - ay = 1 - ax$, $ax = -ay$, i.e. $[0, 0]$. The behavior of the biorthogonal grid system around the conformal point may be classified (Bookstein, 1978, pp. 103-107) by studying the lines through the conformal point that are themselves included in the biorthogonal grid. For projections, there is only one such line, the polar axis itself. For the bilinear map, we can see that there must be two others. Figure 8.4-XIc shows a few positions of the principal cross along a circular path from positive pole to negative pole. At the positive pole, the major strain axis passes through the conformal point. As both strains are rotating clockwise along this counterclockwise circuit, soon the *minor* strain axis must sweep through the conformal point, and shortly afterward the *major* strain must pass through that point again; and, finally, the minor strain again—but by then we have arrived at the negative pole of the form.

This intuitive deduction can be verified analytically without much difficulty. One of the principal strains will pass through the conformal point whenever the affine derivative map at $[x, y]$ leaves orthogonal the line-elements $[x, y]$ and $[-y, x]$ upon the square. The images of these directions upon the kite are

$$\begin{bmatrix} 1 - ay & -ax \\ -ay & 1 - ax \end{bmatrix} \begin{bmatrix} x \\ y \end{bmatrix} = \begin{bmatrix} x - 2axy \\ y - 2axy \end{bmatrix},$$

$$\begin{bmatrix} 1-ay & -ax \\ -ay & 1-ax \end{bmatrix} \begin{bmatrix} -y \\ x \end{bmatrix} = \begin{bmatrix} -y - ax^2 + ay^2 \\ x - ax^2 + ay^2 \end{bmatrix} .$$

The right-hand sides are perpendicular whenever their dot product vanishes: there results a single equation in x and y,

$$(4ax + 1)y^3 + (3a)y^2 - (4ax^3 + 3x^2)y - x^3 = 0 .$$

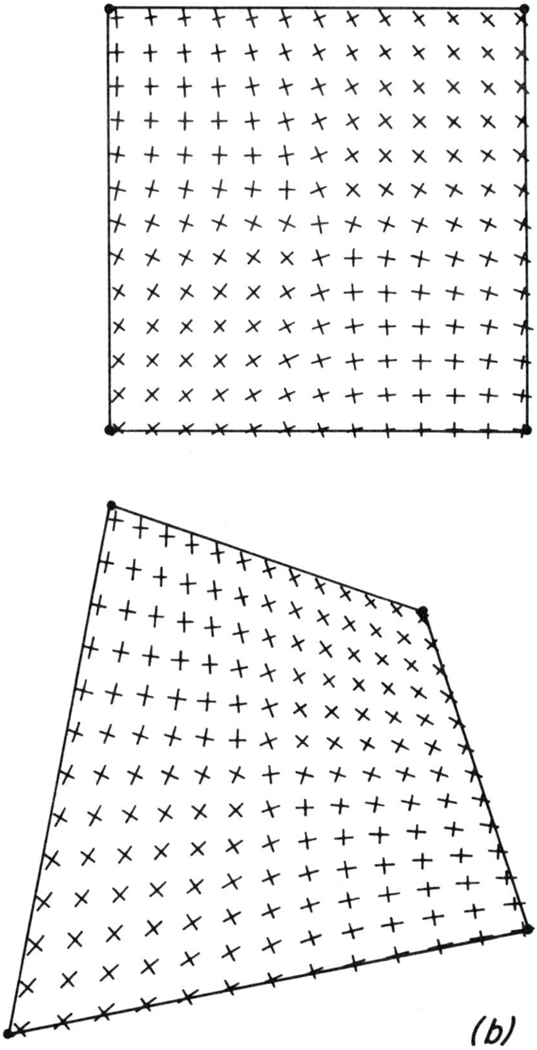

Figure 8.4-XIb. The symmetric tensor field representing this map by its affine derivative.

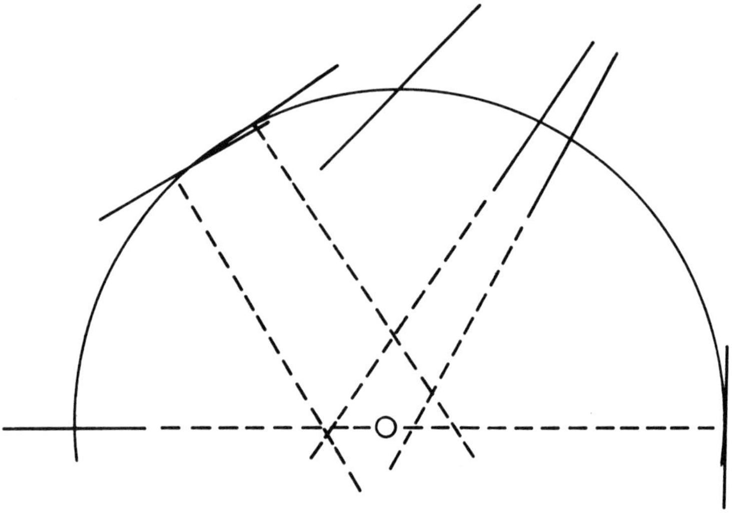

Figure 8.4-XIc. Why there must be three lines through the conformal point that lie on the biorthogonal grid. (See text.)

One root of this equation is $x = y$, the polar axis. The other two roots satisfy

$$(4ax + 1)y^2 + (4ax + 4)xy + x^2 = 0$$

or

$$(4ax + 1)\lambda^2 + (4ax + 4)\lambda + 1 = 0$$

where $\lambda = y/x$, the slope of the direction out of the conformal point.

Near $[x, y] = [0, 0]$, the equation reduces to $\lambda^2 + 4\lambda + 1 = 0$ with roots $-2 \pm \sqrt{3}$. But these values are the tangent and cotangent of $105°$. These roots are at exactly $\pm 60°$ to the root $\lambda = 1$ already located along the polar axis, so that near our conformal point the biorthogonal grids have perfect hexagonal symmetry. As ax moves away from 0, the orientation of these roots rotates slightly. All this is visible in the empirically computed biorthogonal grids for this bilinear mapping, Figure 8.4-XId.

The general bilinear map of the square, whether or not purely inhomogeneous, corresponds to the map of some starting square according to the

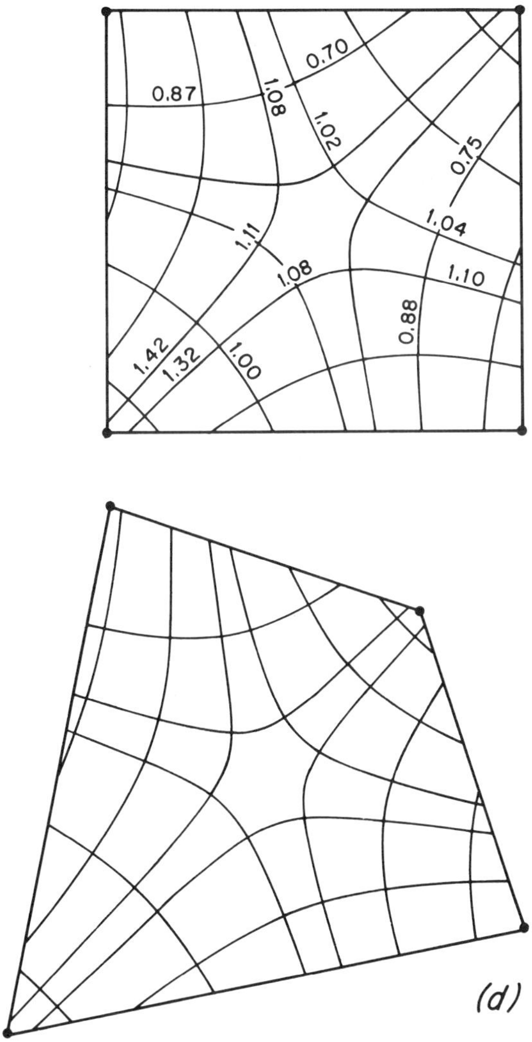

Figure 8.4-XId. The biorthogonal grids for the mapping in Figure 8.4-XIa, with some principal strains indicated.

Cartesian grid of Figure 8.4-XIe and common biorthogonal grid of Figure 8.4-XIf. Purely inhomogeneous maps of various parameters a correspond to squares of various sizes about the common conformal point [0, 0] of the mapping; the smaller squares have smaller shift parameters. Transforms with a homogeneous component shift the starting square away from the conformal point [0, 0]; if the shift is great enough, the conformal point will pass outside the boundaries of the starting square, and the grids that result will have the simpler topology of the distorted rectangle with one polarity always dominant, Figure 8.2-IIc. For the general bilinear mapping, which does not begin with a starting form precisely square, the form of the grids near the singularity may be shown to be exactly the same. Away from the singularity, there is an additional twisting, or spiral turning of the system.

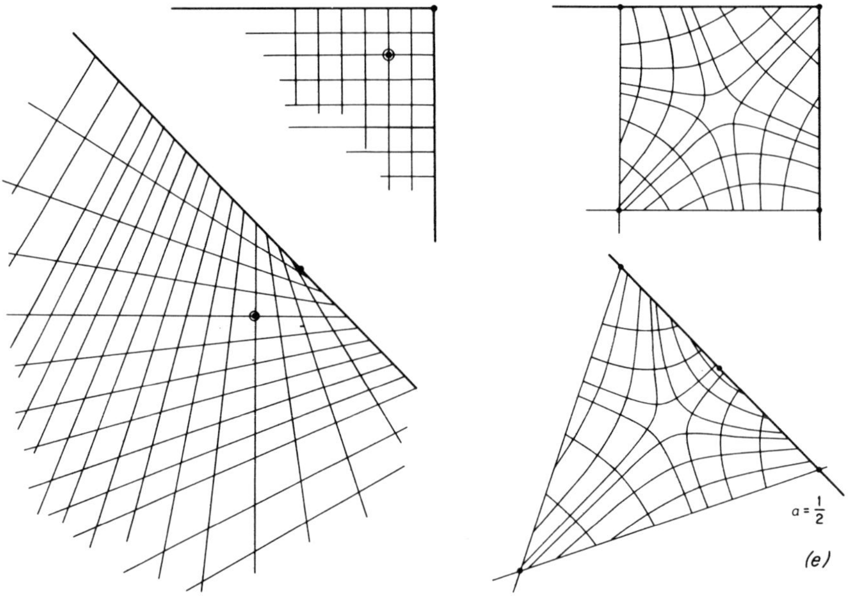

Figure 8.4-XIe. The full bilinear tableau for the entire half-plane (cf. Figure 8.4-IXc), and its single biorthogonal grid.

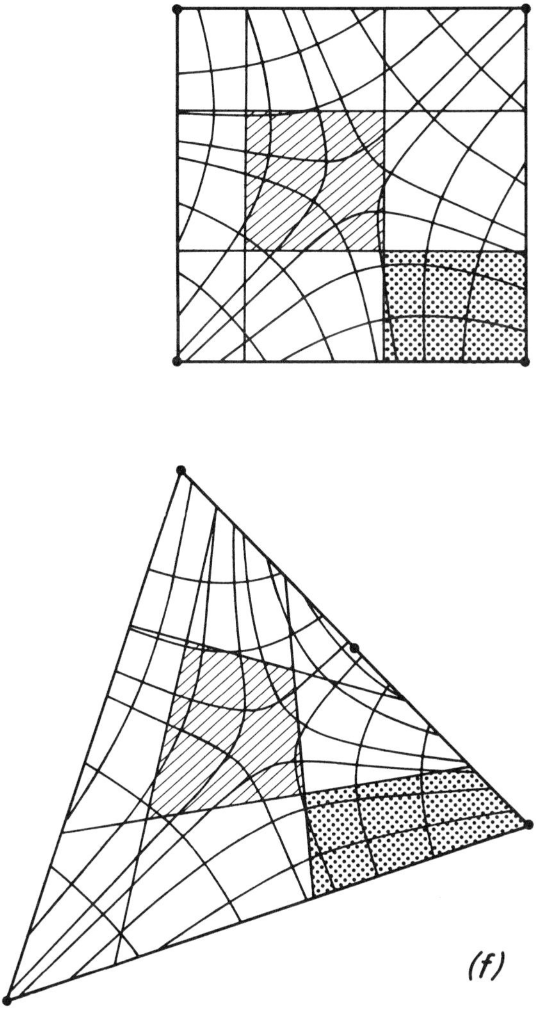

Figure 8.4-XIf. In the mapping from square to quadrilateral, the presence of a homogeneous component displaces the conformal point from the center of the square. Changes in the value of a alter the scale at which this single grid bears the transformation in question.

The Generic Difference between these Mappings

The qualitative distinction between these two classes of grid behavior around the conformal point is stable against changes of form of either the starting or the finishing quadrilateral and stable, also, against changes of detail in the interpolated homology map. When the direction of the major principal strain *co*rotates in the passage from the positive pole to the negative pole of the form, Figure 8.4-VIIIb left, the conformal point bears a singularity with a single strain trajectory through it. In its vicinity, the grid system resembles a confocal system of ellipses and hyperbolas near one focus—which, in turn, resembles the parabolic coordinates of classical mathematical physics, Figure 8.1-Ib. When the major principal strain *counter*rotates as we pass from the positive to the negative pole, Figure 8.4-VIIIb right, the conformal point supports three strain trajectories. (For the projection the additional two were imaginary: the isotropic lines $\lambda = y/x = \pm \sqrt{-1}$.)

To these different sense of the rotation of the principal major strain correspond different behaviors of the mapping along the boundary. The bilinear mapping is uniformly linear on each edge separately; the projection mapping is nonlinear along each edge, graded from positive to negative as one passes away from the positive pole and from negative to positive as one passes away from the negative pole.

8.5 Tensor Fields and Morphogenetic Explanation: Some Speculations

Putative morphogens based on reaction-diffusion models are relative chemical concentrations, scalar fields having a well-determined value or values, in principle, at every point of a region. The only singularities they are likely to have owe to zeros in the denominators of ratios. Gradients, along with related concepts such as phase, are vector fields having a direction and magnitude at every point; singularities arise where the magnitude is zero, for there the direction is undefined. Symmetric tensor fields, as depicted by a biorthogonal grid, bear *two* perpendicular directions and *two* magnitudes at every point, and their singularities are of a different topology than those of the lower-order fields. While at the singularity of polar coordinates *no* direction is defined, at the singularity of a symmetric tensor field we see preferred directions defined in a geometrically new way. While polar coordinates have only the one topological form of singularity, the symmetric tensor field has two different forms: one with a single preferred direction, the other with three.

I believe it is time for a major rethinking of the whole notion of morphogenetic field. Current metaphors, such as are used in the discussions of compartment boundary formation elsewhere in this volume, tacitly assume morphogens which are either scalar or vector fields. Discussions of singularities are thereby impoverished and explanations of experimental findings made defective to an unknown extent. For tensors are as real as scalars and vectors: a cell or tissue has mechanical integrity, after all, so that a structurally encoded orientation could as easily be a strain axis (tensor) as a gradient (vector) or chemical concentration (scalar).

I have no evidence in hand that this possibility is in fact the case—I know of no attempts to measure a real tensor field and match its singularities to organizing centers for ensuing morphogenetic phenomena. But I would nevertheless draw the reader's attention to four possible applications of tensors in morphogenetic explanation: the symmetrical bifurcation of axial gradients of growth; the orientation and anatomical meaning of certain shape-change grids observed empirically; the emergence of the three-root singularity from juxtaposition of two one-root singularities; and the emergence of reorganization from a one-parameter model mixing homogeneous and inhomogeneous components of change.

Bifurcation of Axial Gradients

Conventional compartment models locate axes by resonances or catastrophes within the interior shape; axes modeled so have no natural polarity. But in the study of growing systems, which are subject at all times to real, material strains, the model of singularities of tensor fields may be much more appropriate to an explanation of the physical branching structures which result. For instance, in regeneration of the newt limb (Connelly and Bookstein, 1983) a longitudinal axis bifurcates, Figure 8.5-XIIa, as successive digits appear. The resulting bifurcated fields still point "forward"; the original axial polarity is maintained in spite of the bifurcation. This phenomenon suggests a model bearing two channels of information, one for direction and one for spatial gradient independent of direction; the model of a strain trajectory seems well-suited, as this construct may have a singularity of direction while its numerical magnitude is behaving quite properly. The bifurcation of a pair of digits, then, might be viewed as the splitting of the positive-right trajectory of Figure 8.5-XIIb in the vicinity of the parabolic singularity.

An alternate model for this same bifurcation is available using the other generic singularity, the one with triple symmetry. Under the bilinear mapping, two adjacent edges of the boundary grow relatively more than the other two—in effect, grow *over* the other two while the width of the system is maintained constant. This resembles the actual situation in the regenerating newt limb: while the limb maintains a fairly constant width after amputation, and actively increases in length, a cap of epithelium at the end, the *apical cap,* soon ceases to grow. In the literature of this phenomenon, the cap has been considered an inducer of differentiation slightly proximal to itself. But it could as well induce the bifurcating tensor field of Figure 8.5-XIIc, so that condensing protocartilage finds itself with multiple principal directions rather than the previous single axial orientation.

The Singularity of Human Calvarial Growth

In the normal growth of the human cranium viewed laterally, Figure 8.5-XIII,

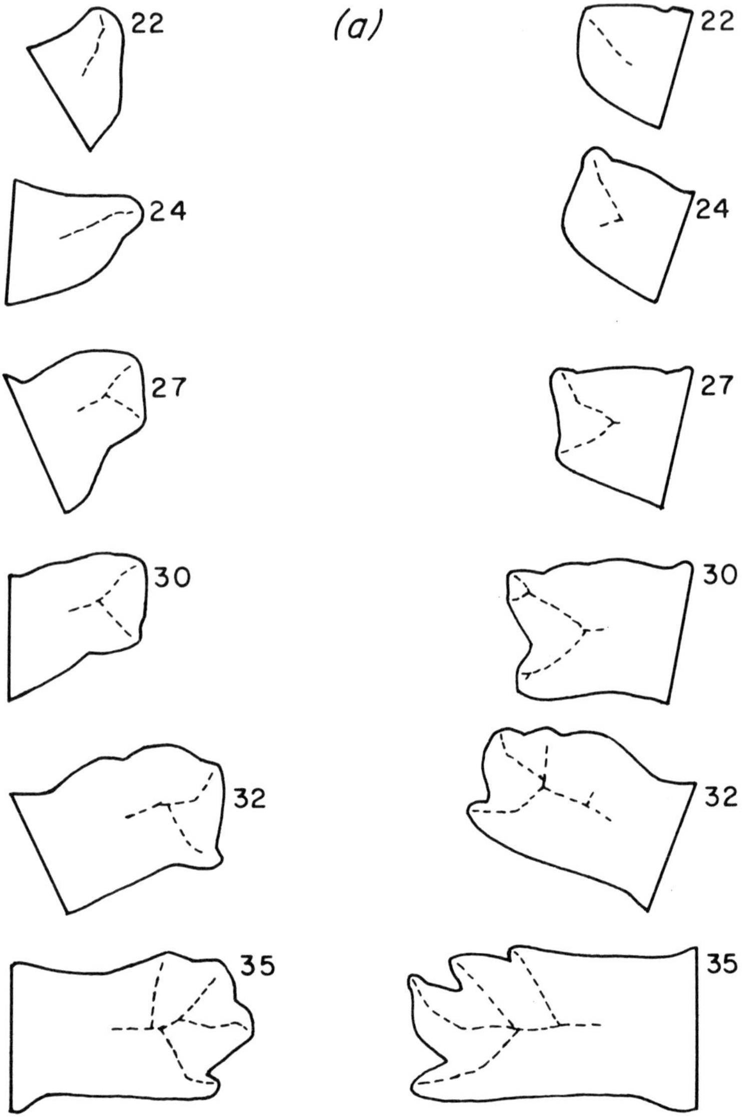

Figure 8.5-XII. Bifurcation of the developing vertebrate limb skeleton.
Figure 8.5-XIIa. Typical findings (after Connelly and Bookstein, 1983). The outlines shown are silhouettes of a single pair of newt forelimbs at various stages of regeneration after amputation. The dashed lines represent *medial axes* of the forms, loci of points equidistant from the boundary in two distinct directions.

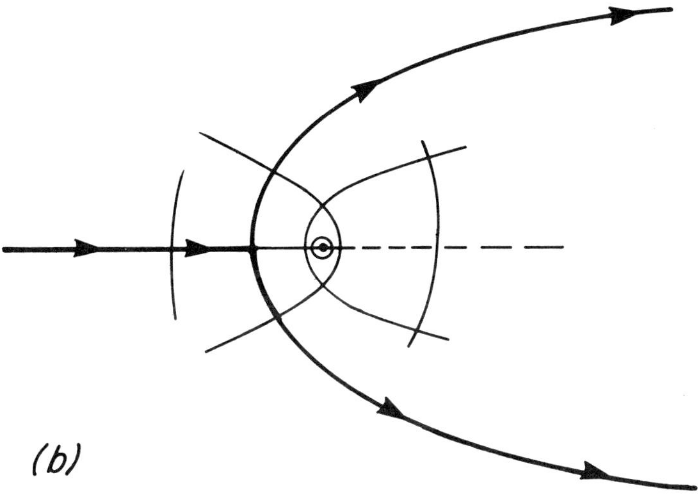

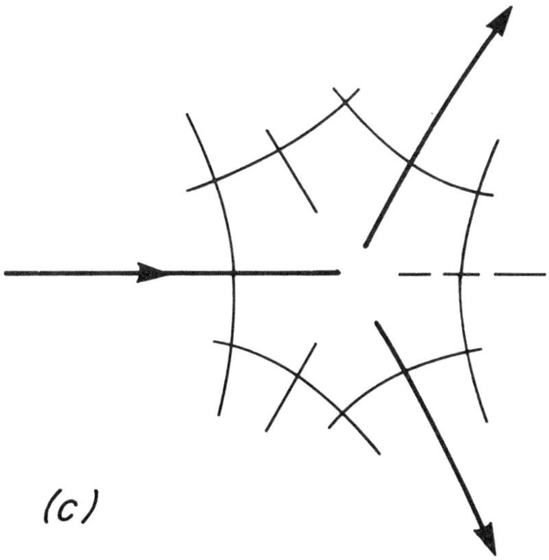

Figure 8.5-XIIb. One model: jump to another branch of the projective singularity.
Figure 8.5-XIIc. Another model: development along all axial polarities of the three-root singularity.

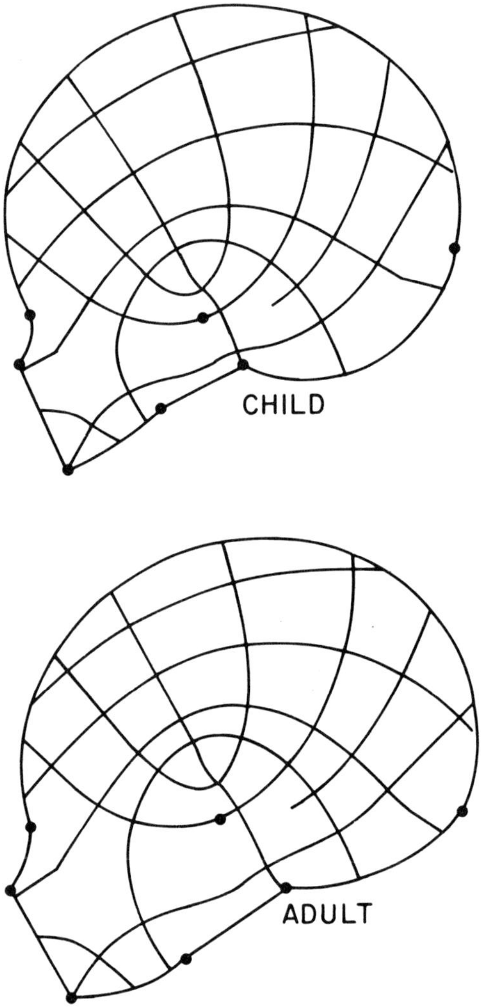

Figure 8.5-XIII. Biorthogonal grids for the observed growth of the human calvarium. (From Bookstein, 1978, Figure VII-10.) Note the one-root singularity.

there is an apparent singularity in the vicinity of the sella turcica (pituitary fossa)—an apparent "growth center." The same pattern is seen in comparisons of juvenile with adult shapes throughout the anthropoid primates (Bookstein, 1978). Using another mathematical model, Moss et al. (1981) located the same center for the head as a so-called *allometric center,* a sort of mean conformal point based in measurement of angles at a distance rather than in the local structure of a tensor field. The same authors later rejected this notion (Moss et al., 1983), in view of its statistical imprecision. In my view the retraction was in error. The locus they extracted was of morphogenetic significance; however, their mathematical model lacked geometrical language for discriminating the tensor interpretation from the vector.

The traditional explanation of this general change in form is the positive allometry of the jaws with respect to the brain. Notice, however, that the invariant axis of the singularity is not oriented anteroposteriorly, along that gradient, at all. The shape change along this axis bears a substantial component of shear, owing to the "orthocephalization" of cranial growth, the movement of the jaws forward, out from under the brain-case. The principal strain trajectory through the singularity instead runs more or less vertically, from a point in the forehead down to the insertion of the spinal cord. This suggests additional, rather different explanations of the regulation of the change, explanations perhaps associated with maintaining the orientation of the visual system or the mechanical advantage of the muscles of mastication. In any case, the detailed geometry of deformation of the cranium carries more information than a mere scalar summary of relative rates of increase of a size measure in the various regions.

Emergence of a Three-Root Singularity from the Juxtaposition of Two One-Root Singularities

Figure 8.5-XIVa models the two basic singularities of the tensor field, that from the projection and that from the bilinear mapping, using a notion of "centers" under conditions of horizontal polarity and vertical constraint. The location of the singularity is at the open circle; the "constraint" (which is merely our measure of scale) is noted by arrows to the two landmarks held at (relatively) constant spacing. We may then sketch the principal axes of the grid corresponding to the projection as a system of polarities radiating approximately from the singularity at the center. The axes of the grid corresponding to the bilinear mapping, on the other hand, appear to radiate instead from the positive and negative poles separately.

When we juxtapose two of these parabolic organizing fields, as in Figure 8.5-XIVb, in-between the two attracting centers emerges the aspect of the bilinear field (with its direction reversed). There develop three positive gradients rather than two. In other words, *from the abutting of two fields, each with a single privileged direction, there emerges a new field with three.* Figure 8.5-XIVc demonstrates this explicitly, averaging the two centered maps by the elastic algorithm of Bookstein (1978).

There is a classic set of morphogenetic experiments to which this observation is directly relevant. When one abuts two same-sided embryonic chick wings on the same side of the embryo, or grafts two newt limb stumps to the same regeneration site, there are often observed not *two* wings in the adult animal but *three,* and the middle one is reversed in polarity (a left wing between two right wings, or vice versa). These experiments have long been used to support the model of intercalation of polar coordinates. But on the tensor model, the appropriate kinematic model is not intercalation of a coordinate, but irreversible bifurcation of a polarity.

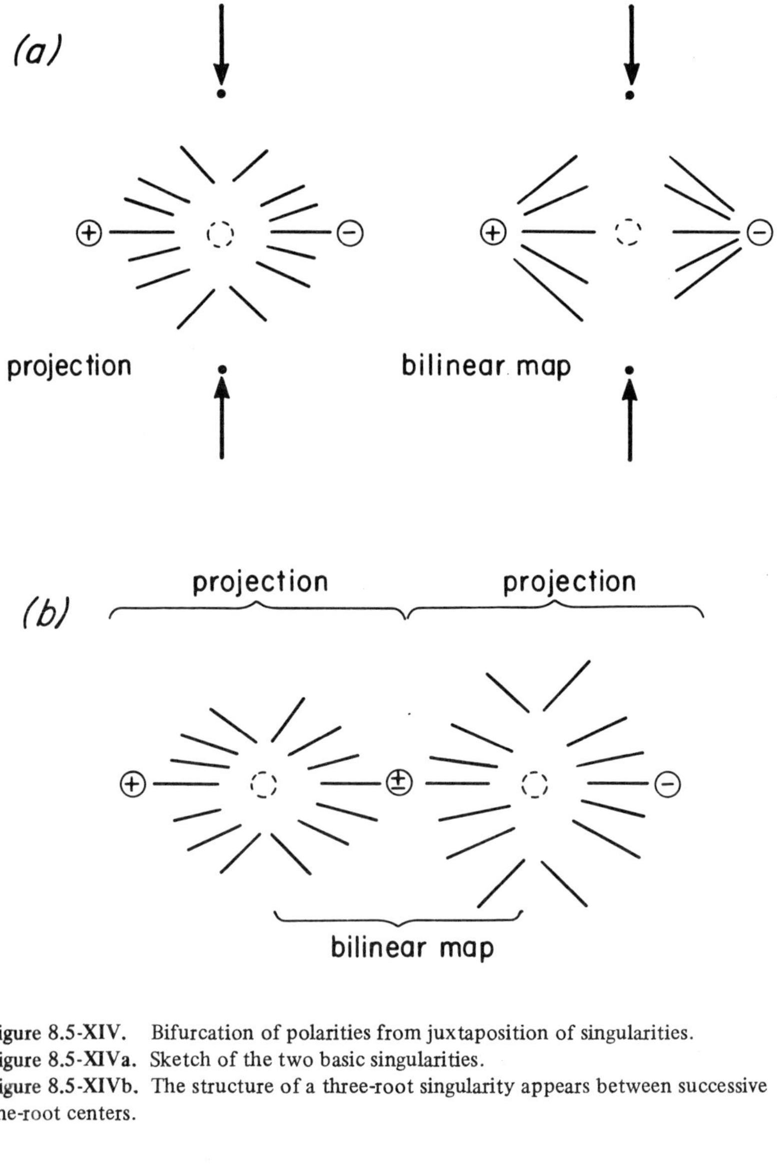

Figure 8.5-XIV. Bifurcation of polarities from juxtaposition of singularities.
Figure 8.5-XIVa. Sketch of the two basic singularities.
Figure 8.5-XIVb. The structure of a three-root singularity appears between successive one-root centers.

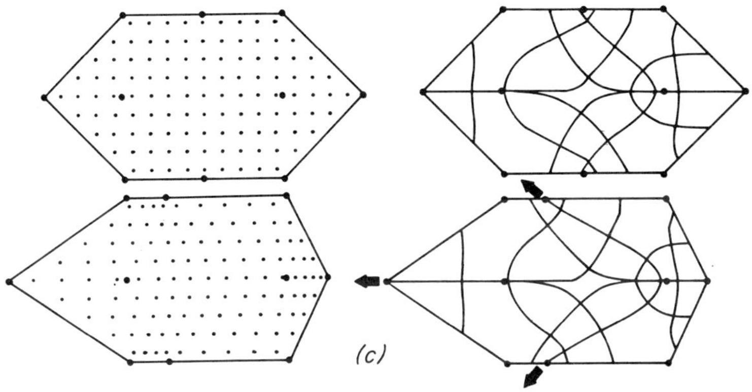

Figure 8.5-XIVc. Simulation of this bifurcation by relative displacement of eight homologous points. Left, the mapping; right, its grids.

If a full axis of these parabolic polar centers is established (perhaps as the result of some resonance process—I am not a tensor chauvinist), the summated morphogenetic field will induce a symmetrical pair of positive gradients at 120° for every initial center. This seems to me a most suggestive metaphor for the production of bilateral structures by segmentation, a process for which there is currently no satisfactory mathematical model.

Timing of Reorganization as Admixture of Components

I conclude this set of speculations by recalling from Section 8.3 the two components, homogeneous and inhomogeneous, of quadrilateral transformations. Under either bilinear mapping or projection, the purely inhomogeneous transform of a square manifested a conformal point precisely at its center. Displacement of the conformal point is equivalent to augmenting the transformation by a homogeneous component. When that component is sufficiently strong, the conformal point is displaced quite outside the boundary of the region we are studying, so that the influence of inhomogeneity is seen only in the gentle curving of the grid lines, Figure 8.2-IIc, and the shallow gradients along them: nonlinearity rather than polarity.

I suggest a time-dependent admixture of this homogeneity as an alternative to current explanations of crucial staging events in morphogenesis. Catastrophe models presume a scalar parameter that is varied past the point at which solutions of a variation equation change their topological properties. In the tensor model, the critical points represent not bifurcation of global minimizing criteria but spatial competition of gradient patterns. Bifurcation can arise directly from the sum of two tensor fields, each constant over time, if the ratio of their relative strengths varies. Biological processes often proceed by independent time scales, suggesting a likely site for a putative "master morphogen" regulating the combination. Variation of the mixture of fields over time is equivalent to a drift of the conformal point in geometrical space. When it encounters tissue, then and there geometric ramifications of structure may begin. This model is at least as congenial as those requiring nonlinear regulation of the entire system for generation of global critical points, resonances, and the like.

Acknowledgement

The writing of this essay was supported by N.I.H. grants DE-05410 to Fred L. Bookstein and DE-03610 to Robert E. Moyers. The Fortran program BIORTH which produced the symmetric tensor fields and the biorthogonal grids summarizing them was underwritten by the preceding grants and also by N.S.F. grant SOC 77-21102 to Fred L. Bookstein. Section 8.5 benefited by conversations with Thomas Connelly of the Department of Anatomy at the University of Michigan.

References

Bôcher, Maxime. 1894. *Über die Reihenentwicklungen der Potentialtheorie.* Teubner, Leipzig.

Bookstein, Fred L. 1978. *The Measurement of Biological Shape and Shape Change.* Springer-Verlag, Berlin.

Bookstein, Fred L. 1981. Coordinate systems and morphogenesis. In T.G. Connelly et al., eds., *Morphogenesis and Pattern Formation,* pp. 262-282. Raven Press, New York.

Bookstein, Fred L. 1984. A statistical method for biological shape change. *Journal of Theoretical Biology,* 107: 475-520.

Bookstein, Fred L., B. Chernoff, R. Elder, J. Humphries, G. Smith, and R. Strauss. *Morphometrics for the Systematist. The Geometry of Size and Shape Change, Especially in Fishes.* Completed manuscript.

Connelly, T.G., and Fred L. Bookstein. 1983. Method for three-dimensional analysis of patterns of thymidine labeling in regeneration and developing limbs. In J.F. Fallon and A.I. Caplan, eds., *Limb Development and Regeneration,* pp. I: 525-536. Alan R. Liss, Inc.

Jacobson, A.G., and R. Gordon. 1976. Changes in the shape of the developing vertebrate nervous system analyzed experimentally, mathematically, and by computer simulation. *Journal of Experimental Zoology,* 197: 191-246.

Moss, M.L., R. Skalak, L. Moss-Salentijn, G. Dasgupta, H. Vilmann, and P. Mehta. 1981. The allometric center. The biological basis of an analytical model of growth. *Proceedings of the Finnish Dental Society,* 77: 119-128.

Moss, M.L., R. Skalak, M. Shinozuka, L. Moss-Salentijn, and H. Vilmann. 1983. Statistical testing of an allometric centered model of craniofacial growth. *American Journal of Orthodontics,* 83: 5-18.

Schüepp, O. 1966. *Meristeme.* Birkhauser Verlag, Basel.

Thompson, D'A.W. 1961 (1917, 1942). *On Growth and Form,* ed. abr. J.T. Bonner. Cambridge University Press, Cambridge.

Winfree, A.T. 1980. *The Geometry of Biological Time.* Springer-Verlag, New York.

THE HISTORICAL RECORD

9

PRINCIPES DE BIOLOGIE MATHEMATIQUE (PRINCIPLES OF MATHEMATICAL BIOLOGY)

PROF. VITO VOLTERRA

TRANSLATED BY P.M. ANTONELLI (1983)

PART I The Bases of the Theory of the Struggle for Existence

9 A.1 Population and the *quantité de vie*

1. We have, since 1926, devoted several works to the study of the struggle for existence.

In this paper we shall leave aside all the details and applications for which we refer you to the works just mentioned (Volterra, 1931; Volterra and D'Ancona, 1935) and we shall consider only the general laws and principles.

In our former works we considered two cases: 1) the one where several species compete for the same food and 2) the one where the individuals of various species, living together, devour one another.

The first case is the easiest to deal with. The laws that are deduced from the general equations have been very satisfactorily verified by several biologists, among whom, quite recently, M. Gause.

The second case is harder to deal with. It is the one we will consider in this article.

2. Let us name $N_1, N_2, ..., N_n$ the number of individuals of n species living in the same environment, i.e. a biological association. We will suppose these numbers are such as can be considered as quantities varying in a continuous manner and to which one can apply infinitesimal methods. This is what has mostly been done in statistical theories. We shall call these numbers the populations of the different species. But other numbers, that we shall now define, must be added to these.

Let us refer to a graphic representation.

Through a vertical strip corresponding to a given environment, let us drive equal and consecutive horizontal strips corresponding to years following one another and starting from a strip that is the origin of time.

Let us drive vertical segments, each of which starts from the initial year or from the year when the life of an individual of a given species living in the environment considered begins and goes on through all his life, stopping at the year where its existence ends or at the last considered year. Each individual's life is measured by the number of horizontal strips encountered by each segment, and, at the end of a given time, the total number of these encounters can be considered to express the quantity of life of the species from the origin of time to the year one has stopped at.

But the number of encounters of a horizontal strip with the vertical segments gives the population of the species for the corresponding year. Consequently, the sum total of the populations of the different years from the origin of time to a given year is equal to the sum total of the encounters last considered. It can then be considered as the measure of the quantity of life of the species over the same lapse of time.

If we call N_h the population of the year h

$$\sum_{1h}^{m} N_h$$

will express the quantity of life of the species from the origin of time to the year m.

Switching from the discontinuous to the continuous and calling $N(t)$ the population at the t time

$$X = \int_0^t N(t)dt$$

will be the *quantité de vie* of the species over the period of time $(0, t)$. In statistical calculations, N or X, can be independently considered as those quantities interlinked by the relations.

$$X = \int_0^t N(t)dt, \quad N(t) = \frac{dX}{dt}$$

9A.2 The Fundamental Equations

1. Let us suppose that the n species living together feed on one another. To deal with this case we shall apply the technique that can be called the *encounters-method.*

We can suppose that each encounter between two individuals of different species brings about a result beneficial to one of the species and damaging to the other; or that the result is null for one and the other. Let us consider the r and s species.

If N_r is the number of individuals making up the r species and N_s that of the s species, the probability for one individual of one of these species to meet with one of the other species will be proportional to N_rN_s; in other words, over a unit of time, the number of encounters may be represented by $m_{rs}N_rN_s$, m_{rs} being a constant coefficient.

Let us suppose that, for each encounter, P_{rs} (this number will of course be a fraction) individuals of a species, of the r species for example, die; then, over the unit of time, there will be $m_{rs}P_{rs}N_rN_s$ individuals of that species doomed to die.

In order to calculate the effect of this destruction on the number of the individuals of the other species, let us call $\beta_1, \beta_2, \ldots \beta_n$ the average weights of the individuals in each of the n species and $P_1, P_2, \ldots P_n$ the total weight of the individuals making up these species.

This is why the number of the individuals of each species will be expressed by

$$N_1 = \frac{P_1}{\beta_1}, \quad N_2 = \frac{P_2}{\beta_2}, \ldots \quad N_n = \frac{P_n}{\beta_n}.$$

And if an individual of the r species is devoured by an individual of the s species, the P_r weight will become $P_r - \beta_r$, while the P_s weight will become $P_s + \beta_r$ and, consequently, the numbers of the individuals of the two species will become (considering things roughly, as we suppose that the total weight of the devoured individual is added to the devourer's weight).

$$\frac{P_r - \beta_r}{\beta_r} = N_r - 1, \qquad \frac{P_s + \beta_r}{\beta_s} = N_s + \frac{\beta_r}{\beta_s}.$$

Still very roughly, we shall then determine the decrease in the number of individuals of the r species over a unit of time by

$$m_{rs}p_{rs}N_rN_s$$

and the increase of the number of individuals of the s species over the same time by

$$m_{rs} p_{rs} N_r N_s \frac{\beta_r}{\beta_s} .$$

As we already mentioned, we stick to the hypothesis that the living substance of the r species is transformed, as soon as it is devoured, into substance of the s species. Of course, this hypothesis is far from corresponding to the observations of biology, but an approximation is sufficient for our calculations.

2. If we state

$$m_{rs} p_{rs} \beta_r = a_{rs}$$

we shall have, as a decrease of the individuals of the r species

$$\frac{1}{\beta_r} a_{rs} N_r N_s$$

and, as an increase of the individuals of the s species

$$\frac{1}{\beta_s} a_{rs} N_r N_s .$$

Supposing $a_{rs} = - a_{sr}$ we shall be entitled to say that, over one unit of time and as a consequence of the encounters of the individuals of the r and s species, the numbers of the individuals of the r species and of the s species will increase, respectively,

$$\frac{1}{\beta_r} a_{sr} N_r N_s , \quad \frac{1}{\beta_s} a_{rs} N_r N_s .$$

So that, over a dt time they will increase by

$$\frac{1}{\beta_r} a_{sr} N_r N_s dt, \quad \frac{1}{\beta_s} a_{rs} N_r N_s dt .$$

We shall reach the same result for any other species.

In other words, we have adopted the $1/\beta_1$, $1/\beta_2$, ..., $1/\beta_n$ as the equivalents of the individuals of each species: For admitting that $1/\beta_r$ individuals of the r species can be transformed into $1/\beta_s$ individuals of the s species means that $1/\beta_r$ individuals of the r species are equivalent to $1/\beta_s$

individuals, of the s species.

As a first very rough approximation, we have then taken the inverse average weights as equivalents. But if we just admit, through a hypothesis, the existence of equivalent numbers even if they do not coincide with the inverse average weights, we get the same result as the one just reached.

3. Now, if we call ε_r the coefficient of increase of the r species when it exists by itself, then, when n species live together, the variation according to time of the N_r numbers will occur according to the differential equations

$$\frac{dN_r}{dt} = \left(\varepsilon_r + \frac{1}{\beta_r} \sum_{1}^{n}{}_{s}\, a_{sr} N_s\right) N_r \quad (r = 1, 2, \ldots, n) \tag{9A.2-1}$$

that is

$$\beta_r \frac{dN_r}{dt} = \left(\varepsilon_r \beta_r + \sum_{1}^{n}{}_{s}\, a_{sr} N_s\right) N_r \tag{9A.2-2}$$

where

$$a_{sr} = - a_{rs}, \qquad a_{rr} = 0 .$$

We shall call ε_r the *intrinsic increase coefficients* and

$$\varepsilon_r + \frac{1}{\beta_r} \sum_{1}^{n} a_{sr} N_s = I_r$$

the effective increase coefficients or demographic coefficients.

4. It is easy to see that, once the N_r^0 initial values for $t = 0$ have been given, we shall be able to integrate the (9A.2-2) equations through regular analytic functions. We just have to compare through majoration the (9A.2-2) equations with the equation

$$\frac{dN}{dt} = (a + nbN)N$$

where $a > |\varepsilon_r|$, $b > |a_{sr}/\beta_r|$, integrating through series of powers of t. If the initial values of N_r are positive, they will remain so, for

$N_0 = N_r^0 e^{\int_0^t I_r dt}$. They will always be finite and, because of the preceding relation, they will not be annulled for any of the values of t. We can state the theorem that *if a species exists at a certain moment, it will always exist and will have always existed.* This is the theorem of *the conservation of species.*

It may happen that a few of the N_r tend to zero for $t = \infty$. We then say that the corresponding species are becoming exhausted (see Section 9A.5, No. 1).

5. By introducing the X_r elements, we shall be able to replace the (9A.2-2) equations by the equations

$$\beta_r \frac{d^2 X_r}{dt^2} = \left(\varepsilon_r \beta_r + \sum_{1s}^{n} a_{sr} \frac{dX_s}{dt}\right) \frac{dX_r}{dt} . \qquad (9A.2\text{-}3a)$$

We shall see that the passage from the (9A.2-3a) equations to the (9A.2-2) ones is far from being as trivial as it may seem at first.

For the (9A.2-3a) equations can be reduced to depend upon an equation of the calculation of variations in exactly the same way as happens to the equations expressing most of the natural physical phenomena (see Section 9A.3, No. 2).

Stating $dX_r/dt = X_r'$, $d^2X_r/dt^2 = X_r''$ the (9A.2-3a) equations are written

$$\beta_r X_r'' = \left(\varepsilon_r \beta_r + \sum_{1s}^{n} a_{sr} X_s'\right) X_r' . \qquad (9A.2\text{-}3b)$$

When the N_r quantities vary with time, we shall say there are biological variations.

9A.3 Equilibrium Cases

1. The $N_1, N_2, \ldots N_r$ will be constants when

$$\frac{dN_1}{dt} = \frac{dN_2}{dt} = \ldots = \frac{dN_n}{dt} = 0 \qquad (9A.3\text{-}1a)$$

that is when

$$\left(\varepsilon_r \beta_r + \sum_{1}^{n} {}_s\, a_{sr} N_s\right) N_r = 0 . \qquad (9A.3\text{-}1b)$$

The species are then in a stationary state. The N_r will also have to be positive. If one or several of these quantities are null, they will stay null always because of the principle of the conservation of species. It comes to suppressing some species of the association and to considering an association made up of a lesser number of species.

The (9A.3-1b) equations will be satisfied if we have

$$\varepsilon_r \beta_r + \sum_{1}^{n} {}_s\, a_{sr} N_s = 0 . \qquad (9A.3\text{-}2a)$$

These conditions are sufficient and not necessary, as the (9A.3-1b) can be satisfied without all the conditions being verified. But then some of the N_r will have to be null which brings us back, as we just mentioned, to considering a residual association made up of a lesser number of species. In the same way, it may happen that some of the roots of the (9A.3-2a) equations are null. In these cases too, the state will be stationary but the association may then be reduced. Which is why, *when we study the stationary states, we shall only consider the case where the (9A.3-2a) equations have positive roots.*

These are the cases we shall call *the equilibrium cases of the association.* We shall neglect all the others. The equilibrium will be preserved through the mutual actions of the different species without our having to consider some species as being exhausted (see Section 9A.2, No. 4, and Section 9A.5, No. 1).

2. We shall call the (9A.3-2a) equations, the fundamental equations. We shall say the solution is positive if all the roots are positive.

The positive solutions will give equilibrium cases.

The determinant of the (9A.3-2a) equations

$$D = \begin{vmatrix} 0, & a_{21} & \cdots\cdots & a_{n1} \\ a_{12}, & 0 & \cdots\cdots & a_{n2} \\ \cdots & \cdots & \cdots & \cdots \\ a_{1n}, & a_{2n} & \cdots\cdots & 0 \end{vmatrix} \qquad (9A.3\text{-}3)$$

will be called *the fundamental determinant.*

It is *skewsymmetrical,* i.e. the elements of the diagonal are null and the elements symmetrical in relation to the diagonal of equal absolute value and opposite sign. By the theory of determinants, it will be a square if n is an even number and null if n is an odd number.

3. Let us have $q_1, q_2, \dots q_n$ as solutions of the (9A.3-2a) equations; because of the $a_{sr} = -a_{rs}$ relations, we shall have

$$\sum_1^n \varepsilon_r \beta_r q_r = 0 . \tag{9A.3-4}$$

Let us suppose $q_1, q_2, \dots q_n$ are positive.

We deduce right away from the preceding relation the theorem:

If all the intrinsic increase coefficients are of a same sign, equilibrium is not possible.

4. We shall now examine closely the two cases of the fundamental determinant being different from or equal to zero.

First Case Where the Fundamental Determinant Is Not Null

This case can only come about if the number of species is even. The fundamental equations will always have one and only one solution and, if the equilibrium equations solutions are positive, there will be only one case of equilibrium.

In this case, let us change the ε_r into $\varepsilon_r + \Delta\varepsilon_r$ and call Δq_r the variations of the q_r such that $q_r + \Delta q_r$ (solutions of the (9A.3-2a) equations) remain positive. We shall have

$$\beta_r(\varepsilon_r + \Delta\varepsilon_r) + \sum_1^n {}_s\, a_{sr}(q_s + \Delta q_s) = 0$$

and consequently

$$\beta_r \Delta\varepsilon_r + \sum_1^n {}_s\, a_{sr} \Delta q_s = 0 \tag{9A.3-5a}$$

so that

$$\sum_1^n {}_r\, \beta_r \Delta\varepsilon_r \Delta q_r = 0 . \tag{9A.3-5b}$$

Consequently, *if we increase (or change) simultaneously all the intrinsic coefficients of increase, the equilibrium populations of a few species will increase and that of other species will decrease,* for because of (9A.3-5a), all the Δq_r will not possibly be null and, because of (9A.3-5b) those that are not null will not possibly have the same sign.

If all the a_{sr} quantities $(s = 1, 2, \dots r - 1, r + 1, \dots n)$ are positive or null, the r species will not be devoured by any of the other ones; so, because of the (9A.3-5a) equation, if $\Delta\varepsilon_r$ is negative (positive) the Δq_s will not possibly be all negative (positive) or null; and, consequently, if the increase coefficient of the rth species decreases (increases), a few of the equilibrium populations of the other species will increase (decrease).

Similarly, if all the a_{sr} quantities $(s = 1, 2 \dots r - 1, r + 1, \dots n)$ are negative or null we shall have that the r species does not devour the others and, consequently, if $\Delta\varepsilon_r$ is negative (positive) the q_s of the (9A.3-5a) equation will not possibly be all positive (negative) or null; and there will consequently be a consequent opposite to the one we have just found.

Second Case. Null Fundamental Determinant

5. This is always the case when the number of species is odd. But it may also occur when the number of species is even.

If the fundamental equations have a solution, there is an infinity of them.

6. Let us assume that the intrinsic increase coefficients are null, then the fundamental equations always have solutions different from zero. The positive ones correspond to equilibrium cases. If there is a positive solution, there will be an infinity of them. Let us call

$$q_1^{(h)}, q_2^{(h)} \dots\dots q_n^{(h)} \qquad (h = 1, 2 \dots g)$$

all the independent solutions. We shall have

$$\sum_{1}^{n}{}_s \, a_{sr} q_s^{(h)} = 0 \,. \qquad \text{(9A.3-6a)}$$

All the solutions will be given by

$$\sum_{1}^{g}{}_h \, I_h q_1^{(h)} , \quad \sum_{1}^{g}{}_h \, I_h q_2^{(h)} , \dots \sum_{1}^{g}{}_h \, I_h q_n^{(h)}$$

where $I_1, I_2, \dots I_k$ are arbitrary constants.

7. Let us now assume that the intrinsic coefficients of increase are not all null and that there is a solution $q_1^0, q_2^0 \dots q_n^0$ to the fundamental equations.

Then

$$q_1^0 + \sum_{1\,h}^{g} I_h q_1^{(h)},\quad q_2^0 + \sum_{1\,h}^{g} I_h q_2^{(h)},\ \dots\ q_n^0 + \sum_{1\,h}^{g} I_h q_n^{(h)} \qquad (9A.3\text{-}7)$$

will also be solutions. Those formulas will give all possible solutions. For if there are two of them, their differences will satisfy the fundamental equations in the case of null intrinsic increase coefficients.

If one of the solutions is positive, there will be an infinity of them and, consequently, when the fundamental determinant is null, if there is one state of equilibrium, there will be an infinity of them.

8. Let us suppose that, when the ε_r are not all null, the fundamental equations have the $q_1^0, q_2^0 \dots q_n^0$ solution. We shall have

$$\beta_r \varepsilon_r + \sum_{1\,s}^{n} a_{sr} q_s^0 = 0\,. \qquad (9A.3\text{-}2b)$$

But the (9A.3-6a) equations can be written, as $a_{sr} = - a_{rs}$

$$\sum_{1\,s}^{n} a_{rs} q_s^{(h)} = 0 \qquad (9A.3\text{-}6b)$$

so we shall get

$$\sum_{1\,r}^{n} \beta_r \varepsilon_r q_r^{(h)} = 0, \qquad (h = 1, 2 \dots g). \qquad (9A.3\text{-}8)$$

So those equations have to be satisfied for the (9A.3-2b) to be verified. It only takes one unsatisfied (9A.3-8) for the equilibrium not to be possible.

9. The theorems we gave in the first case (non-null determinant) (see Section 9A.3, No. 4) can be extended to the case of the null determinant, but we then have to take into account the fact that the $\Delta\varepsilon_r$ cannot be arbitrarily chosen.

9A.4 Integrals of the Fundamental Equations

1. From the $a_{sr} = - a_{rs}$ condition we get the (9A.2-3b) equations

$$\sum_{1}^{n}{}_r\,\beta_r X_r'' = \sum_{1}^{n}{}_r\,\varepsilon_r \beta_r X_r'$$

and, integrating,

$$\sum_{1}^{n}{}_r\,\beta_r X_r' - \sum_{1}^{n} \varepsilon_r \beta_r X_r + C = 0 \qquad (9A.4\text{-}1a)$$

C being a constant quantity.

2. Then (9A.2-3a) equations give us

$$\beta_r \frac{d}{dt} \log X_r' = \varepsilon_r \beta_r + \sum_{1}^{n}{}_s\, a_{sr} X_s' \qquad (9A.2\text{-}3c)$$

and, integrating,

$$\Theta_r = \beta_r \log X_r' + \sum_{1}^{n}{}_s\, a_{rs} X_s - \varepsilon_r \beta_r t - C_r = 0 \qquad (9A.4\text{-}2a)$$

the C_r being constants. We deduce from it the integral

$$\Theta = \sum_{1}^{n}{}_r\, X_r' \Theta_r = \chi + Z - (\sum_{1}^{n}{}_r\, \varepsilon_r \beta_r t + C_r) X_r' = 0\,, \qquad (9A.4\text{-}2b)$$

naming Z the bilinear form

$$Z = \sum_{1}^{n}{}_r \sum_{1}^{n}{}_s\, a_{rs} X_r' X_s \qquad (9A.4\text{-}3a)$$

and χ the expression

$$\chi = \sum_{1}^{n}{}_r\, \beta_r X_r' \log X_r'\,. \qquad (9A.4\text{-}3b)$$

Combining the (9A.4-l) and (9A.4-2b) integrals, we find

$$\begin{aligned}\Omega &= \Theta + \sum_{1}^{n}{}_r\, \varepsilon_r \beta_r X_r - \sum_{1}^{n}{}_r\, \beta_r X_r' - C \\ &= \sum_{1}^{n} \beta_r X_r' \log X_r' + \sum_{1}^{n}{}_r \sum_{1}^{n}{}_s\, a_{rs} X_s X_r' - \sum_{1}^{n} \varepsilon_r \beta_r t X_r' \\ &\quad - \sum_{1}^{n}{}_r\, C_r X_r' + \sum_{1}^{n}{}_r\, \varepsilon_r \beta_r X_r - \sum_{1}^{n}{}_r\, \beta_r X_r' - C = 0\,. \end{aligned} \qquad (9A.4\text{-}2c)$$

3. Let us suppose the (9A.3-2a) equations have

$$q_1, q_2 \cdots q_n, \qquad (9A.4\text{-}4)$$

roots. We shall then have identically

$$\varepsilon_r \beta_r = -\sum_{1\,s}^{n} a_{sr} q_s$$

and the (9A.2-2) equations may then be written

$$\beta_r \frac{dN_r}{dt} = \sum_{1\,s}^{n} a_{sr}(N_s - q_s)N_r \,.$$

Considering $a_{sr} = - a_{rs}$, we shall get

$$\sum_{1\,r}^{n} \frac{\beta_r}{N_r} (N_r - q_r) \frac{dN_r}{dt} = 0$$

and consequently

$$\sum_{1\,r}^{n} \left(\beta_r \frac{dN_r}{dt} - q_r \beta_r \frac{1}{N_r} \frac{dN_r}{dt} \right) = 0$$

and integrating

$$\sum_{1\,r}^{n} (\beta_r N_r - q_r \beta_r \log N_r) = C' \qquad (9A.4\text{-}5a)$$

C' being a constant.

We shall now have a close look at the different cases*.

4. *First Case. Non Null Fundamental Determinant*

1a) The ε_r are not all null. Then there is only one (9A.4-4) solution. The roots will not be all null and we shall have the (9A.4-5a) integral.

1b) If the ε_r are null, the (9A.4-4) roots will all be null and the (9A.4-5a) integral will be reduced to

*We have used for this exposition an interesting remark of M.B. Levi (1931).

$$\sum_{1}^{n}{}_r \beta_r N_r = C'', \tag{9A.4-5b}$$

C'' being a constant.

5. *Second Case. The Fundamental Determinant is Null*
The values different from zero

$$q_1^{(h)}, q_2^{(h)}, \ldots q_n^{(h)} \qquad (h = 1, 2 \ldots g)$$

will exist and, because of the (9A.2-1a), (9A.2-1b) and (9A.3-6b) equations we shall have

$$\sum_{1}^{n}{}_r \beta_r q_r^{(h)} \frac{dN_r}{N_r} = \sum_{1}^{n}{}_r \beta_r \varepsilon_r q_r^{(h)} dt$$

and integrating

$$\sum_{1}^{n}{}_r \beta_r q_r^{(h)} \log N_r = \sum_{1}^{n}{}_r \beta_r \varepsilon_r q_r^{(h)} t + c^{(h)}, \tag{9A.4-5c}$$

$c^{(h)}$ being the constants.

2a) All the (9A.3-8) relations are not satisfied (the ε_r not being all null) and, consequently, the fundamental equations do not have solutions. Consequently, the (9A.4-5a) integral will not exist and the (9A.4-5c) integrals will still be extant, and in one of them at least the t coefficient is not null.

2b) The ε_r are null and then the (9A.4-5c) integrals become

$$\sum_{1}^{n}{}_r \beta_r q_r^{(h)} \log N_r = c^{(h)}. \tag{9A.4-5d}$$

Moreover, we shall have the integral

$$\sum_{1}^{n}{}_r \beta_r N_r = C''. \tag{9A.4-5a}$$

2c) The ε_r are not null and there are $q_1^0, q_2^0 \ldots q_n^0$ solutions to the (9A.3-2a) equations. In this case the equations (see Section 9A.3, No. 8)

$$\sum_{1}^{n}{}_r \beta_r \varepsilon_r q_r^{(h)} = 0 \qquad (h = 1, 2, \ldots g)$$

must be satisfied and the (9A.4-5a) integral and (9A.4-5c) integrals exist (the coefficient of t being null), that is

$$\sum_1^n (\beta_r N_r - \beta_r q_r^0 \log N_r) = C' \tag{9A.4-5e}$$

$$\sum_1^n{}_r \beta_r q_r^{(h)} \log N_r = c^{(h)} \qquad (h = 1, 2, \dots g) \tag{9A.4-5d}$$

that is

$$N_1^{\beta_1 q_1^{(h)}} N_2^{\beta_2 q_2^{(h)}} \dots N_n^{\beta_n q_n^{(h)}} = \text{const.} \tag{9A.4-5f}$$

Moreover the (9A.3-2a) will also have the solutions (see Section 9A.3, No. 7)

$$q_1 = q_1^0 + \sum_1^g{}_h I_h q_1^{(h)}, \quad q_2 = q_2^0 + \sum_1^g{}_h I_h q_2^{(h)}, \dots q_n = q_n^0 + \sum_1^g{}_h I_h q_n^{(h)}$$

which is why the (9A.4-5a) integral will be written

$$\sum_1^n{}_r (\beta_r N_r - \beta_r q_r^0 \log N_r) - \sum_1^g{}_h I_h \sum_1^n{}_r q_r^{(h)} \beta_r \log N_r = C'.$$

But the I_r are arbitrary. We have again the (9A.4-5e) and (9A.4-5d) integrals.

6. To conclude, we have obtained the (9A.4-1a) and (9A.4-2a) integrals and must now deduce from these consequences, from which will come in turn the general laws that rule the struggle for existence phenomena and biological fluctuations. This is what the following sections are devoted to.

9A.5 Consequences of the Integrals

1. We first have to establish some definitions that we shall make use of herein. If N(t) denotes the number of individuals in a species and if it is always bounded between two fixed positive numbers, we shall say that the species has *variations bounded between two positive numbers.*

N(t) can only tend to zero when $t = \infty$. In this case the species is said to have become exhausted (see Section 9A.2, No. 4). Moreover, we shall say that it is becoming exhausted in a regular manner. But if N can, without tending to zero, take arbitrarily small values when $t \to \infty$ (that is when its lower limit is null), we shall say that the species has become exhausted in an irregular manner, in opposition to the other case. To this convention of mathematical language corresponds a biological postulate, that of the actual disappearance of the species in all the cases where the N corresponding function has a null limit inferior (see Section 9A.5, No. 7).

If $N(t)$ is bounded between two positive numbers and if $N(t)$ has maxima and minima for the values of $t > t_0$ whatever t_0 is, we shall say that the species experiences *fluctuations*.

We shall also say that the *fluctuations are damped* when the oscillation (or difference between the upper and lower limits) for $t > t_0$ becomes and stays as small as we want if we increase t_0 sufficiently. In this last case, and in this case only, the fluctuations will allow N to tend to a determined and finite limit when n tends to ∞.

Finally, we shall say that $N(t)$ *asymptotically tends* to the q limit when $N(t)$ does not have fluctuations and tends to the determined and finite q limit for $t = \infty$. When the average of the $N(t)$ number over the (t_0, t) interval tends to a limit for $t = \infty$, we shall call this limit the *asymptotic average* of N.

2. Let us suppose that the (9A.3-2a) fundamental equations have positive roots $q_1, q_2 \ldots q_n$.

There will be an equilibrium state corresponding to these roots and consequently, we shall have the integral (see Section 9A.4, No. 3)

$$\sum_{1}^{n}{}_{r}\ \beta_r(N_r - q_r \log N_r) = C' . \qquad (9A.4\text{-}5a)$$

Switching from the logarithms to the numbers themselves we shall write*

$$\left(\frac{e^{N_1}}{N_1^{q_1}}\right)^{\beta_1}\left(\frac{e^{N_2}}{N_2^{q_2}}\right)^{\beta_2}\cdots\left(\frac{e^{N_n}}{N_n^{q_n}}\right)^{\beta_n} = c' \qquad (9A.5\text{-}1)$$

c' being a constant.

Stating

*We could do without this transformation and work on the logarithmic formulas directly (see Section 9A.5, No. 6).

$$n_r = \frac{N_r}{q_r}$$

$$c = c' q_1^{\,q_1\beta_1} q_2^{\,q_2\beta_2} \ldots q_n^{\,q_n\beta_n}$$

(9A.5-1) will become

$$\left(\frac{e^{n_1}}{n_1}\right)^{\beta_1 q_1} \left(\frac{e^{n_2}}{n_2}\right)^{\beta_2 q_2} \ldots \left(\frac{e^{n_n}}{n_n}\right)^{\beta_n q_n} = c\,. \qquad \text{(9A.5-2)}$$

As $q_1, q_2 \ldots q_n$ are positive, $n_1, n_2 \ldots n_n$ are positive too and, consequently,

$$\frac{e^{n_r}}{n_r} \geqslant e$$

and consequently

$$\left(\frac{e^{n_r}}{n_r}\right)^{q_r\beta_r} \leqslant \frac{c e^{q_r\beta_r}}{e^{\sum_1^n q_h\beta_h}} = K e^{q_r\beta_r}$$

where

$$K = \frac{c}{e^{\sum_1^n q_h\beta_h}}\,.$$

We deduce from it

$$\frac{e^{n_r}}{n_r} \leqslant e \sqrt[q_r\beta_r]{K}\,.$$

It shows that n_r must be bounded between two fixed positive numbers one of which is bigger and the other smaller than unity. For let us consider a curve, the equation of which is

$$y = \frac{e^x}{x} \qquad \begin{array}{l} 0 < x < \infty \\ e \leqslant y < \infty . \end{array}$$

The ordinate will take its e minimal value for $x = 1$ and will become infinite for $x = 0$ and $x = \infty$. Let us drive a parallel to the x axis at the $y_0 > e$ distance. It will intersect the curve in two points, having the x^0 and x' abscissa.

If $e^x/x < y_0$ we shall have

$$0 < x^0 < 1 < x' < \infty, \qquad x^0 < x < x'.$$

So, n_r^0 and n_r' being two positive numbers between which is unity, and stating

$$n_r^0 q_r = N_r^0, \qquad n_r' q_r = N_r'$$

we shall have

$$N_r^0 < N_r < N_r',$$

N_r^0 and N_r' being two positive numbers between which the q_r quantity is bounded.

We gather from all this the following theorem:

If there exists an equilibrium state and if the initial state is different from that state, the numbers of the individuals of each species will have variation limited between positive numbers.

3. Let us now suppose that the fundamental determinant is not null. Then one can prove the theorem above this way:

If there is a state of equilibrium and if we start from a state different from that of equilibrium, we have fluctuations of species that are not damped.

come up: 1) the $N_1, N_2, \ldots N_n$ tend to determined and finite limits; 2) these numbers experience oscillations of an amplitude larger than some positive number.

Now, $N_1, N_2, \ldots N_n$ cannot tend to $q_1, q_2, \ldots q_n$, i.e. $n_1, n_2, \ldots n_n$ cannot tend to 1, for the smallest value of the c constant is

$$m = e^{q_1 \beta_1 + q_2 \beta_2 + \ldots + q_n \beta_n}$$

a value that corresponds to $n_1 = n_2 = \dots n_n = 1$.

At one given moment, it only takes one of these quantities being positive and different from unity, to render $c > m$. This is the reason why, if, in the initial state, $n_1, n_2, \dots n_n$ are not equal to unity, one must have $c > m$. But if $n_1, n_2, \dots n_n$ tend to 1, the first number of (9A.5-2) should tend to m, while staying constantly equal to $c > m$. Thus, $N_1, N_2, \dots N_n$ cannot tend towards $q'_1, q'_2, \dots q'_n$ numbers different from $q_1, q_2, \dots q_n$ in part or totally. For in this case the quantities $\beta_r(dN_r/dt)$ should tend towards determined and finite limits:

$$\left(\varepsilon_r \beta_r + \sum_1^n{}_s\, a_{sr} q'_s\right) q'_r \,.$$

But if N_r and dN_r/dt tend to determined and finite limits, then the derived dN_r/dt must tend to zero and, consequently,

$$\varepsilon_r \beta_r + \sum_1^n{}_s\, a_{sr} q'_s = 0 \,.$$

Resolving these equations in correlation with the q's, we have

$$q'_s = q_s \qquad (s = 1, 2, \dots n)$$

for the fundamental determinant is not null, which is in contradiction with the hypothesis we required for the q's.

So it is necessary for some of the $N_1, N_2, \dots N_n$ to maintain nondamped fluctuations while time grows indefinitely. Q.E.D.

4. Let us go back to the (9A.2-2) equations and integrate between t_0 and t. We shall have

$$\frac{\beta_r}{T} \log \frac{N_r}{N_r^0} = \varepsilon_r \beta_r + \sum_1^n a_{sr} \frac{1}{T} \int_{t_0}^{t} N_s dt \qquad \text{(9A.5-3a)}$$

N_r^0 being the value of $N_r(t_0)$ and $T = t - t_0$. Let us state

$$\frac{1}{T} \int_{t_0}^{t} N_r dt = n_r \,,$$

n_r will be the average of N_r over the interval (t_0, t).

So the (9A.5-3a) equations will be written

$$\frac{\beta_r}{T} \log \frac{N_r}{N_r^0} = \varepsilon_r \beta_r + \sum_1^n {}_s\, a_{sr}\, n_s \,. \qquad (9A.5\text{-}3b)$$

Taking a large enough T, we shall be able to make the moduli of the first members of the preceding equations as small as we want and, consequently, to make the n_r as close to the q_r as we want, from which we deduce

$$\lim_{t\to\infty} n_r = q_r \,.$$

So the q_r *are the asymptotic average of the* N_r *and, consequently, they are independent of the initial values of the* N_r.

5. If we take the initial values of $n_1, n_2, \ldots n_n$ close enough to unity, we shall be able to make c as close to

$$e^{\sum_1^n {}_h\, q_h \beta_h}$$

as we want and, consequently, K as close to 1 as we want. But we have

$$e \leqslant \frac{e^{n_r}}{n_r} \leqslant e\, \sqrt[q_r \beta_r]{K} \,.$$

So we shall be able to keep the n_r as close to 1 as we want, from which is deduced the following proposition:

> *If the initial state is close enough to the equilibrium state, the biological association will stay as close to that state as we want.* This is why *the equilibrium state will be stable.*

6. Let us consider the case where fundamental equations have null or negative roots. Let us call $r_1, r_2, \ldots r_h$ the species that correspond to the positive roots that we shall call $q_{r_1}, q_{r_2}, \ldots q_{r_h}$; and let us call $s_1, s_2, \ldots s_h$

the species corresponding to null roots. Let us first suppose that there are no negative roots.

The (9A.4-5a) integral will be written

$$\sum_{1}^{h}{}_{i} \beta_{r_i} (N_{r_i} - q_{r_i} \log N_{r_i}) + \sum_{1}^{k}{}_{i} \beta_{s_i} N_{s_i} = C' .$$

And the smallest value of

$$\beta_{r_i} (N_{r_i} - q_{r_i} \log N_{r_i}) \tag{9A.5-4}$$

will be achieved if $N_{ri} = q_{ri}$ and will equal

$$\beta_{r_i} q_{r_i} (1 - \log q_{r_i}) .$$

Consequently, the first of the preceding sum-totals will be larger or equal to

$$\sum_{1}^{h}{}_{i} \beta_{r_i} q_{r_i} (1 - \log q_{r_i}) = c_1 .$$

So

$$\sum_{1}^{k}{}_{i} \beta_{s_i} N_{s_i} \leqslant C' - c_1 .$$

This proves that the N_{s_i} are limited. Moreover, the N_{r_i} must stay bounded between positive numbers, for the (9A.5-4) expressions grow indefinitely when the N_{r_i} get close to zero or increase indefinitely.

7. Let us now suppose that the fundamental determinant is not null and see what happens to the N_{s_i} when t grows indefinitely. If they too stayed bounded between positive numbers, the (9A.5-3b) formulas would be valid, which is why the asymptotic averages of all the N_i should tend towards the roots of the fundamental equations, which is impossible because, among these roots, some are null, while the asymptotic averages should be bounded between positive numbers.

So, for a sufficiently large t, a few of the N_{s_i} will have to be smaller than any arbitrarily chosen positive number. As is easily seen, it does not prove that there are N_s tending to zero, but that there must be species that will become exhausted whether in regular fashion or not (see Section 9A.5, No. 1).

If the $s_1, s_2, \ldots s_k$ species correspond to null roots, and if to the $v_1, v_2, \ldots v_g$ species correspond the negative roots $-p_1, -p_2, \ldots -p_g$, the (9A.4-5a) integral will become

$$\sum_{1}^{h}{}_i\, \beta_{r_i}(N_{r_i} - q_{r_i} \log N_{r_i}) + \sum_{1}^{k} \beta_{s_i} N_{s_i} + \sum_{1}^{g} \beta_{v_i}(N_{v_i} + p_i \log N_{v_i}) = C' .$$

The N_{s_i} and N_{v_i} can be seen to be limited, but one cannot assert that all $N_1, N_2, \ldots N_n$ will be bounded between two positive numbers.

8. Let us consider the case where the fundamental determinant is null.

1) If the fundamental equations have solutions constituted of positive roots, the theorem stated at the end of Section 9A.5, No. 2 will be verified.

2) If the roots are positive or null, the result of Section 9A.5, No. 6 is valid.

3) If the fundamental equations have null and negative roots it can be concluded, through reasonings analogous to those of Section 9A.5, No. 7, that the species corresponding to these null and negative roots will be limited.

4) If the $\varepsilon_1, \varepsilon_2, \ldots \varepsilon_n$ are not null and if all the (9A.3-8) relations are not satisfied, there will not be any solution to the fundamental equations. Consequently, there will be $q_1^{(h)}, q_2^{(h)}, \ldots q_n^{(h)}$ values such that

$$N_1^{\beta_1 q_1^{(h)}} N_2^{\beta_2 q_2^{(h)}} \ldots N_n^{\beta_n q_n^{(h)}} = Ce^{zt}$$

C being a constant

$$z = \sum_{1}^{n}{}_r\, \beta_r \varepsilon_r q_r^{(h)} \gtrless 0 .$$

When t grows indefinitely, the second member tends to 0 or ∞. *So, in this case, it is impossible for the numbers of each species to be bounded between two positive numbers.*

5) The intrinsic coefficients of increase being null we shall have the (9A.4-5b) and (9A.4-5c) integrals, which is why the species are limited. If one of the $q_1^{(h)}, q_2^{(h)}, \ldots q_n^{(h)}$ solutions is positive, that is if an equilibrium

state exists, we shall have the $N_1, N_2, \dots N_n$ bigger than a positive number and, consequently, all these numbers will be bounded between two positive numbers.

We can give a significant enough example of this case. Let there be three species, the numbers of individuals of which can be considered as the cartesian coordinates of a space.

The fundamental equations may be written

$$\beta_1 \frac{dN_1}{dt} = (cN_2 - bN_3)N_1$$

$$\beta_2 \frac{dN_2}{dt} = (aN_3 - cN_1)N_2$$

$$\beta_3 \frac{dN_3}{dt} = (bN_1 - aN_2)N_3$$

$a = a_{32} = -a_{23}$, $b = a_{13} = -a_{31}$, $c = a_{21} = -a_{12}$ being the constants of one sign. The states of equilibrium will correspond to the straight line

$$\frac{N_1}{a} = \frac{N_2}{b} = \frac{N_3}{c}$$

and the (9A.4-5b) and (9A.4-5f) integrals to the plane

$$\beta_1 N_1 + \beta_2 N_2 + \beta_3 N_3 = C_2 \qquad (9A.4\text{-}5b_1)$$

and the surface

$$N_1^{\beta_1 a} N_2^{\beta_2 b} N_3^{\beta_3 c} = C_3 . \qquad (9A.4\text{-}5f_1)$$

The intersection of the plane with this surface is a closed curve which will give us the cycle of the values of N_1, N_2, N_3. The cycle being closed, one can see that the phenomenon is a periodic one.

PART II The General Laws of the Struggle for Existence

9B.1 The Principle of the Conservation of Demographic Energy

1. Let C_0 be a constant equal to the limit superior of $\sum_1^n \beta_r \varepsilon_r X_r$. The (9A.4-1a) integral will be written

$$L + M = \sum_1^n{}_r \beta_r N_r + (C_0 - \sum_1^n{}_r \beta_r \varepsilon_r X_r) = \text{const.}$$

From the biological point of view, considering L as *an actual demographic energy* and M as *a potential demographic energy,* we will have one being transformed into the other while their sum stays constant. This proposition is analogous to the theorem of virial forces in mechanics.

2. Let us label $I_1, I_2, \ldots I_n$ the demographic coefficients and suppose that $\delta X_1, \delta X_2, \ldots \delta X_n$ are the infinitesimal changes of the life quantities. One can consider

$$\Sigma_r \beta_r I_r \delta X_r$$

as *the increasing demographic work.* If we take

$$I_r = \varepsilon_r + \frac{1}{\beta_r} \sum_1^n{}_s a_{sr} N_s$$

and is we suppose that the δX_i are the natural increases taking place during the time dt we will have

$$\delta X_r = N_r \, dt$$

and consequently the demographic work will be

$$\Sigma_r \beta_r \varepsilon_r N_r \, dt$$

i.e. the work corresponding to the increases caused by the reciprocal actions of the different species will be null and the total work will be reduced to the auto-increases only.

3. In certain cases, the life spent by one species in a certain environment modifies its properties so that the population increase rates change.
For example, this is what happens when catabolic products are emitted by

the different individuals. For those products are, in some cases, capable of poisoning the environment. The variation of the environment thus produced by the different species instantaneously, is the result of the activity of the individuals in the preceding instants. If every species' activity remains constant it may be considered at every instant as proportional to the life quantity of the species and, consequently, the change of every r species' increase rate may be given by an expression

$$\frac{1}{\beta_r} \sum_{1}^{n}{}_s\, c_{rs} X_s$$

where the c_{rs} are constant quantities.

When the action between the different species is reciprocal we shall have $c_{rs} = c_{sr}$ and consequently the above expression will be written

$$\frac{\partial}{\partial X_r} \left[\frac{1}{2} \sum_{1}^{n}{}_r \sum_{1}^{n}{}_s\, c_{rs} X_r X_s \right] .$$

Stating

$$P = \sum_{1}{}_r\, \beta_r \varepsilon_r X_s + \frac{1}{2} \sum_{1}^{n}{}_r \sum_{1}^{n}{}_s\, c_{rs} X_r X_s \qquad \text{(9B.1-1)}$$

the fundamental equations will become

$$\beta_r X_r'' = \left(\frac{\partial P}{\partial X_r} + \sum_{1}^{n}{}_s\, a_{sr} X_s' \right) X_r' \qquad \text{(9A.2-3c)}$$

and, consequently,

$$\sum_{1}^{n}{}_r\, \beta_r X_r'' = \frac{dP}{dt} .$$

Integrating, one obtains

$$\sum_{1}^{n}{}_r\, \beta_r N_r = P - c_0 , \qquad \text{(9A.4-1b)}$$

c_0 being a constant.

In order to use a nomenclature as close as possible to that of dynamics, we shall call P *the demographic potential.* The potential demographic energy will be

$$c_0 - P = n$$

and we will have as before

$$L + n = \text{const.}$$

The demographic work done in the time dt will be

$$dP.*$$

In this section we have introduced a demographic potential having second degree terms along with first degree ones. In what is to follow, and unless we explicitly say the contrary, we shall suppose that the demographic potential only has first degree terms.

9B.2 The Three Laws of Biological Fluctuations

1. We saw that in the case of an odd number of species, the fundamental determinant is null and that there are generally no equilibrium states and that the numbers of individuals of some species must increase indefinitely or become smaller than any positive number. So in the case of an odd number of species, chances are that the association will not be preserved and that it will finally become an association made of an even number of species.

2. Let us then especially examine the case of an even number of species, the fundamental determinant not being null. Let us moreover suppose that all the roots of the fundamental equations are positive and that there is a state of equilibrium

In this case we shall be able to formulate the three fundamental laws of fluctuation.

First Law. Law of the Conservation of the Fluctuations
The number of individuals of different species are bounded between positive numbers and there always are fluctuations that are not damped (see Section 9A.4, No. 3).

* In the book *The Biological Associations from a Mathematical Point of View* (Hermann, Paris, 1935), taking into account the variations of catabolic actions with the time, integro-differential equations were found. The reciprocity of the c_{rs}, c_{sr} coefficients would not be verified in the experiment of Mr. Regnier and Miss Lambin.

Second Law. Law of the Conservation of Averages
If one takes as the averages of the individuals of the different species the averages for infinitely long lapses of time (asymptotic averages) these averages are constants independent of the initial values of the numbers of the individuals of the species (see Section 9A.5, No. 5).

Third Law. Law of the Perturbation of Averages
Should all the species be uniformly destroyed proportionally to the numbers of the individuals, there will always be species that will benefit from it (i.e. whose averages will increase) and species that will be damaged (i.e. whose averages will decrease).*

Among the first ones, there will be one at least of the ones that are devoured by others and among the second ones there will be one at least of the ones that devour others.

In order to demonstrate this law, let us recall that the averages (asymptotic averages) equal the equilibrium populations. Consequently, what is said of the latter apply to the former. And we know that if $\Delta\varepsilon_1, \Delta\varepsilon_2, \dots \Delta\varepsilon_n$ have the same sign, a few at least of the $\Delta q_1, \Delta q_2, \dots \Delta q_n$ are not null and have different signs (see Section 9A.3, No. 4).

Suppose $\Delta q_1 > 0$. There are two possible hypotheses: either some of the $a_{21}, a_{31}, \dots a_{n1}$ are negative and then the 1 species will be devoured by others; or all those quantities are positive or null, and then the 1 species will devour others while none will devour it. In that second hypothesis (see Section 9A.3, No. 4), by taking $\Delta\varepsilon_1, \Delta\varepsilon_2, \dots \Delta\varepsilon_n$ negative we will find among the species $2, 3, 4, \dots n$ a few whose equilibrium population will increase.

There will be at least one species that is devoured by others the population of which will increase while diminishing the $\varepsilon_1, \varepsilon_2, \dots \varepsilon_n$.

We supposed that Δq_1 was positive. Consequently, there must be at least one negative one among $\Delta q_2, \Delta q_3, \dots \Delta q_n$. Suppose $\Delta q_2 < 0$, then either there will be positive ones among the $a_{12}, a_{32}, \dots a_{n2}$ and the 2 species will be devoured by others without devouring any; consequently, if $\Delta\varepsilon_1, \Delta\varepsilon_2, \dots \Delta\varepsilon_n$ are negative there will be some species devouring others whose averages will decrease.

Thus is the third law demonstrated.

3. In order to make things even clearer we can give other detailed explanations of the third law.

*We suppose that the destruction is mild enough to allow the possibility of a stationary state.

For three categories of species can be distinguished in a biological association: 1) the ones that devour others without being devoured by any; 2) the ones that are devoured by others without devouring any; 3) the ones that are devoured by other species and also devour others.

All categories may exist, or two of them. If there is only one it must be of the third kind.

Now, the second part of the former wording of the third law can be replaced by the following words:

Among the first ones there will be one at least belonging to the second or third category and among the second ones there will be one at least belonging to the first or third category.

Let us suppose that the only existing categories are the first and second ones*, then, among the species that benefit from a decrease of the intrinsic coefficients of increase, there will be one at least belonging to the second category/ and, among the damaged ones, there will be one at least belonging to the first category.

This was the particular form I had used to state the third law in my previous works, limiting its signification to either devoured or devouring species. The terms we have just used to express the third law give it a new extension by suppressing any restriction.

We obviously follow analogous reasoning if, instead of decreasing ε_1, ε_2, ... ε_n, we increase them.

4. Let us now take a particular version of the general case by considering the case of two species, the first one devouring the second one. The (9A.5-1) integral will become

$$\left(\frac{e^{N_1}}{N_1^{q_1}}\right)^{\beta_1}\left(\frac{e^{N_2}}{N_2^{q}}\right)^{\beta_2} = \text{const.}$$

and, by considering N_1 and N_2 (q_1 and q_2 being positive) as the cartesian coordinates of the plane we will have a closed cycle. So the phenomenon will be periodical and the asymptotic averages will be the averages during one period.

The three laws of biological fluctuations will then be:

*In this case the number of species of the first category must equal the number of those of the second one.

First Law. Law of Periodic Cycle
The fluctuations of the two species are periodic.

Second Law. Law of the Conservation of Averages
The averages of the numbers of the individuals of the two species over a period of time are constant and do not depend upon initial values.

Third Law. Law of the Perturbation of Averages
If the two species are uniformly destroyed proportionally to the number of their individuals, the average of the number of individuals of the devoured species increases and the average of the number of individuals of the devouring species decreases.

9B.3 The Variational Principle

1. We said in Section 9A.2, No. 5, that it is possible to bring the equations (9A.2-3a) back to depending upon a question of variational calculus. We are now going to realize that assertion and show too that it is true even if we consider instead of the (9A.2-3a) equations the (9A.2-3c) equations.

The tendency to reduce natural problems to problems of minima has always been there, for we think nature in its manifestations, tends to spare the greatest possible part of what it spends accomplishing the various phenomena. This was Maupertuis's starting point when, in a famous book, he thought he established one of nature's fundamental principles, which he called the principle of least action and which was to become the basis of dynamics.

The philosophical principle Maupertuis went by was that nature always selects the simplest means to act. Resting on that concept, he endeavored to draw all laws of rest and movement from one single metaphysical concept. Descartes had tried to take as a starting point the principle of the *quantité de movement* and Leibnitz that of virial forces. Maupertuis first gave a mathematical definition of his action quantity and proceeded to try to deduce solutions of natural problems from the principle of least action. He applied it to the collision of hard elastic bodies. He also reminds us in his book of Fermat's famous principle of light refraction, connecting it with his own principle.

Maupertuis's principle is obviously a lot more comprehensive than Descartes or Leibnitz's, for the latter only give the integrals of the dynamics equations while Maupertuis's principle is equivalent to the very equations.

Lagrange rested the dynamics on another basis and proved as a consequence of his equations the principle of least action. He thus brought mechanics into the variational calculus that he had helped to create.

Hamilton was the one to develop the question next and allow it to progress, first by establishing the so-called Hamilton's principle, then by developing the action principle.

And finally, Jacobi systematized the general theory through his reaching an equation which has recently had a great extension and which proves more extensive every day.

We shall see that the same succession can be observed with the biological problem and show that struggle for existence equations can be put in the canonical form and can further lead to a Jacobian type of equation.

2. While studying the integrals of fundamental equations (see Section 9A.4) we found the expression

$$\chi = \sum_{1}^{n} {}_{r}\, \beta_r X_r' \log X_r' , \qquad (9A.4\text{-}3b)$$

N_r being the population of a species, dN_r/N_r is its *infinitesimal relative increase* and $\int_1^{N_r} (dN/N)$ the total relative increase necessary to obtain the actual value N_r starting from a single individual.

Now, dX_r being the species's life quantity infinitesimal increase, we have

$$N_r = \frac{dX_r}{dt} = X_r' .$$

One can call *infinitesimal vital action* the quantity

$$\beta_r \log N_r \, dX_r = \beta_r N_r \log N_r \, dt = \beta_r X_r' \log X_r' \, dt,$$

trying to adopt an analogue to that used in mechanics (by I/β_r one means the equivalent of the r species; Section 9A.2, No. 2).

So during the time interval (0, t) the total action will be

$$\int_0^t \beta_r N_r \log N_r \, dt = \int_0^t \beta_r X_r' \log X_r' \, dt$$

and, if one considers the association of n species, the vital action will be

$$A = \int_0^t \sum_{1}^{n} {}_{r}\, \beta_r X_r' \log X_r' \, dt = \int_0^t \sum_{1}^{n} {}_{r}\, \beta_r N_r \log N_r \, dt . \qquad (9B.3\text{-}1)$$

3. We have also found (see Section 9A.4) the bilinear form

$$Z = \sum_{1\,r}^{n} \sum_{1\,s}^{n} a_{rs} X'_r X_s \tag{9A.4-3a}$$

and, in Section 9B.1, the *demographic potential*

$$P = \sum_{1\,r}^{n} \beta_r \varepsilon_r X_r + \frac{1}{2} \sum_{1\,r}^{n} \sum_{1\,s}^{n} c_{rs} X_r X_s \tag{9B.1-1}$$

Let us form

$$\Phi = \chi + \frac{1}{2} Z + P = \sum_{1\,r}^{n} \beta_r X'_r \log X'_r + \frac{1}{2} \sum_{1\,r}^{n} \sum_{1\,s}^{n} a_{rs} X'_r X_s + P \tag{9B.3-2a}$$

We shall have

$$\frac{\partial \Phi}{\partial X'_r} = \beta_r \log X'_r + \beta_r + \frac{1}{2} \Sigma_s a_{rs} X_s \tag{9B.3-2b}$$

$$\frac{\partial \Phi}{\partial X_r} = \frac{1}{2} \Sigma_s a_{sr} X'_s + \frac{\partial P}{\partial X_r} \tag{9B.3-2c}$$

Let us now consider

$$U = \int_{t_0}^{t} \Phi \, dt \, . \tag{9B.3-3a}$$

By writing

$$\delta U = 0 \tag{9B.3-3b}$$

the Euler equations will be

$$\frac{d}{dt} \frac{\partial \Phi}{\partial X'_r} - \frac{\partial \Phi}{\partial X_r} = 0 \, , \tag{9B.3-3c}$$

that is

$$\frac{d}{dt} (\beta_r \log X'_r) = \frac{\partial P}{\partial X_r} + \sum_{1\,s}^{n} a_{sr} X'_s \tag{9B.3-3d}$$

which are the (9A.2-3c) equations introduced in Section 9B.1 and which are reduced to the (9A.2-2a) equations of Section 9A.2 when the demographic potential is reduced to first degree terms.

We have then reduced the fundamental equation of the struggle for life to a problem of variational calculus.

9B.4 Canonical Equations

1. Having put the fundamental equations (in cases where the demographical potential has second degree terms) into the Lagrangian form, we shall now be able to obtain the canonical form.

Let us put

$$p_r = \frac{\partial \Phi}{\partial X_r'} = \beta_r \log X_r' + \beta_r + \frac{1}{2} \sum_1^n {}_s\, a_{rs} X_s \qquad (9B.3\text{-}2d)$$

from which we deduce

$$X_r' = e^{(1/\beta_r)(p_r - \beta_r - \frac{1}{2} \sum_1^n {}_s\, a_{rs} X_s)} . \qquad (9B.3\text{-}2e)$$

By taking

$$H = \Phi - \sum_1^n {}_r\, p_r X_r' = P - \sum_1^n {}_r\, \beta_r X_r'$$

$$= P - \sum_1^n {}_r\, \beta_r e^{(1/\beta_r)(p_r - \beta_r - \frac{1}{2} \sum_1^n {}_s\, a_{rs} X_s)} , \qquad (9B.4\text{-}1a)$$

we shall have

$$\frac{\partial H}{\partial p_r} = -e^{(1/\beta_r)(p_r - \beta_r - \frac{1}{2} \sum_1^n {}_s\, a_{rs} X_s)} = - X_r' = - \frac{dX_r}{dt} . \qquad (9B.4\text{-}1b)$$

But from (9B.4-1a), (9B.3-2c), (9B.3-3c), and (9B.3-2d) we have

$$\frac{\partial H}{\partial X_r} = \frac{\partial P}{\partial X_r} + \frac{1}{2} \sum_1^n {}_s\, a_{sr} X_s' = \frac{\partial \Phi}{\partial X_r} = \frac{d}{dt} \frac{\partial \Phi}{\partial X_r'} = \frac{dp_r}{dt} . \qquad (9B.4\text{-}1c)$$

We thus come to the canonical equations

$$\frac{dp_r}{dt} = \frac{\partial H}{\partial X_r}, \quad \frac{dX_r}{dt} = -\frac{\partial H}{\partial p_r} \qquad (9B.4\text{-}2a)$$

where H is given by the expression

$$H = P - \sum_{1\,r}^{n} \beta_r\, e^{(1/\beta_r)(p_r - \beta_r - \frac{1}{2}\sum_{1\,s}^{n} a_{rs}X_s)} \qquad (9B.4\text{-}1a)$$

2. The canonical equations have the integral

$$H = \text{const.}$$

This integral is the one of the demographic energy conservation that we gave in Section 9B.1, No. 3.

We also found (see Section 9A.4, No. 3), when the demographic potential has only first degree terms, the integral

$$\sum_{1\,r}^{n} \beta_r(N_r - q_r \log N_r) = \sum_{1\,r}^{n} \beta_r(X_r' - q_r \log X_r') = \text{const.} \qquad (9A.4\text{-}5a)$$

We can write

$$K = \sum_{1\,r}^{n} \left[\beta_r e^{(1/\beta_r)(p_r - \beta_r - \frac{1}{2}\sum_{1\,s}^{n} a_{rs}X_s)} - q_r(p_r - \beta_r - \frac{1}{2}\sum_{1\,s}^{n} a_{rs}X_s)\right]$$
$$= \text{const.}$$

It is easy to check, through very simple calculations, that the Poisson bracket (H, K) is null.

3. The Jacobian equation deduced from the (9B.4-2a) canonical equations will be

$$C = \sum_{1\,r}^{n} \beta_r \left[\varepsilon_r X_r - e^{(1/\beta_r)(\frac{\partial V}{\partial X_r} - \beta_r - \frac{1}{2}\Sigma_s a_{rs}X_s)}\right] \qquad (9B.4\text{-}2b)$$

and the integrals of the canonical equation will be

$$\frac{\partial V}{\partial a_r} = b_r, \quad \frac{\partial V}{\partial X_r} = p_r,$$

$V(X_1, X_2, \ldots X_n; a_1, a_2, \ldots a_n)$ being a complete integral of the equations with partial derivatives (9B.4-2b) and $a_1, a_2, \ldots a_n; b_1, b_2, \ldots b_n$ being constants.

9B.5 Commuting Integrals

1. The (9A.2-2b) equations have the integrals (see Section 9A.4, No. 2)

$$\beta_r \log \frac{dX_r}{dt} + \sum_{1s}^{n} a_{rs} X_s - \varepsilon_r \beta_r t = \text{const.} \qquad (9A.4\text{-}2a)$$

Thanks to (9B.2-2d) equations they can be written

$$\frac{p_r + \frac{1}{2} \sum_{1s}^{n} a_{rs} X_s}{\varepsilon_r \beta_r} - t = \text{const.} \qquad (9B.5\text{-}1)$$

Putting

$$\frac{p_r + \frac{1}{2} \sum_{1s}^{n} a_{rs} X_s}{\varepsilon_r \beta_r} = H_r$$

and eliminating the t time, we shall find the integrals of the canonical equations (9A.3-2a) (P having only first degree terms) to be

$$H_r - H_i = H_{ri} = \text{const.}$$

It is easy to verify that

$$(H, H_r) = 1, \quad (H_r, H_h) = \frac{a_{hr}}{\varepsilon_h \beta_h \varepsilon_r \beta_r} \qquad (9B.5\text{-}2)$$

and consequently,

$$(H, H_{ri}) = 0 \qquad (9B.5\text{-}3)$$

$$(H_{rh}, H_{g\ell}) = \frac{a_{gr}}{\varepsilon_g \beta_g \varepsilon_r \beta_r} + \frac{a_{\ell h}}{\varepsilon_i \beta_i \varepsilon_h \beta_h} + \frac{a_{r\ell}}{\varepsilon_r \beta_r \varepsilon_\ell \beta_\ell} + \frac{a_{hg}}{\varepsilon_h \beta_h \varepsilon_g \beta_g} \qquad (9B.5\text{-}4)$$

Let us put

$$- H - K = L - \sum_1^n{}_i\, q_i \beta_i \qquad (9B.5\text{-}5)$$

we shall have

$$L = \sum_1^n{}_r\, \varepsilon_r \beta_r q_r H_r = \sum_1^n{}_r\, \varepsilon_r \beta_r q_r H_{ri}$$

because of the (9A.3-4) relation (Section 9A.3, No. 3)

$$\sum_1^n{}_r\, \varepsilon_r \beta_r q_r = 0 .$$

Consequently,

$$(L, H_h) = \sum_1^n{}_r\, \varepsilon_r \beta_r q_r (H_r, H_h) = \frac{1}{\varepsilon_h \beta_h} \sum_1^n{}_r\, a_{rh} q_r = -1$$

and

$$(L, H_{rh}) = 0 .$$

2. So the integrals of the canonical equations

$$H, L, H_{rh}$$

are independent and commuting.

Every linear combination of $H_{12}, H_{13}, \ldots H_{1n}$ commutes with the H and L functions. If one could find $n - 2$ independent commuting ones and independent from L too, the problem would be reduced to quadratures, for we would have n independent commuting integrals.

But unless one states some limitation on the constant quantities $a_{rs} = -a_{sr}, \beta_r, \varepsilon_r$, this is impossible, for one could get $H_{12}, H_{13}, \ldots H_{1n}$ from it, linearly expressed by commuting functions and, consequently, themselves commuting, which would be in contradiction with the (9B.5-4) equations But for particular values of the $a_{rs} = -a_{sr}, \beta_r, \varepsilon_r$ constants, things can go differently.

3. Let us imagine n is any n and point out that, in order for H_{1h}, H_{1g} to commute, it is necessary and sufficient to have (see (9B.5-4))

$$\frac{a_{gh}}{\varepsilon_g \beta_g \varepsilon_h \beta_h} + \frac{a_{1g}}{\varepsilon_1 \beta_1 \varepsilon_g \beta_g} + \frac{a_{h1}}{\varepsilon_h \beta_h \varepsilon_1 \beta_1} = 0 .$$

Consequently, if

$$a_{rs} = \varepsilon_r \beta_r \varepsilon_s \beta_s (m_s - m_r) \qquad (9B.5\text{-}6)$$

then, $m_1, m_2, \ldots m_n$ being constants, $H_{12}, H_{13}, \ldots H_{1n}$ will commute. They will, moreover, be independent, for each of them contains a p_r that the others do not contain.

So, in cases where the a_{rs} have the (9B.5-6) form, the problem is reduced to quadratures.

4. One can verify this result directly from the (9A.2-2) equations (see Section 9A.2, No. 3). For, if we state

$$\Sigma\, \varepsilon_s \beta_s N_s = N, \qquad 1 - \Sigma_s\, \varepsilon_s \beta_s m_s N_s = M$$

they become

$$\frac{1}{\varepsilon_r} \frac{d}{dt} \log N_r = m_r N + M$$

and, if M and N are eliminated,

$$\frac{\frac{1}{s_1}\frac{d}{dt}\log N_1 - \frac{1}{s_2}\frac{d}{dt}\log N_2}{m_1 - m_2} = \frac{\frac{1}{\varepsilon_1}\frac{d}{dt}\log N_1 - \frac{1}{\varepsilon_3}\frac{d}{dt}\log N_3}{m_1 - m_3}$$

$$= \ldots\ldots = \frac{\frac{1}{\varepsilon_1}\frac{d}{dt}\log N_1 - \frac{1}{\varepsilon_n}\frac{d}{dt}\log N_n}{m_1 - m_n}$$

which immediately gives us $n - 2$ integrals independent of time.

But in the (9A.2-2) equations, eliminations occur and we get $n - 1$ equations having a multiplier, which proves that the integration is reduced to quadratures.

9B.6 The Principle of Least Action in Biology

1. We have found (see Section 9A.4, No. 2) the integral

$$\Omega = \sum_{1}^{n}{}_r X_r' \Theta_r + \sum_{1}^{n}{}_r \varepsilon_r \beta_r X_r - \sum_{1}^{n}{}_r \beta_r X_r' - C = 0 \qquad (9A.4\text{-}2c)$$

of the (9A.2-2a) equations

$$\beta_r \frac{d^2 X_r}{dt^2} = \left(\varepsilon_r \beta_r + \sum_{1}^{n}{}_s a_{sr} \frac{dX_s}{dt}\right) \frac{dX_r}{dt}\,. \qquad (9A.2\text{-}2a)$$

If we vary the X_r, quantities so that $\delta t = 0$ (isochronous variation), we shall find

$$\delta\Omega = \sum_{1}^{n}{}_r \delta X_r' \Theta_r + \sum_{1}^{n}{}_r \delta X_r \left(\sum_{1}^{n}{}_s a_{sr} X_s' + \varepsilon_r \beta_r\right)$$

and, supposing the (9A.4-2a) equations are verified by the $X_1, X_2, \dots X_n$ we shall have

$$\delta\Omega = \sum_{1}^{n}{}_r \delta X_r \left(\sum_{1}^{n}{}_s a_{sr} X_s' + \varepsilon_r \beta_r\right).$$

So, if the isochronous variation is such that

$$\sum_{1}^{n}{}_r \delta X_r \left(\sum_{1}^{n}{}_s a_{sr} X_s' + \varepsilon_r \beta_r\right) = 0$$

we shall have $\delta\Omega = 0$ that is we will be able, through an isochronous variation, to conserve the integral, (9A.4-2c).

Consequently, the two conditions

$$\sum_{1}^{n}{}_r \delta X_r \left(\sum_{1}^{n}{}_s a_{sr} X_s' + \varepsilon_r \beta_r\right) = 0 \qquad (9B.6\text{-}1a)$$

$$\delta\Omega = 0 \tag{9B.6-1b}$$

are equivalent for a same isochronous variation.

Obviously, we can arbitrarily take $n-1$ of the $\delta X_1, \delta X_2, \ldots \delta X_n$ and the nth will be determined by the (9B.6-1a). Consequently, we shall be able to simultaneously nullify $\delta X_1, \delta X_2, \ldots \delta X_n$ at the bounds 0, t.

2. We saw that the (9B.6-1a) and (9B.6-1b) conditions are equivalent. The second one expresses the fact that the isochronous variation conserves the (9A.4-2c) integral.

Let us now try to interpret the first condition.

In this regard, let us call that the demographic coefficient of the various species existing at one certain moment being (see Section 9B.1, No. 2) $I_1, I_2, \ldots I_n$ and $\delta X_1, \delta X_2, \ldots \delta X_n$ being virtual variations of the life quantities, we can consider

$$\Sigma_r \beta_r I_r \delta X_r$$

as the demographic virtual work. Given that demographic coefficients are expressed by

$$\varepsilon_r + \frac{1}{\beta_r} \sum_{1s}^{n} a_{sr} N_s = \varepsilon_r + \frac{1}{\beta_r} \sum_{1s}^{n} a_{sr} X'_s$$

then

$$\sum_{1r}^{n} \left(\varepsilon_r \beta_r + \sum_{1s}^{n} a_{sr} X'_s \right) \delta X_r$$

will be the demographic virtual work for the variations $\delta X_1, \delta X_2, \ldots \delta X_n$. This is the reason why the (9B.6-1a) condition expresses that the virtual demographic work is null.

3. We have found that the struggle for existence laws are dependent on a question of variational calculus, i.e. that when the biological fluctuation equations are satisfied, the integrand is stationary for all the infinitesimal variations of the parameters which characterize uniquely the successive states of a biological association (Section 9B.3).

We are now going to demonstrate that the natural passage from one to another state of the association does correspond, under given conditions, to a minimum of the expression we called vital action (Section 9B.3, No. 2).

What comes out of it is a principle analogous to the principle of least action in mechanics.

Let us first write

$$2\Phi - \Omega = \sum_1^n{}_r \beta_r X_r' \log X_r' + \frac{d}{dt}\left[\sum_1^n{}_r \varepsilon_r\beta_r t X_r + \sum_1^n{}_r (C_r + \beta_r)X_r\right] + C$$

and then

$$\chi = \sum_1^n{}_r \beta_r X_r' \log X_r'$$

$$= 2\Phi - \Omega - \frac{d}{dt}\left[\sum_1^n{}_r \varepsilon_r \beta_r tX_r + \sum_1^n{}_r (C_r + \beta_r) X_r\right] - C$$

and consequently

$$A = \int_0^t \chi dt$$

$$= \int_0^t (2\Phi - \Omega)dt - \left[\sum_1^n{}_r \varepsilon_r \beta_r tX_r + \sum_1^n{}_r (C_r + \beta_r) X_r\right]_0^t - Ct. \tag{9B.3-1}$$

Let us choose $X_1, X_2, \dots X_n$, so that the (9A.2-3a) equations are verified, i.e. so that we have

$$\frac{d}{dt}\frac{\partial\Phi}{\partial X_r'} - \frac{\partial\Phi}{\partial X} = 0.$$

Let us moreover suppose that the isochronous variations $\delta X_1, \delta X_2, \dots \delta X_n$ satisfy the (9B.6-1b) condition and suppose that those quantities are null at the bounds 0, t. We shall have, through the (9B.3-1) equation

$$\delta A = 0.$$

4. We are getting to the final result. For we get from the (9B.3-1) equation

$$\delta^2 A = \delta^2 \int_0^t \chi\, dt = \int_0^t \sum_1^n{}_r \beta_r \frac{\delta X_r'^2}{X_r'} dt$$

and the X_r' are positive, consequently,

$\delta^2 A > 0.$

This goes to show that *any infinitely small variation of the* $X_1, X_2, \dots X_n$ *conserving the* (9A.4-2c) *integral, causes an increase of the vital action.* So we have a minimum. This proves the principle of least vital action.

5. One could use the (9B.6-1a) relation directly, instead of the equivalent (9B.6-1b) relation, to obtain the same result, but there is an easier, more direct way of reaching the result above, and which will make it more precise, through the steps we shall now show.

Let us change in

$$\chi = \sum_{1r}^{n} \beta_r X'_r \log X'_r = \sum_{1r}^{n} \beta_r N_r \log N_r$$

X_r into $X_r + \Delta X_r = X_r + \xi_r$ and, consequently, X'_r into $X'_r + \xi'_r = N_r + \gamma_r$ where $\gamma_r = \xi'_r$.

Supposing that $\gamma_r > -N_r$ we shall have $N_r \log N_r$ turn into

$$(N_r + \gamma_r) \log(N_r + \gamma_r) = N_r \log N_r + \gamma_r(\log N_r + I) + N_r f\left(\frac{\gamma_r}{N_r}\right)$$

stating that

$$f(x) = (1 + x) \log (1 + x) - x.$$

Consequently $A = \int_0^t \chi \, dt$ will become

$$\int_0^t \chi \, dt + \int_0^t \sum_{1r}^{n} \beta_r(\log N_r + I)\xi'_r \, dt + \int_0^t \sum_{1r}^{n} \beta_r N_r f\left(\frac{\gamma_r}{N_r}\right) dt.$$

If the ξ_r are null at the $0, t$ limits we shall have

$$\int_0^t \sum_{1r}^{n} \beta_r (\log N_r + I) \xi'_r \, dt = - \int_0^t \sum_{1r}^{n} \beta_r \frac{N'_r}{N_r} \xi_r \, dt \qquad (9B.6\text{-}2)$$

and if the equations

$$\sum_{1r}^{n} \xi_r \left(\sum_{1}^{n} a_{sr} N_s + \varepsilon_r \beta_r\right) = 0$$

are verified, we shall find that, because of the (9A.2-2) equations, the (9B.6-2) expression is null. Consequently, $A = \int_0^t \chi \, dt$ will grow by

$$\int_0^t \sum_1^n{}_r \beta_r N_r f\left(\frac{\gamma_r}{N_r}\right) dt .$$

a quantity that will be positive for $\gamma_r > - N_r$ and null only when all the γ_r are null.

For $(df(x))/dx = \log(1 + x)$ and consequently $f(\gamma_r/N_r)$ will decrease, while γ_r varies between $-N_r$ and 0, will be equal to 0 for $\gamma_r = 0$ and will increase for the positive values of γ_r.

We shall then have the following proposition, taking into account the fact that the population cannot be negative.

Let us isochronously modify the natural passage of a biological association from one state to another, by varying the populations of different species.

The vital action will increase if the quantities of life at the initial point in time and at the final point in time do not change and if the demographic virtual work is null at every point in time (see Section 9B.6, No. 4).

So we have an actual minimum of vital action, which is the principle of least action in biology.

6. We have thus constituted a *biological dynamics* that is like *the dynamics of material systems.* For the variational principle (see Section 9B.3) can be compared to *Hamilton's* principle and we have also obtained in biology the principle corresponding to the least action one. The passage from one to the other, either in the biological field or in the field of material systems, can be effected in an analogous fashion, considering for mechanics moving systems through which no mechanical work is executed, as one considers for biology vital action variations without demographic work.

We may point out that in biology we have indeed a minimum of the action, which is not always true for material systems mechanics.

That circumstance must not surprise us as the general principles we have just compared, while being apparently analogous, differ from one another because of the functions which, on one hand, express mechanical action, on the other, vital actions.

Arricia (Roma), September 1936.

References

Gause, G.F. 1934. *The Struggle for Existence.* The Williams & Wilkins Company, Baltimore. S. i-x, 1-163.

Levi, B. 1931. *Bull. Un. Mat. It.,* Anno X, No. 4.

Regnier et Lambin. 1934. Étude d'un cas d'antagonisme microbien *(B. Coli. Staphylococcus aurens),* avec des remarques par Vito Volterra. Comptes rendus de l'Acad. des Sciences. December 1934.

Volterra, Vito. 1931. *Leçons sur la théorie mathématique de la lutte pour la vie.* Gauthier-Villar, Paris.

Volterra, Vito et Umberto D'Ancona. 1935. *Les associations biologiques au point de vue mathématique.* Hermann et Cie, Paris.

10

PRINCIPES DE PHILOSOPHIE ZOOLOGIQUE (PRINCIPLES OF ZOOLOGICAL PHILOSOPHY)

J. VON GOETHE

TRANSLATED BY B. TAYLOR (1980)

On the 22nd February of this year there took place at a sitting of the French Academy an important event which must have highly important consequences. In this sacred place of science where everything customarily proceeds in the most restrained manner in the presence of a large audience, where one encounters the moderation, indeed dissimulation, of well brought-up people; where one replies to differences of opinion only with moderation, where what is doubtful is set aside rather than disputed, a dispute over a scientific point is taking place which threatens to become personal, but which, looked at closely, is of great significance.

What has come to the surface here is the never-ending conflict between the two ways of thought, into which the scientific world has long been divided and which was always present among our neighboring scientific researchers but which has broken unusually violently on this occasion.

Two excellent men, the permanent secretary of the academy, Baron Cuvier, and a worthy member, Geoffrey de Saint-Hilaire, are the opponents; the former is of great renown to everyone, the second to scientific researchers. Colleagues at the same institution for thirty years, they have taught natural history at the Jardin des Plantes, both of them [have been] most earnestly active in this enormous field, at first working together communally, but gradually divided through a difference of views and preferring to avoid each other.

Cuvier works untiringly as a differentiator, describing exactly what is at hand and is winning mastery over an immeasurable breadth. Geoffrey de Saint-Hilaire, on the other hand, is quietly concerned with the analogies of creatures and their hidden relationships. The former proceeds from the particular to the whole which is postulated but never regarded as recognizable.

The latter cherishes the whole in its [or his] inner sense and continues to live in the conviction that the details can be worked our gradually. But it is important to notice that much which the latter is succeeding in proving clearly and distinctly in experiments [experience] is gratefully accepted by the former. Likewise the latter in no way despises whatever comes to him from the other that decides single details, and so they agree on many points without admitting a [mutual] reaction on this account. For the dissector, differentiator, depending on experimentally gained facts will not admit a preconception, a premonition of the detail in the whole. To want to recognize and know that which one does not see with the eyes, which one cannot represent tangibly he declares positively to be a presumption. The other one, on the contrary, holding to certain basic principles, yielding to a higher leadership does not wish to acknowledge the authority of that method of approach.

After this introductory "essay" no one will hold it against me if I repeat what was said above: two different ways of thought are in play here, which, in human kind, are usually divided and are split up so much that, as everywhere, even in the scientific realm, they are difficult to find united, and as they are divided, perhaps cannot be united. Indeed it goes so far that if the history of science and our own long experience can be of use, then it might be feared that human nature can scarcely be saved from this division. I will expand on what I have said.

The differentiator employs so much exact observation that he needs an uninterrupted attentiveness, a skill that pursues things to their smallest detail, which notes the variations of forms, and finally likewise needs the definite gift of the mind of naming these differences, to the extent that one cannot very well suspect him if he is proud of this and may esteem this method of approach as the only one that is well-founded and correct.

If he sees resting on this [method] the fame that is accorded him, he may not find it easy to force himself to share his recognized advantages with another [person] who has apparently made it easier for himself to reach a goal, where, in fact, the wreath [of honor] should only be awarded for diligence, hard work and perseverance.

Indeed, the one who starts out from the idea [IDEE] of being able to imagine something, the one who knows how to seize on a main concept to which experience is gradually subordinated, who lives in secure confidence that he will certainly again meet in single cases what he has found here and there and already expressed in general. We have to attribute to a man who is so inclined a kind of pride, a certain inner feeling of his advantages when he does not yield from his side, but can at least bear a certain depreciation which is often displayed toward him by the opposing side, even of a gentle, moderate kind.

But what makes the dispute irreconcilable may well be as follows. As the differentiator is concerned throughout with the tangible, can prove what he achieves, demands no special views, never brings forth what might appear as a paradox, then he must win a larger, indeed, a general public, while on the other hand the latter finds himself more or less a hermit who is not always able to agree even with those who feel duty bound to him. This antagonism has often occurred already in science, and the phenomenon must repeatedly recur, since as we have just seen, the [disparate, opposing] elements for it continue to grow divided, and, wherever they touch, always cause an explosion.

Now this mostly occurs when individuals of different nations, different ages or some other separated circumstances, affect each other. However, in the present case there appears the remarkable circumstance that two men, equally advanced in age, colleagues at the same institution, busy for so long in different directions in the same field, avoiding each other, tolerating each other, each continuing in his own direction, practicing the finest way of life, are yet finally subject to and subordinated to an outbreak [of hostilities] and a mutual dislike that is finally [made] public.

Having spent some time dealing in generalities, it is now apposite to approach more closely the work of St. Hilaire.

Since the beginning of March the Parisian daily newspapers have entertained us [with details] of such an occurrence by applauding one side or the other. The quarrel continued in several following sessions [of the French Royal Academy] until finally Geoffrey de Saint-Hilaire found the circumstances suited to the removal of those discussions from the circle [i.e., the Academy] and through a publication of his own to bring the matter before a larger public.

I have read and studied that publication and had to overcome many a difficulty, thus deciding [to bring out] this present essay so that many who take up the script I have mentioned might give me friendly thanks for helping them with some introduction. So may the chronicle of these latest French academic disputes be accepted as part of the work in question.

15th February 1830

Geoffrey de Saint-Hilaire brings out a report about an essay in which some young people bring out observations concerning the organization of molluscs; indeed with a particular preference for the approach which one calls a priori, and where the unity of organic composition is praised as the true key to observations of nature.

22nd February

Baron Cuvier comes up with his speech in rebuttal and fights against the pretentious single principle, declares it to be a subordinate one and expresses another [principle] which he declares to be higher and more fruitful.

At the same sitting
Geoffrey de Saint-Hilaire improvizes an answer in which he expresses his confession of faith even more strongly.

Sitting of 1st March
Geoffrey de Saint-Hilaire reads an essay in the same tone, in which he seeks to represent the theory of analogies as new and extremely useful.

Sitting of 22nd March
He [Saint-Hilaire] undertakes to apply usefully the theory of analogies to the organization of fishes.

At the same session
Baron Cuvier attempts to destroy the arguments of his opponent by connecting his assertions to the Os hyoides, which had been mentioned.

Sitting of 29th March
Geoffrey de Saint-Hilaire defends his views on the Os hyoides and adds some final observations.

The journal "le Temps" dated 5th March publishes a resumê favorable to Geoffrey de Saint-Hilaire under the headline, "On the teaching of philosophical harmonies in living beings." In its edition of 22nd March, the "National" does the same.

Geoffrey de Saint-Hilaire decides to take the matter out of the circle of the Academy, gets printed what has already happened and writes a foreword to it, "On the theory of analogies" and dates it 15th April.

By this means he now adequately makes clear his conviction so that he fortunately meets our wishes of bringing forth the situation in as generally understandable a form as possible, as he also maintains in a supplement the necessity of dealing [with the subject] in print, as in oral discussions what is right and what is wrong is heard equally.

Quite favorably to foreigners he mentions with contentment and agreement what the Germans and Edinburghers have achieved in this subject and acknowledges himself to be their ally, and from this the scientific world can promise itself meaningful advantages.

But here let us first add several alternating remarks in our manner of alternating from the general to the particular, so that there may be the most benefit for us.

If, in the history of states as well as scholars, we meet many examples that some particular, often small and chance event takes place which sets the previously hidden parties in open opposition, then we find here the same case, but this one has the unfortunate peculiarity that the very cause which has called forth these disputes is of a quite particular kind and leads the situation into areas where it is threatened by boundless confusion, in that the scientific points which are mentioned can neither arouse particular interest nor can they become clear to the large majority of the public. It must therefore be profitable to take the dispute back to its first elements.

But as everything that happens among men in their higher senses, looked at from an ethical point of view, must be considered and judged, but above all the personality, the individuality of the persons in question must especially be considered, so we wish first of all to make ourselves acquainted, if, only in most general terms with the life history of the two men we have mentioned.

Geoffrey de Saint-Hilaire, born in 1772, was employed as Professor of Zoology in 1793 at the time the Jardin du Roi was designated a public teaching institution. Soon afterwards Cuvier was called to this institution; both men worked harmoniously together, as well intentioned young men do, unconscious of their inner difference.

In 1798 Geoffrey de Saint-Hilaire became a member of the greatly problem. plagued expedition to Egypt and was thereby to some extent alienated from his teaching career; but his sense of proceeding from the general to the particular was constantly strengthened, and, after his return, during his participation in the great work on Egypt, he found the most desired opportunity to employ and use his method.

The trust that his views as well as his character had gained for him, showed itself subsequently when in 1810 the government sent him to Portugal in order to organize studies there, as they expressed themselves. He came back from this ephemeral undertaking and enriched the Paris Museum with much of importance.

As he now continued to work untiringly at his subject he was recognized by the nation also as an honourable man and in 1815 was elected a deputy [M.P.]. But this was not the showplace for him to shine; he never ascended the podium.

He clearly expressed at last the principles by which he observed nature in a work published in 1818. He expressed his main thought: Its organization of animals is subordinated to a general plan, modified only here and there, from which one concludes the differentiating [of animals].

Let us now turn to his opponent:

Georg Leopold Cuvier was born 1769 in what was at that time still a part of Wurttemberg; from this he gained a fairly exact knowledge of German language and literature. His decided inclination to natural history brought him into contact with the excellent Kielmeyer, which [contact] was later continued from abroad. I remember seeing, in 1793, earlier letters from Cuvier to the above mentioned researcher [and these were] remarkable for the characteristically masterfully drawn anatomies of lower organisms researched in the text.

During his stay in Normandy he worked on the Linnean class of worms, but stayed in touch with colleagues in Paris, and Geoffrey de Saint-Hilaire persuaded him to come to the capital. They joined in publishing several works for didactic purposes, and especially sought an ordering of mammals.

The qualities of such a man did not go unnoticed; in 1795 he was employed at the Central School in Paris and accepted as a member of the Institute in its first class. For the use of that school he published in 1798 "Tableaux élémentaires de l'histoire naturelle des animaux. 8."

He received the position of Professor of Comparative Anatomy and through his incisiveness gained a broad, clear overview, and through a clear, shining lecture [style] won the most widespread and loud applause. After Daubenton's departure he took his place at the College de France, was recognized by Napoleon, and entered the Department of Public Instruction. As a member [of the Department] he travelled through Holland and a part of Germany, through the parts of the [German] Empire which were then incorporated as *départements* [of France] in order to inspect schools and teacher training institutions; the report he submitted is available. I knew provisionally that he did not neglect in it to set out the advantages of German schools over the French ones.

Since 1813 he was called to higher state affairs, in which he was confirmed after the return of the Bourbons, and he continues to the present to be involved in public as well as scientific activity.

His works cannot be taken in at a glance; they encompass the whole realm of nature and his explanations also serve our knowledge of objects and as an example of [scientific] treatment.

He did not think only of researching and ordering the boundless kingdom of living organisms; even long extinct species owe their scientific resurrection to him.

How well he knows the whole world of mankind and is able to penetrate into the characters of his excellent coworkers is shown in the memorials which he erects to dead members of the Institute, where his extensive overview of all scientific areas may be recognized.

May the reader forgive the sketchiness of those biographical attempts; it was not a question here of instructing the necessary participants, to put something new before them, but only to remind them of what must have long been known about the two dignitaries.

But now one may well ask: what cause, what necessity does a German have to gain closer knowledge of this quarrel, perhaps even of joining one side or the other? If, however, one maintains that any scientific question, wherever it is discussed, will interest any cultured nation—as one may regard the scientific world perhaps as one single body—so one must prove here that we are particularly summoned on this occasion.

Geoffrey de Saint-Hilaire names several Germans as being of the same opinion as himself; Baron Cuvier, on the other hand, seems to have formed the most unfavourable ideas of our German efforts in this field. He expresses himself in a statement of 5th April [page 24 in the memorandum] as follows: "I know well, I know that in certain minds, behind this theory of analogies at least in some confused way another very old theory may lie concealed, which, already long refuted, has been brought up again by some Germans in order to favor the pantheistic system which they call natural philosophy [NATURPHILOSOPHIE] ." To comment on this utterance word for word, to make its sense clear, to set out the pious innocence of German thinkers about nature, would probably take an octavo volume; I want to try in succession to reach my purpose in the shortest way.

The situation of a researcher of nature like Geoffrey de Saint-Hilaire is indeed of the sort that it must give him satisfaction to be informed of the efforts of German researchers to some extent, to be convinced that they cherish similar views to his own and that he can therefore expect informed applause from their side and, if he demands it, sufficient support. As then, especially in modern times, it has never been to our western neighbors' disadvantage when they gained some knowledge of German research and effort.

The German researchers of nature which are named on this occasion are: Kielmeyer, Meckel, Oken, Spix, Tiedemann, and immediately one has to admit thirty years of our participation in these studies. But I may well maintain it has been more than fifty years that we have seen ourselves chained with true inclination to such investigations. Scarcely anyone but me remembers those beginnings, and may it be granted to me to mention here those young researchers by means of whom some light may even be shed on present quarrels.

"I'm not teaching; I'm telling." (Montaigne)

Second Section

"I'm not teaching; I'm telling." is how I closed the first section of my observations on the work in question. Now, however, I find it advisable, in order to determine more closely the standpoint from which I would like to be judged, to quote the words of a Frenchman which might express better than anything else the means by which I will try to make myself understood.

"There are intelligent men who have their own way of lecturing. They begin, following their mode, speak at first about themselves and only unwillingly get away from their own personality. Before they lay before you the results of their pondering, they feel a need to recount first where and how those observations came to them."

Grant me therefore without further presumption in that sense the handling, indeed only in the most general terms, of course of the history of those sciences to which I have dedicated my years.

Here one should mention therefore how early [in my life] an echo of natural history, undefined but persistent, had its influence on me. In the very year of my birth, 1749, Count Buffon published the first part of his "Histoire Naturelle" and aroused great sympathy [agreement] amongst those Germans who were then very susceptible to French influence. The volumes followed yearly, and so the interest of a cultured society paralleled my growth without my being aware of more than the name of this important man as well as the names of his eminent contemporaries.

Count Buffon [was] born in 1707. This excellent man had a joyous, free overview, lust for life and joy in all life in existence; he was gladly interested in everything there was. Man of life, of the world, he had throughout all [his life] the wish to please by his teaching, to enter our good graces by means of his instruction. His presentations are more depictions than descriptions; he presents the creature in its entirety, doing this especially willingly with regard to a human being; on account of which he makes the domestic animals follow the latter immediately. He uses all that is known; he does not only know how to use researchers of nature; he also understands how to make use of the results of all travellers. He is seen in Paris, the great center of knowledge, as superintendent of the already important royal cabinet, favored abroad, wealthy, raised to the rank of Count, and still treating his readers as noble and dignified.

From this standpoint he knows how to form the all-encompassing from the single details, and even if—which first affects us here—in Volume 2, page 555 Buffon writes, "The arms of a human being in no way resemble the forelegs of animals, just as little as they do the wings of birds"—he there speaks in the sense of the crowd who, looking naturally, accept things as they are.* But inwardly it is developed better; for in Volume 4, page 379 he says, "There is an original and general design which can be followed a long way." Thereby he fixed the basic maxim of comparative natural philosophy once and for all.

May one forgive me these hasty, almost outrageously hasty words with which we present such a worthy man. It is enough to convince us that, irrespective of the boundless details to which he devotes himself, he did not fail to recognize something all-encompassing. One thing is certain: if we now go through his works we will find that he was conscious of all the main problems with which natural philosophy is concerned and earnestly exercised to solve them, even if he was not always fortunate [in doing so]. This in no way diminishes the respect we feel for him, when one sees that we latecomers

*Goethe is referring to Buffon's *Histoire Naturelle Générale et Particuliere,* 15 Vols. (Paris, 1749-67).

boast only all too early as if we had already solved many of the questions he raised. Irrespective of all that we must admit that when he sought to gain a higher view he did not despise the help of imagination. Because of this, the world's applause increased noticeably, but he then to some extent removed himself from the real element on which science should be based and seemed to transfer these concerns to rhetoric and dialectic.

Let us seek to become ever clearer in such an important matter:

Count Buffon was employed as chief overseer of the Royal Garden; he is supposed to have based his working out of natural history on it. His learning is towards the whole of what is alive, interacts, and has particular relationships to human beings.

For the details he needed a helper and called on Daubenton, a compatriot.

The latter seized the opportunity [but] from the opposite side and is an exact incisive anatomist. This subject owes him a lot, but he stuck so much to the detailed that he did not join together even the most closely related.

Unfortunately this quite different approach caused an irreparable division even among two men. As to how it may have been decided, suffice it to say that since 1768 Daubenton has taken no further part in Buffon's Natural History, but has continued to work diligently on his own, and after Buffon left at a ripe old age, the likewise aged Daubenton remained at his post and attracted a younger colleague in Geoffrey de Saint-Hilaire. The latter wanted a journeyman and found him in Cuvier. Surprisingly enough the same difference quietly developed in these two likewise gifted men, but at a higher level. Cuvier keeps decidedly to detail and in a systematically ordering sense for a wider overview leads to and necessitates a method of arrangement. Geoffrey, according to his way of thinking, seeks to penetrate inside the whole, but not, like Buffon, into what is at hand, existing already developed, but into what is functioning, becoming, developing. And so, secretly, the repeated conflict grows and remains hidden longer than the older one, while high, sociable culture, certain conveniences, silent forbearance postpone the outbreak year by year, until finally a small cause reveals the electricity of the Leidner bottle artificially separated, the secret conflict, and there is a powerful explosion.

But let us now continue with our observations on those four often mentioned men of science, even if we are to repeat ourselves to some extent, for it is they who, without detriment to any others, shine as the founders and developers of French natural history and form the seed from which so much that is desirable is fortunately developing, heading the important institution for almost a century, expanding it, using and furthering natural history by all possible means, and representing the synthetic and analytical approach of science. Buffon takes the external world as he finds it as entity that hangs together in all its manifold details with mutual relationships in contact. Daubenton, as an anatomist, constantly employed in dividing and parting, keeps away from connecting with anything else anything he has found on its own. He carefully sets down everything side by side, measures, and describes everything on its own account.

Cuvier works in the same sense only with more freedom and circumspection. He has been given the gift of noticing boundless details, of differentiating, of comparing, placing and ordering, and thus making great gains.

But even he has a certain apprehension of a higher method, which he, however, has himself not dispensed with, and, although unconscious of it, yet uses this method and so he represents Daubenton's qualities in a higher sense. Likewise we might say that Geoffrey to some extent points back towards Buffon. For when Buffon effects that great synthesis of the empirical world and absorbs it, at the same time making himself acquainted with and uses all the pointers which are offered for the purpose of differentiating, so Geoffrey approaches the great abstract unity that the former only suspected. He is not afraid of this unity and knows while he assimilates it how to use its divisions to his advantage.

In the history of knowledge and science the situation will never arise again that for such a long time a science would be advanced by such highly important men in constant opposition in the same place and position, with reference to the same objects, following their office and duty, and which, instead of summoning them to a mutual working out together even from different viewpoints through the unity of the task laid upon them, puts them into hostile opposition not because of the object but by the way of looking at it. Such a remarkable case must serve us all, and science itself, for the best! So may each of us say at this opportunity that SEPARATING and CONNECTING are two indivisible acts of life. Perhaps it is better to say that it is indispensable, whether one likes it or not, to proceed from the whole to the detail and from the detail to the whole, and the more vitally these two functions of the mind are related, like inhaling and exhaling, the better will be the outlook for the sciences and their friends.

We leave this point, to return to it later, when we have spoken of those men who in the seventies and eighties of the last century took us further along their particular path.

Petrus Camper, a man with a singular ability to observe and to connect which was tied to a remarkable gift for observation and imitation [–drawing] and so, through reproducing what he had experienced aroused that in himself and knew how to sharpen his meditations with spontaneity.

The great gains he made are generally recognized: I mention here only the facial line by means of which the protrusion of the forehead as the housing of the intellectual organ became more noticeable above the lower, more animal-like, shape and became more suited for thinking.

Geoffrey gives him this magnificent testimonial [Page 149 of his Note*],

*From the daily proceedings of the French Royal Academy.

"A wide-encompassing mind, highly developed and ever pondering, he has such a lively, deep feeling for the agreement of organic systems, that he preferred to seek out extraordinary cases where he could find cause to occupy himself with problems, an opportunity to use sharpness of mind, in order to bring so-called anomalies back to rule." And what could I add if more than a hint were to be given here!

This might well be the place to note that the researcher of nature first and most easily learns to recognize this way the value, the worth of a law or rule. If we continually see only that which is fixed we come to think it must be so, that it has always been so decided and is therefore stationary. But if we see the anomalies, misformations, gigantic deformities, then we recognize that the rule is indeed firm and eternal but is at the same time alive, that creatures do not grow from the rule but can within it transform themselves into the misshapen, but at any time, as if held back by reins, must acknowledge the unavoidable majesty of the law.

Samuel Thomas Sömmerring was stimulated by Camper. A highly capable mind that was awakened to observing, noting, and thinking. His work on the brain and his most thoughtful saying that man is principally distinguished from animals by the fact that the mass of his brain is to a large degree heavier than the complex of the other nerves, which is not the case with other animals, was of the greatest consequence.

And what attention was given in that receptive era to the yellow spot in the center of the retina! How much, as a consequence, do the sense organs, the eye and the ear, owe to his insight and his drawing!

His manner, [through] a written relationship with him was thoroughly enlivening and encouraging. A new fact, a fresh view, a deeper consideration were shared and effectiveness aroused. Everything that was budding developed quickly and youthfulness did not suspect the difficulties which lay in its path.

Johann Heinrich Merck, employed as military paymaster in Hesse-Darmstadt, deserves to be mentioned here in every way. He was a man of untiring intellectual activity which did not reveal itself through important results only because, as a talented dilettante, he was drawn and driven from all sides. He also shared a lively interest in comparative anatomy where the talent in drawing which expressed itself clearly and definitely fortunately helped him.

But the real inducement was given by the remarkable fossils to which until that time scientific attention had not been directed; these [fossils] were being dug up in the river area of the Rhine in large numbers and variety. With covetous fondness he acquired many excellent examples, the collection of which was placed in the ducal Hesse museum after his death, arranged and there carefully preserved and increased by the discerning curator von Schleiermacher.

My inner relationship to both men, at first through personal acquaintance then through a continued correspondence increased my inclination toward these studies. Therefore I sought, in accordance with my inborn disposition, above all for a clue, or as one might also term it, a point from which one could start, a maxim to which one could hold, a circle from which one could not wander.

If, in our field nowadays, there are striking differences [of opinion], then there is nothing more natural than that they had to occur in greater numbers and more often in those days, because everyone, setting out from his own standpoint, was concerned with using everything for his own purposes.

In comparative anatomy in the broadest sense, in so far as it was to establish a morphology, one was then always concerned as much with the differences as with the similarities. But I soon noticed that one had toiled until now only with regard to breadth; one compared, just as they happened to be at hand, animal with animal, animals with animals, animals with people, and from this there developed an incomprehensible diffuseness in that it was partly suitable for all cases, but partly completely failed to work out.

Then I laid the books aside and went directly to nature, to an animal skeleton that was visible at a glance. The stance on four feet was the most decisive and I began to examine it in order from head to tail.

Here the intermaxillary bone was above all the most striking, and so I observed it through a great variety of kinds of animals.

But quite other observations were thriving at that time. The close relationship of apes to human beings forced the researcher of nature to painful ponderings. The excellent Camper thought he had found the difference between apes and men in that the former was endowed with an intermaxillary bone but that such a bone was missing in humans.

I can not express what a painful feeling it was for me to find myself in decided opposition to a man to whom I owed so much, whom I hoped to approach [in fame, wisdom, etc.], to acknowledge myself as his pupil, from whom I hoped to learn everything.

Anyone who would intend to become acquainted with my efforts of that time will find whatever has been printed in the first volume of what I have supplied to morphology. What trouble has been taken, even in the pictures—on which everything depends—to depict the different, varying forms of that bone, can all be seen in the proceedings of the Imperial Leopold-Carolin Academy of Natural Researches, where, both the text has been reprinted and where the pictures, after remaining hidden for long years, have had a friendly reception. Both are found in the first section of the fifteenth volume.

But, before we open up that volume, I have something else to recount, to note and to admit, which, even if it were of no great importance, may yet serve to the benefit of our struggling followers.

Not only quite callow youth but also the already experienced man, as soon as a pregnant, coherent thought occurs to him will want to share it and awaken a similar way of thought in other people.

Thus I did not notice the misconception when I was thoughtlessly good-natured enough to send to Peter Camper the treatise which one will now find, translated into Latin with part outlined, part completed drawings. I then received a very detailed, benevolent answer in which he praised highly the attentiveness I had given those objects, did not object to the drawings, even though such objects are better taken [directly] from nature. He dispensed good advice and pointed out several advantages. He even seemed a little astonished at this effort and asked if perhaps I wanted to have this pamphlet printed, pointed out in detail the difficulties with the engravings as well as how to overcome them. In short, he took all fitting interest in the project as a father and patron.

But there was not the slightest sign that he had noticed my purpose of opposing his own opinion and seeing something else as a program. I replied modestly and received even more detailed and favorable letters, which, looked at carefully, dealt with the material contents but in no way concerned my purpose, to the extent that, as this connection could further nothing, I quietly let it lapse, without, as I perhaps should have done, creating the meaningful experience that one cannot convince a mastermind of his error, because it was accepted as part of his mastery and thus legitimized.

Unfortunately, together with so many other documents, those letters have been lost and they would have brought home very clearly the diligent outlook of that man as well as my youthful and faithful deference.

But yet another misfortune befell me: an excellent man, Johann Friedrich Blumenbach, who fortunately had devoted himself to natural philosophy, and had begun particularly to work through comparative anatomy, in his compendium of comparative anatomy, came out on Camper's side and denied that human beings had an intermaxillary bone. My embarrassment was now greatly increased by this, because a valuable textbook, a reliable teacher was thoroughly setting aside my conclusions and views.

But such an intelligent, continually researching, and thinking man could not always stick to a preconceived opinion forever, and I am indebted to him that, through cordial relationships, he taught me sympathetically about this and many other points, by informing me that the intermaxillary bone is divided from the upper jaw in hydrocephalic children and also shows itself as a weakness in a double wolf jaw.

Now, however, I can call forth those works which then were rejected with protests and have remained in obscurity for so many years and can request some attention now be given to them.

I must first devote my attention to the above mentioned illustrations for the sake of complete clarity, still more, however, to point to d'Alton's great osteological work, from which one may gain a far greater, freer, and thorough overview.

During all this, however, I have the purpose of requesting the reader to regard all that has been said up to now and what is still to be said as directly or indirectly relevant to the quarrel between those two excellent French nature-researchers of whom one presently speaks constantly.

Then I may now presuppose that one will take three drawings I have just noted and be inclined to go through them with us.

As soon as one talks of illustrations it is understood that one is talking about form [Gestalt] ; but in the present case we are directed to the function of the parts directly, for form stands in relationship to the whole organization [organism] , to which each part belongs and thereby also to the external world, of which the completely organized being [WESEN] must be regarded as a part. In this sense we now go to work without further pondering.

On the first illustration we see that bone which we recognize as the foremost [bone] of the whole animal structure is formed in various ways. A closer observation makes us see that through it the most necessary nourishment is suited to the animal: thus the more varied the nourishment the more varied is the formation of this organ. In the deer we find a light, toothless bony semi-circle for plucking grass-stalks and leafy boughs reasonably. In the ox we see roughly the same form [Gestalt] , only broader, coarser, and stronger in proportion to the needs of the creature. In the third drawing we have the camel, which, sheeplike, displays a certain, almost monstrous, indecision so that the intermaxillary bone can scarcely be distinguished from the upper jawbone or the incisor tooth from the canine tooth.

On the second diagram the horse is shown with a significant intermaxillary bone containing six blunted incisor teeth; in a young specimen, the here undeveloped canine tooth is fully ascribed to the upper jawbone.

It is worthy to note on the second picture of the same plate the upper jawbone of the *Sus babirussa* seen from the side; here one sees the wonderful canine tooth quite certainly contained in the upper jawbone, while its alveolus scarcely skims the piglike toothed intermaxillary bone and shows not the slightest influence on it.

On the third plate* we devote our attention to the third drawing, the wolf's jaw. The protruding intermaxillary bone, equipped with six stout, sharp incisor teeth is clearly distinguished by a suture from the upper jawbone in

*The third, fourth, and fifth plates referred to here by Goethe appeared originally in Joseph Guillaume Edouard D'Alton, *Vergleichende Osteologie,* Volume 12 (Bonn, 1821-28). Plates 3 and 5 were reproduced in *Goethe's Morphologische Shriften,* W. Troll, ed. (Jena, n.d.), as Plate XXV on page 376 and Plate XVIII on page 288, respectively.

Drawing b, and one notes, although very protuberant, the close relationship with the canine tooth. The lion's jaw, more compact, stronger in the tooth and more powerful, shows that differentiation and relationship even more exactly. The same fore-jaw of the polar bear, powerful but clumsy, coarse, a characterless formation, in every respect less fitting for seizing than for crunching, the Canales palatini broad and open, but yet no trace of that suture which, however, can be drawn in one's mind and its course shown.

On the fourth plate, Trichechus rosmarus gives cause for many kinds of observations. The great preponderance of the canine teeth forces the intermaxillary bone to retreat, and the contrary creature thus receives an appearance similar to a human being. Figure 1, a reduced size drawing of an adult animal lets the divided intermaxillary bone be seen clearly; one also notices how the powerful root, fixed into the jawbone, during continued struggle upwards has produced a kind of swelling on the surface of the cheek. Figures 2 and 3 are drawn from a young animal of the same size. In this specimen the intermaxillary bone is completely separated from the upper jawbone, as there the canine tooth remains undisturbed in its alveolus which belongs only to the upper jawbone.

After all this we may boldly maintain that the great elephant tooth is likewise rooted in the upper jawbone. During this we have to remember that during the tremendous strain which is here placed on the upper jawbone the neighboring intermaxillary bone should produce a lamina for the strengthening of the gigantic alveolus if not to its formation.

I believe we have found out this much by careful examination of several specimens even though the skull diagram in the twelth volume are not decisive.

For it is here that we need the genial spirit of analogy standing at our side so that we do not get an inaccurate picture of a truth that has been tested in many examples through a single doubtful case, but even these show due honor to the law where it might be withdrawn from us the phenomenon.

On the fifth plate, ape and human are juxtaposed. In what concerns the latter enough has already been shown clearly by a special preparation, division, and dissolving of the said bone. Perhaps both figures, the goal of the whole treatise, could have been depicted more frequently and more clearly and set up side by side. But just then, in that most fruitful time, inclination and activity were blocked in that subject, so that we must thankfully recognize [the fact] when a highly honored society of scientific researchers honor these fragments of their attention and wish to preserve the memory of honest endeavors in the indestructible corpus of its minutes.

Once more, however, we must request the continued attention of our readers, for spurred on by Mr. Geoffrey [de Saint-Hilaire] himself, we have another organ to consider in the same sense.

Nature remains eternally respectable, for ever recognizable up to a certain point, eternally at the disposal of a reasonable man. She turns many sides to us; what she conceals she at least hints at; to the observer as well as to the thinker she gives many sided encouragement, and we have reason to disdain no means by which her external features may be noted more clearly and her inner features may be researched more thoroughly. Therefore, without further ado, we turn to FUNCTION for our purposes.

Properly understood, function is thought of as the being in action, and so we are concerned, summoned by Geoffrey himself, with the human arm and the forelegs of an animal.

Without wishing to appear learned, we begin with Aristotle, Hippocrates, and Galen according to the latter's report. The serene Greeks ascribed to nature a very dear power of reason. She has arranged everything so well that one will have to find the whole always perfect. She gave the powerful animals claws and horns, the weaker ones delicate legs. But the human being is especially provided with his dexterous hand by means of which he can produce swords and spears instead of horns and claws. The reason why the middle finger is longer than the others is a merry one to hear.

However, if we continue further in our way we must put before us the great work of d'Alton and take from its richness the proofs for our observations.

We assume as universally known the human forearm, its connection with the hand and the wonders achieved by it. There is nothing intellectual in what falls within this realm.

If, after this, one observes predatory animals and [notes] how their claws and talons are busy fitted only for gaining food and how, except for some instinct to play, they remain subordinated to the intermaxillary bone and are servants of the eating mechanism.

In horses the five fingers have been enclosed in a hoof; we see this in an intellectual view, even if through some monstrosity the divisibility of the hoof into fingers did not convince us. This noble creature needs no violent acquisition of his food; an airy, not too moist meadow furthers his free existence, which really only seems suited to an endless movement of its roaming, comfortable disposition; a natural tendency which man knows only too well how to use for useful and passionate purposes.

If we now attentively examine this part [of the animal] through the most diverse species then we find that its perfection and that of its function increases or decreases according to whether PRONATION and SUPINATION can more or less easily and completely be practiced. Many animals possess such an advantage to a greater or lesser degree; but, as they necessarily use the forearm for standing and moving, they exist most of the time in pronation and as in this mode the radius is turned inwards to the thumb, with which it is organically connected, then the same [the radius], as indicating the real point of emphasis, becomes more important according to the set of circumstances, indeed finally almost exclusively in its place.

Amongst the most mobile forearms and most skilled hands we can perhaps count those of the squirrel and related rodents. Their light body, insofar as it more or less achieves an upright stance and the hopping movement do not let the forepaws become coarse. There is nothing more graceful to watch than a squirrel peeling a fir-cone: the center pillar is thrown away quite clean, and it would perhaps be worthwhile to observe if these creatures did not gnaw away and obtain the seeds in the spiral sequence in which they developed. We can here fittingly think of the two protruding gnawing teeth of this family, which, contained in the intermaxillary bone, have not been presented in our plates but which are produced the more numerously in d'Alton's notebooks.

It seems to be extremely remarkable that, through a secret agreement, with a more advanced activity of the hand the front teeth at the same time achieve greater facility. For while these teeth in other animals are used to seize food, here food is brought to the mouth in a skilled manner by the hands, whereby the teeth are now devoted solely to chewing and this thus to some extent becomes technical.

But here we are led into the temptation not so much to repeat the above expressed Greek dictum as to alter it progressively. "Animals are subject to the tyranny of their limbs," we might say, while, indeed, they use them for the extension and continuation of their existence without any further ado; as, however, the activity of every such determination, even needlessly, always continues, so, therefore, rodents, when they have satisfied [their hunger], must begin to destroy until finally this tendency produces through the beaver an analogue of reasonable archictectonics.

But we may not continue in this way because we would lose ourselves in the boundless, so we will summarize briefly.

As an animal feels itself ever further determined for standing and walking, the more the radius will increase in strength, so that [the ulna] finally almost disappears and only the olecranon remains as the most necessary articulation with the upper arm. If one goes through d'Alton's illustrations one will make basic observations on this subject and always finally catch sight, at this part and others, of the existence which is distinguished through the form in a living and inter-related function.

But now, however, we must think of the case where ample indication of the organ is left and just at the point where any function ceases completely which enables us to penetrate the secrets of nature from a new direction.

One should take the notebook of d'Alton the younger representing ostrich-like birds and observe from the first to the fourth plate, from the skeleton of the ostrich to that of the cassowary of New Holland, and should notice how the forearm is progressively drawn together and simplified.

Although this organ which really makes man a man, a bird a bird, finally appears most peculiarly abbreviated so that one could express it [the organ] as a chance malformation, yet the collected single masses of a limb can be distinguished by it; the analogue of their form can not be mistaken, just as little, however far they extend, where they fit together, and although the foremost decline in number, those that remain do not give up their determined relationship.

This important point, which one must keep in view when examining the osteology of higher animals, has been seen perfectly correctly and forcefully expressed by Geoffrey: that one can discover any special bone which seems to be hidden from us most reliably inside the limits of its neighbors.

He is apprised of another basic truth which is directly connected to it: that, in fact, thrifty nature has prescribed itself a state, a budget, in whose single chapters she preserves for herself the fullest discretion, but in the final analysis remains true to herself, in that, if on one side too much is spent, she draws it from the other and strikes a balance in the most decisive manner. These two sure signposts, to which our Germans have owed so much for so many years, have been recognized by Geoffrey to the extent that, in the progress of his scientific life, they have always yielded him the best services; as they then absolutely do away fully with the sad expedient of final causes.

So much may be enough to indicate that we may not leave out of consideration any kind of manifestation of the labyrinthine organism, if, by viewing the external, we wish to penetrate into the most central.

From what has been said so far one may see that Geoffrey has penetrated to a high level of thought suitable to the idea. Unfortunately, on many points his language does not offer him the right expression, and as his opponent finds himself in the same situation, the quarrel becomes unclear and confused for this reason. We want to try to illuminate this circumstance modestly. For we may not delay in making this opportunity noticeable, as a dubious use of words in French lectures and in the disputes of excellent men gives cause to significant errors. One believes one speaks in pure prose and one is already speaking metaphorically; one uses metaphors in a different way than the other, extends it in a related sense, and so the quarrel becomes unending and the riddle insoluble.

Matériaux. One makes use of this word to express the parts of an organic being which together form either a whole or a subordinate part of the whole. In this sense one would call the intermaxillary bone, the upper jawbone, the palatine bone materials from which the arch of the jaw is constructed; likewise one would regard the upper arm bone, both those of the forearm and the numerous ones of the hand as materials of which the human arm and animal foreleg is composed.

In the most general sense, however, we designate by the word materials unconnected, perhaps even unrelated bodies receiving their relationships through arbitrary determination. Beams, planks, laths are materials of one kind from which one can construct many kinds of buildings, thus e.g. a roof. Tiles, copper, lead, zinc have absolutely nothing in common with them and yet, according to circumstances, are needed for the completion of the roof.

We must therefore put a much higher construction on the French word matériaux than is its due, although it occurs unwillingly because we can foresee the consequences.

Composition. A likewise unfortunate word, related mechanically with the previously mechanical. The French introduced such a word when they began to think and write about the arts; for thus it is the painter composes his painting, the musician is called a composer once and for all, and yet, if both wish to deserve the true name of an artist, they do not put their works together but they develop some inherent picture, a higher harmony fitted to nature and art.

Just as in art, this expression is demeaning when one is talking about nature. Organs do not get composed as if they were something previously finished, they develop from and by each other to a necessary existence that reaches towards the whole. So there may be talk of function, form, color, measurement, mass, weight, or other factors, whatever they may be called, all is admissible in observation; the living organism goes its way undisturbed, continues its growth, sways, strives, and finally reaches its perfection.

Embranchment is likewise a technical word of carpentry and expresses the idea of adding beams and rafters in and on each other. A case where this word appears permissible and expressive when it is used to designate the branching of one road into several.

We believe we see here in detail as well as on the whole the after effects of that epoch where the nation was given up to sensualism, accustomed to make use of material, mechanical, atomic expressions; so then inherited linguistic usage reaching even into everyday dialogue but as soon as conversation rises to the intellectual level it clearly strives against the higher view of men of excellence.

We would add yet one more word, the word plan. Because, in order to compose materials well, a certain previously thought-out arrangement is necessary, so they use the word plan but are thereby immediately led to the concept of a house or a town, which, however reasonably [they may be] arranged, can still offer no analogy to an organic being. Yet they use, thoughtlessly, buildings and streets as a comparison; so therefore the expression Unité du plan gives rise to misunderstandings and to rebuttal and contention and the question on which all depends is obscured thereby.

Unité du type would have brought the matter closer to the right path, and this was so close in that they know how to use the word "type" in the context of speèch, as it should really stand at the head and contribute to the neutralizing of the quarrel.

First let us only repeat that Count Buffon even in 1753 has printed that he acknowledged "a general primitive plan—that one can follow a great distance—on which all seems to have been conceived" [trans. from French] Vol. IV p. 379*.

"What further proof is needed?"

But this might be the place to return to the dispute from which we started and to bring forth its consequences in chronological order as far as possible.

One should remember that the volume [or: issue] which caused what is before us now is dated 15th April 1830. The collective daily papers immediately take cognizance of the affair and declare themselves for or against

In June the publishers of the Revue Encyclopédique discussed the matter, not without favor to Geoffrey. They declare it [the matter] to be European, i.e. important within and outside the scientific circle. They extensively quote an essay of the excellent man which deserves to be generally known, where he speaks out briefly and concisely.

One sees how passionately the dispute is treated from the fact that on the 19th of July, when the political ferment had reached a high degree, this far distant scientific and theoretical question was occupying and exciting such minds as these.

Be that as it may, we are directed by this controversy [to look] at the internal, particular relationships of the French Academy of Sciences, for the following may well be the reason why this inner discord did not come to light earlier.

In earlier times the sittings of the Academy were closed, only members were admitted and discussed their experiences and opinions. Gradually friends of science were given friendly admittance; other importunors could consequently hardly be refused and so they gradually found themselves in the presence of a significant audience.

If we carefully observe the way of the world we learn that all public proceedings, be they religious, political, or scientific, sooner or later become thoroughly formal.

The members of the French Academy, as is fitting in good society, therefore stood back from all basic and violent controversy. Lectures were not debated; they were given to commission for study and treated according to their expert judgment, following which one essay or another experienced the honor of being accepted in the memoirs of the Academy. That much is known to us in general.

But now it can be announced in our case that the quarrel, once it has broken out, will have a decided influence on such a tradition.

In the sitting of the Academy of 19th July we can hear an echo of those

*Buffon's *Historie Naturelle Générale et Particuliere,* 15 Vols. (Paris, 1749-67).

differences, and now the two permanent secretaries, Cuvier and Arago, are in conflict.

Until now, as we have heard, it was the custom at each succeeding session to refer only to the titles of previous lectures and then to set the matter aside.

The other permanent secretary, Arago, however, made an unexpected exception just that time and brought up in detail the protest submitted by Cuvier. The latter protested, however, against such innovations, which would cause a lot of time to be expended and at the same time complained about the incompleteness of the resumé that had just been given.

Geoffrey de Saint-Hilaire disagreed, citing the example of other institutes where such things happened with profit.

He was himself then disagreed with, and it was finally considered necessary to give the matter further consideration.

In a sitting on 11th October Geoffrey read an essay on the particular forms of the back of the head of crocodiles and the *Teleosaurus,* here he reproached Cuvier for omission in observation of these parts. The latter stood up, very much against his will, as he assured those present, but he felt impelled to do so by those reproaches, in order not to yield in silence. This is a remarkable example for us, what great damage can occur when a dispute about higher views is spoken about [only] in its details.

Soon afterwards a session was held which we here want to remember in Mr. Geoffrey's own words as given by him in the "Gazette Medicale" of 23rd October.

"The present newspaper and other public prints had spread the news that the dispute that had developed between Mr. Cuvier and myself was to be taken up at the next sitting. People hurried along to hear the developments of my opponent which he had announced in advance [as being] about the temporal bone of crocodiles.

"The hall was more crowded than usual and it was thought that among the listeners could be seen not only those who, inspired by pure interest, came in from scientific fields; there were rather some curiosity seekers to be noticed and opinions from an Athenian-like ground floor of quite varied views to be heard.

"This circumstance, when communicated to Mr. Cuvier, led him to postpone his essay to another sitting.

"Informed by his opening projects I considered myself ready to reply and was very happy [at the chance] of seeing this matter settled.

"For rather than a scientific contest I prefer to lodge my deductions and conclusions with the Academy.

"I had written my essay with the intention, if I had spoken on the spur of the moment about the situation, of trusting it to the academic archives with the condition: ne varietur."

A year has now passed since those events and one can be convinced from what has been said that we have remained attentive to the consequence of such an important scientific explosion even after the great political explosion. But now, so that the above remarks do not age completely, we wish to assert only this much, that we believe we have noticed that our neighbors' scientific observations in this field will, in future, be treated with more liberty and with greater intellectual freedom.

We have found mention of the following names among our German participants: Bojanus, Carus, Kielmeyer, Meckel, Oken, Spix, Tiedemann. If one may now suppose that the findings of these men will now be recognized and used, that the genetic way of thought, which Germans can not now avoid, will win more credit, then we can certainly rejoice in continued sympathetic co-operation from that side [France].